THE RACE WITH NO FINISH LINE

THE RACE WITH NO FINISH LINE

Assessing the Strategy of Regional Great Power Competition

Martin Skold, PhD

Quantico, Virginia
2023

LIBRARY OF CONGRESS CATALOGING-IN-PUBLICATION DATA
Names: Skold, Martin, 1983- author. | Marine Corps University (U.S.). Press, issuing body.
Title: The race with no finish line : assessing the strategy of regional great power competition / Martin Skold.
Other titles: Assessing the strategy of regional great power competition
Description: Quantico, Virginia : Marine Corps University Press, 2023. | Includes bibliographical references and index. | Summary: "The Race with No Finish Line offers a framework for understanding the strategies of states engaged in competition for regional hegemony. Employing insights drawn from business strategy, the author argues for an essentially asymmetric understanding of fundamental policy goals for states engaged in competition for control of a region of the globe, with one state attempting to maintain a dominant position and another attempting, by focusing limited resources, to supplant it"— Provided by publisher.
Identifiers: LCCN 2022024171 (print) | LCCN 2022024172 (ebook) | ISBN 9798985340402 (paperback) | ISBN 9798985340419 (epub)
Subjects: LCSH: Hegemony. | Strategic rivalries (World politics) | Great powers—Foreign relations. | Balance of power. | Strategy. | Geopolitics. | Sea-power—Great Britain—Case studies. | Sea-power—Germany—Case studies. | North Sea—Strategic aspects—Case studies.
Classification: LCC JZ1312 .S58 2022 (print) | LCC JZ1312 (ebook) | DDC 327.101—dc23/eng/20220707 | SUDOC D 214.513:ST 8/3
LC record available at https://lccn.loc.gov/2022024171
LC ebook record available at https://lccn.loc.gov/2022024172

DISCLAIMER
The views expressed in this publication are solely those of the author. They do not necessarily reflect the opinion of Marine Corps University, the U.S. Marine Corps, the U.S. Navy, the U.S. Army, the U.S. Air Force, or the U.S. government. The information contained in this book was accurate at the time of printing. Every effort has been made to secure copyright permission on excerpts and artworks reproduced in this volume. Please contact the editors to rectify inadvertent errors or omissions. In general, works of authorship from Marine Corps University Press (MCUP) are created by U.S. government employees as part of their official duties and are now eligible for copyright protection in the United States; further, some of MCUP's works available on a publicly accessible website may be subject to copyright or other intellectual property rights owned by non-Department of Defense (DOD) parties. Regardless of whether the works are marked with a copyright notice or other indication of non-DOD ownership or interests, any use of MCUP works may subject the user to legal liability, including liability to such non-DOD owners of intellectual property or other protectable legal interests.

MCUP products are published under a Creative Commons NonCommercial-NoDerivatives 4.0 International (CC BY-NC-ND 4.0) license.

Published by
Marine Corps University Press
2044 Broadway Street
Quantico, VA 22134

1st Printing, 2023
ISBN: 979-8-9853404-02
DOI: 10.56686/9798985340402

THIS VOLUME IS FREELY AVAILABLE AT WWW.USMCU.EDU/MCUPRESS

For Christina Goodlander, whose love, patience, and wisdom are in every page of this work.

For my parents, Craig and Mary Ann Skold, who set me on this path long ago.

And for my teachers.

Contents

Illustrations	ix
Foreword	xiii
Preface	xvii
Acknowledgments	xxi
Introduction	3
Chapter One	15
On Competition and Strategy	
Chapter Two	58
Some Principles of Competitive Strategy	
Chapter Three	96
A Framework for Strategic Analysis of Great Power Competition for Regional Hegemony	
Chapter Four	111
The Origins and Aims of the Dreadnought Race	
Chapter Five	196
The Dreadnought Race	
Chapter Six	258
The Dreadnought Race in Strategic Perspective	
Conclusion	365
The Strategy of Regional Great Power Competition	
Appendix: A Note on Data Sources and Methodology	377
Selected Bibliography	385
Index	401
About the Author	407

Illustrations

Figure 1.	The OODA loop, as drawn by Col John R. Boyd, USAF	73
Figure 2.	Ratio of British capital ships to German capital ships, 1875–1915	273
Figure 3.	Ratio of British major combatant fleet to German major combatant fleet, 1875–1915	275
Figure 4.	Ratio of British fleet size to German fleet size, 1875–1915	276
Figure 5.	Ratio of British torpedo-craft fleet to German torpedo-craft fleet, 1894–1914	277
Figure 6.	British total defense expenditures as percentage of gross national product, 1875–1913	282
Figure 7.	British naval estimates in pounds, 1875–1914	283
Figure 8.	British naval estimates as percentage of gross national product, 1875–1914	284
Figure 9.	British naval estimates as percentage of total defense budget, 1875–1914	285
Figure 10.	British naval estimates' growth rate from previous year, 1876–1914	285
Figure 11.	British Regular Army total personnel, 1875–1913	287
Figure 12.	British Army total effective personnel, including reserves, militias, and others, 1900–14	287
Figure 13.	British Army estimates in pounds, 1875–1914	288
Figure 14.	British expenditures per soldier in pounds, 1875–1913	288

Illustrations

Figure 15.	British Army estimates' growth rate from previous year, 1876–1914	289
Figure 16.	British defense budget growth rate from previous year, 1876–1914	290
Figure 17.	Share of British defense budget growth attributable to Royal Navy estimates increase, 1876–1914	290
Figure 18.	British gross national product growth rate from previous year, 1876–1913	291
Figure 19.	British revenue as percentage of gross national product, 1875–1913	292
Figure 20.	British overall expenditures as percentage of gross national product, 1875–1913	292
Figure 21.	British defense expenditures as percentage of overall expenditure, 1875–1913	294
Figure 22.	British national debt as percentage of gross national product, 1875–1913	295
Figure 23.	British national debt in pounds, 1875–1913	296
Figure 24.	Ratio of German capital ships to British capital ships, 1875–1915	301
Figure 25.	Ratio of German torpedo-craft fleet to British torpedo-craft fleet, 1875–1915	302
Figure 26.	Ratio of German torpedo-craft fleet to British torpedo-craft fleet during (and immediately before) the Dreadnought Race, 1895–1915	302
Figure 27.	Ratio of German fleet size to British fleet size, 1875–1915	303
Figure 28.	Ratio of German major surface combatant fleet size to British major surface combatant fleet size, 1875–1915	303
Figure 29.	German Army total personnel, peace footing, 1875–1913	305

Illustrations

Figure 30.	German Army estimated total personnel, war footing, 1875–1913	306
Figure 31.	German peacetime military participation, 1875–1913	306
Figure 32.	German total official defense estimates as percentage of gross national product, 1875–1914	310
Figure 33.	German total defense expenditures as percentage of gross national product, 1893–1913	311
Figure 34.	German naval estimates as percentage of total army and navy estimates, 1875–1914	312
Figure 35.	German naval expenditures as percentage of defense expenditures, 1892–1913	313
Figure 36.	Annual percent increase in total German defense estimates, 1876–1914	313
Figure 37.	Percent increase in German total defense expenditures, 1893–1913	314
Figure 38.	Share of German total defense expenditure growth attributable to navy expenditure increase, 1893–1914	316
Figure 39.	Share of German defense estimates growth attributable to increase in navy estimates, 1876–1914	316
Figure 40.	German revenue as percentage of gross national product, 1875–1914	318
Figure 41.	German overall expenditures as percentage of gross national product, 1875–1914	318
Figure 42.	German national debt in pounds, 1875–1914	319
Figure 43.	German national debt as percentage of gross national product, 1875–1914	319
Figure 44.	German actual defense expenditures as percentage of overall estimates, 1892–1913	321
Figure 45.	German defense estimates as percentage of overall estimates, 1875–1914	321

Foreword

It is a rewarding discovery when one's interest in strategic concepts coincides with a rollicking well-told story about great power rivalry. It is even more rare when that combination produces a real contribution to understanding about competitive strategy. Martin Skold's *The Race with No Finish Line: Assessing the Strategy of Regional Great Power Competition* is not only one of those "make-my-day" great reads but also an intellectual stimulant that keeps on giving long after the last page has been turned. This is a well-written and -researched history of the great battleship rivalry between a challenger, Germany, and a century-long status-quo hegemon-of-the-seas, Great Britain. Beyond this, there are three compelling aspects of Skold's contribution that deserve being highlighted.

First, with the publishing of this work, Skold offers a significant empirical test of the "competitive strategy" construct as applied to one of the most notable arms races in history. Equally important, his analysis transcends simplistic action-reaction assumptions by delving deep into the internal rationale, political debates, economic costs, sequence of decisions, and misperceptions of both sides. It is all too easy to demonize the challenger of a strategic balance, but doing so provides neither insight into the causal chain that produces a buildup nor a behavioral map of leverage points to influence its course. Skold possesses a sophisticated ability to make "out-of-body" evaluations of the contestants' motives, commitments, resources, and delusions, all with an objective eye. There is no shortage of past military competitions waiting to be deconstructed, and this book provides both a prototype case-study method and an exemplary template for other scholars to explore and compare hundreds of these competitions.

Foreword

Second, if a reader is looking for a succinct, comprehensive summary of competitive thinking as related to strategy, they would be hard-pressed to find a better compendium and summary of key concepts than the first three chapters in this book. In chapter 1, Skold's survey of competition and strategy, to include their definitions and constructs, is not locked in to one approach but rather adroitly covers the breadth of the field and demonstrates depth in showing how various concepts both contradict and complement one another. Chapter 2 converts strategic definitions into operational principles. In chapter 3, Skold pulls off an intellectual tour de force in producing a framework for strategic analysis, and his derived seven components for the analysis and comparison of competitive strategies are worth the price of this book alone.

Third, the value of Skold's work is defined not only by its study of the past and its unique conceptualization but also by its relevance to an imminent future. At present, the West is awakening from its interregnum slumber of the post–Cold War period to realize that the "end of history" did not materialize with the collapse of Communism or the consumption of globalism. Once again, the West is surprised and shocked by the "tragedy of great power competition," in which major regional players resentful of Western unipolarity block their attempts at theater hegemony. Unilaterally invented "zones of interest" challenge the principle of national self-determination, and their aggressive pursuit with the threat and use of force challenge the core principles of sovereignty and peaceful resolution of disputes. As such, these regional challengers demonstrate a willingness to supersede international norms with national ambitions, to confront the status quo with imperial claims, and to find allurement, even seduction, in the status, influence, and physical manifestation of perceived power that accompanies joining the high-tech weapons "race with no finish line."

Marine Corps University Press has done yeoman's service in making this work available to professional military officers, government decision makers, and the interested public. Hope-

Foreword

fully, in an impending era of "new world disorder" Skold's work will help reawaken interest in the relevance of competitive strategy. One of the most powerful lessons learned in this book is the observation that despite Great Britain winning the battleship competition with Germany, war still broke out between the two countries, with devastating consequences. As powerful as Skold's presentation is here, and as consequential the concept discussed, the intellectual construction of competitive strategy is a "glass half full." There is still much to explore, debate, and develop.

Phillip Karber
President, The Potomac Foundation
Adjunct Professor, Georgetown University, Washington, DC

Preface

This book was born of a desire to win an upcoming game. As noted elsewhere in this volume, this book began as my doctoral thesis for the degree of doctor of philosophy (PhD) in international relations at the University of St. Andrews in Scotland. It has been long in development, having first been proposed as a thesis idea in 2011. At the time, the United States was primarily preoccupied with the Global War on Terrorism (GWOT, in Pentagon-speak), and in fact the successful elimination of al-Qaeda leader Osama bin Laden was still a few months in the future. Acting on intuition, however, I had decided to pursue a different focus for my studies. As I note in the acknowledgements, I credit the contributions of several individuals with the various components of this focus.

The first contribution was the observation made to me nearly a decade prior during my undergraduate studies at Georgetown University in Washington, DC, by Georgetown's famous (and feared) geography and geopolitics instructor, Professor Charles Pirtle, that the Iraq War represented a dangerous distraction from the geopolitical game and a waste of resources for the United States. This was echoed at one point by my friend and mentor Professor Phillip Karber, who, referring to the GWOT writ large, said something to the effect of "We need to be getting ready for the big game and we're stuck in a back-alley knife fight." Professor Karber is on record regarding the large number of friends he lost in the terrorist attacks of 11 September 2001, so this was no idle statement. The takeaway from observations like these, which never left me, was that the GWOT—the preoccupation of the era—was going to be of transient importance. High geopolitics, involving conventional challengers with arsenals and economic resources far more powerful than al-Qaeda's (or Iraq's, or any number of others'), would return soon enough.

Preface

A second category of thinking augmented this. I had been trained in the "net assessment" school of thought in my studies for my master's degree (again at Georgetown University) by an alumnus of the U.S Defense Department's legendary office by that name, Barry Watts, who introduced me to high-level strategic thinking as it pertained to defense planning. In planning my next move to the PhD level, I took inspiration from this and began to think about how strategy might be applied to high geopolitics—again, on the theory that geopolitical competition would make itself known again soon. As I note in the acknowledgments, some of the initial answers I found to this question appeared in books on strategy—business strategy in particular—that Professor Karber was kind enough to recommend to me. If you are preparing for a big game, you want to know how to win.

The question was which aspect of this nebulous issue—the strategy of great power competition—I should focus on. After a lot of reading and thought, I decided to make it about competition for control of a region of the globe. My reasoning was simple enough: great power competition often manifests as competition over a key area, and—crucially—this was a manageable topic for a single work. A book offering a similar analytic framework to that found here on global hegemonic competition, on the scale of the Cold War, eighteenth-century great power rivalry, or anything similar, has yet to be written, though I hope at some point someone will take up that daunting challenge. After a lot of consultation of business strategy literature in particular, I hit on the idea of an analytic framework: rather than simply prescribe best practices, a strategic work should tell its readers what questions to ask to see who is winning a game and why. To put a long story short, the question became this: "In great power competition for control of a region of the globe, what is it about, and what is important to know about it if you want to win or know who is winning?" I have endeavored to answer this question about questions in this book.

In doing so, I have used every "core competency" at my disposal. Classic military strategy literature offered some broad

Preface

outlines. Business strategy literature provided many applicable concepts. Net assessment writings informed "how I think about this." International relations theory provided a key insight in terms of state goals, expressed as the "three metrics." More avant-garde military strategy literature, such as U.S. Air Force colonel John R. Boyd's OODA (observe-orient-decide-act) loop, provided new ways to think about the problem. This may be an eclectic mix, but using whatever is useful is supposed to be the acme of successful strategy; I can only hope such has been the case here.

I hope this book will be useful to defense planners and high-level policy makers, as well as anyone who wants to keep an eye on their work. But I particularly hope it will be relevant to the study of great power competition as it has recently emerged and will emerge. A few years into my studies, Russia occupied Crimea and refocused the Western policy community on conventional conflict and competition once again. About this same time, numerous authors began alerting the U.S. and other Western publics to the global challenge posed by China. As of this writing, the West is aiding Ukraine in its defense against Russian invasion as part of a broader effort to contain Russia, and debate is ongoing about how best to contain China along the first island chain in the Pacific. I can only hope that this book will be a contribution to this discussion.

In doing so, I suppose I can take some credit, and give my various instructors credit, for one simple prediction: conventional geopolitics has returned, and its study is of obvious and central importance for makers and students of U.S. policy. As an American, it is my hope that in the games ahead, the United States secures its interests and acquits itself well—and wins the races with (seemingly) no finish line yet to come.

Acknowledgments

This is the word that year by year
While in her place the School is set
Every one of her sons must hear,
And none that hears it dare forget.
This they all with a joyful mind
Bear through life like a torch in flame,
And falling fling to the host behind—
"Play up! play up! and play the game!"
   ~ Sir Henry Newbolt[1]

Writing a book is inevitably a solitary endeavor, but it has elements of a team sport. In the course of burying oneself in one's work (and cutting out social distractions), it is all too easy to forget the aid one has received along the way. Lest I forget, I must take the opportunity here to recognize those who have helped me to complete the seeming "race with no finish line" that this project has been, and their many contributions.

This study began life as my PhD thesis at the University of St. Andrews in Scotland. For that reason, I thank the University of St. Andrews for taking a chance on me and on this project years ago, and for an excellent education and experience as a student. I must thank my supervisor, Dr. William Vlcek, for all his advice, encouragement, and leadership in producing the dissertation that became this book. Always patient and generous with his time, Dr. Vlcek offered valuable critiques and commentary at every step of the writing process. For everything from long hours spent in rambling discussions of ideas about history and international politics; to insightful guidance and criticism on the substance, structure, and organization of this thesis; to frequent literature recommendations; to a close eye for detail that extended to the smallest grammatical errors and rose to

Acknowledgments

the most intricate philosophical arguments; to advice on the writing process; to simple confidence in the project when its ideas were not yet fully formed; to patience and understanding when required chapters failed to materialize on time—Bill, I am profoundly grateful.

I am grateful as well to my examiners, Professor Sir Hew Strachan of the University of St. Andrews and Professor David Betz of King's College London, for their patience and consideration with earlier drafts of this work. There was indeed much honor in being examined by such a distinguished committee, and much improvement was made by their involvement. I thank Professor Strachan especially for subsequent career advice and encouragement.

I must also, for the same reason, thank Professor Andrew Williams at the University of St. Andrews, who very graciously gave me his time in reading and revising the original work and offered much-needed encouragement along the way.

This book took final form thanks to the intervention of four people in particular. The first suggestion that I consider submitting what was then a raw dissertation to Marine Corps University Press (MCUP) came from Dr. Faye Donnelly at the University of St. Andrews, to whom I am grateful for helping me look up options and for planting the seed of the idea. When I initially floated the idea, Marine Corps lieutenant colonel Brian Kerg first saw potential in the project and referred it to MCUP. Without Lieutenant Colonel Kerg's generosity of time and spirit, this work would have remained unpublished, and I owe him a great debt of gratitude. I am in turn exceedingly grateful to MCUP director Angela Anderson, not only for taking the project on but for advice and encouragement (and good humor) along the way, as well as for firm guidance and a steady hand at the tiller when it came to steering it toward completion. I am likewise grateful to MCUP managing editor Christopher Blaker, whose patience and perseverance through an arduous editing process in the midst of major real-world developments (notably, the Russian invasion of Ukraine) can only be called

Acknowledgments

commendable, and who has likewise been a true pleasure to work with and an inspiration. Thanks to all of you, a somewhat rough-and-ready dissertation received the polish it needed to become a scholarly book for a distinguished audience.

I am also grateful to two anonymous peer reviewers for MCUP, whose input turned a crude manuscript into what I hope is a worthy finished product. I am grateful in particular for one reviewer's insights into U.S. Air Force colonel John R. Boyd's OODA (observe–orient–decide–act) loop and into late-nineteenth-century ship numbers and classifications.

As I note in the preface, I must also acknowledge several people who, sometimes unknowingly, led me to this project. First, I must thank Professor Emeritus Charles E. Pirtle of Georgetown University. A fearsome classroom presence and legendary taskmaster during my undergraduate years, Professor Pirtle first introduced me to leadership long-cycle theory, the intersection of naval power and geopolitics, and many of the strategic concepts contained in this work. It is a tribute to his rigorous and vigorous instructional methods that I have remembered these concepts (and a good many books) all these years.

Second, I thank Georgetown University adjunct professor Barry Watts, who instructed me in my studies for my master's degree in security studies at Georgetown, for helping plant the seeds of this project some years ago. It was Professor Watts, a former colleague of the legendary Pentagon strategic analyst Andrew W. Marshall, who, in his class on net assessment, introduced me to the problem of holistically and comprehensively understanding international competitive relationships and the larger problems of national security strategy. I am also grateful for his recommending me to the University of St. Andrews, without which this book would not have happened.

The greatest contribution to the genesis of this project came from Georgetown University adjunct professor Phillip Karber, whose definition of strategy I employed in this study. I am profoundly grateful to Professor Karber for his mentorship during the many years since I first took his security studies class

Acknowledgments

at Georgetown in 2004. Professor Karber was instrumental in helping formulate the proposal for this project, as well as in recommending me to the University of St. Andrews. I must thank him in particular for the use of his library (and some fine whiskey) on several cold, late nights in January 2011, where, after a crash course in the literature of business strategy, I first wrote the proposal for this study. He was also kind enough to take time from his busy schedule to provide me with several literature recommendations along the way. For all this and more, I am profoundly grateful.

I have dedicated this work in part to my teachers, all and sundry, whose instruction at so many points in my life led me to the opportunity to submit a thesis for consideration for a PhD at one of Scotland's (and Europe's) most distinguished universities. There are too many to name, but those who have played a part in my education up to now know who they are, and they have my sincerest thanks. It all came together in the end. I must thank in particular retired U.S. Air Force captain Robert S. Hopkins III, PhD, whose instruction many years ago set me on the path to the U.S. national security community and, by this way, to writing this book (and whose correspondence has been invaluable during its completion).

I must also thank my students in first- and second-year international relations at the University of St. Andrews, whose enthusiasm for the subject and original ideas in classroom discussions reminded me why I set out to do a PhD in the first place, and who offered me a chance to learn some of the international relations discipline's more intricate theoretical concepts in the best way possible: by teaching them. There is some of each of their thinking in this book.

Several people were kind enough to read the first complete draft of this work and offer advice on its completion. My old Georgetown colleague Matthew Merighi graciously took the time to read the first draft and offered many helpful edits along the way. My brother, Dana Skold, not only read the first draft and offered editorial assistance but also spent several

Acknowledgments

hours with me discussing its theoretical merits and demerits. I regret that space and time considerations prevented me from implementing some of his more ingenious suggestions, but his thoughts and arguments helped greatly in the final formulation of this thesis. Though, for a particular insight that need not be discussed in detail, he has the dubious honor and right of saying, "I told you so." And if discussion of one's ideas is half the fun of a project such as this, a major part of that fun I owe to him.

However, wherever errors are found, then, to borrow and adapt yet another quotation from the great poet of imperialism whose poetry adorns the above chapters: "Let not thy wrath befall them, but deal to *me* the blame."[2]

I must also thank numerous correspondents whose thoughts on the current round of great power competition also informed the final organization of this work, both through direct suggestions and the insight that comes from discussion: these have included Alex Almeida; Elbridge Colby; Adam Elkus; Kerry Ellard; Daniel Foubert; Samuel Goldman; James Griffin; Marine Corps lieutenant colonel Nathaniel Lauterbach; Damir Marusic; John Ryan McGehee; Patrick Porter; Nick Prime; Aris Roussinos; Mark Safranski; Michael Thadani; Farooq Tirmizi; Jonathan Ward; Avi Woolf; and many others. I thank Adam Elkus in particular for singular insight into John Boyd's strategic theories. I thank all these people in addition for promoting my thinking on statecraft and international relations theory in various online fora while this book was coming together.

In writing a book, as in war, the moral factor is paramount, and I am grateful in particular for the support of those friends who, in the preparation of this work, despite seeing me at far less than my best, offered advice and care. These have included, at various times, Alexander Blackford, Richard Booth, Alison Fitzgibbon, and Mary Stamp. As with all such things, the value of the small stuff can only be appreciated by those involved, and it was paramount here. Good friends, like many state competitive goals, have more value than can be quantified; I could not have asked for better.

Acknowledgments

My family—my parents, Craig and Mary Ann Skold; my sister, Whitney; and my brother, Dana—aided and abetted my obsession with international politics for many years and put up with my eccentricities along the way—all of which in the end led to this book. My love and deepest thanks to them all.

Most importantly of all, I must offer my most profound love and gratitude to my wife and soulmate, Christina Goodlander, who traveled with me to Britain and was with me every step of the way. For her patience with me through long and odd hours of study, for her encouragement in dark moments, for the major sacrifices she made on my behalf so that I might complete this book, for countless hours spent discussing international politics in all its many forms, for advice on numerous theoretical and practical matters, and above all for the constant and boundless love she showed me throughout—and for all the joys of marriage which are too many to name—I can only say that I have been blessed. This work, at the end of it all, was, as it should be, a Team Skoldlander production.

Acknowledgments

Endnotes
1. Henry Newbolt, "Vitaï Lampada," in *Admirals All and Other Verses* (London: Elkin Mathews, 1898), 21.
2. Rudyard Kipling, "Hymn Before Action," in *The Seven Seas* (London: Methuen, 1896), 103.

THE RACE WITH NO FINISH LINE

Introduction

Since the dawn of time, men have competed with each other—with clubs, crossbows, or cannon, dollars, ballots, and trading stamps. Much of mankind, of course, abhors competition, and these remain the acted upon, not the actors. . . . Anyone who says there will be no competition in the future simply does not understand the nature of man. . . . [E]ven on a lesser scale the game can be lost, or won. . . . We can lose the game not only because of the nature of our enemies, but because of our own. We understand we cannot ignore the competition, and realize with frustration that we cannot end it by putting our competitor out of business with a bang, but we will not willingly face the fact that we may walk along the chasm, beset by tigers, for many years to come.
~ T. R. Fehrenbach[1]

How do you win a race with no finish line?
~ Attributed to U.S. President Ronald W. Reagan, regarding the Cold War[2]

Around the year 1260 BCE, a great king of the Hittite Empire wrote to a king in Greece, demanding the extradition of a rebel vassal whom the Greeks had given asylum. The Greeks had been aiding this rebel in his effort to establish an independent kingdom in the west of Asia Minor, in what had once been Hittite territory. In the process, the Hittite king alluded to "the matter of Wilusa, over which we were in enmity," which had somehow been resolved. It is known from archaeology that for several decades preceding, both sides had sought control of the strategic coastal territory of western Asia Minor; both sought

Introduction

the submission and loyalty of its smaller kingdoms; and no final settlement could be reached. The final outcome of this prehistoric diplomatic exchange is not known, but the rivalry no doubt continued.

Wilusa, it is now believed, was the old name for the kingdom and citadel that the Greeks later came to call Wilios, and then Ilios: the fabled city that was the object of the legendary Trojan War, in real life a vassal state of the Hittite Empire. It is believed by archaeologists to have been destroyed about the time that the two kings exchanged messages. But although a province could be warred over, and perhaps even annihilated, rivalry between the greater states went on unabated. Their leaders were forced to make the best of such rivalry, managing it to their states' respective advantages as best they could.[3]

Great power competition—stopping short of major war, but observing few other constraints—is as old as Troy.

Three thousand years later, the leader of another great power, frustrated with a seemingly endless competition with a powerful adversary, asked his advisors, "How do you win a race with no finish line?" The answer: "Get the other runner to quit!"[4]

Such competition is the subject of this book, which offers a framework for analysis of the strategies of states that find themselves in enmity but do not immediately go to war. While, as will be noted, several studies have devoted themselves to answering various aspects of the question of why states compete, exceedingly little has been done up to now to answer the question of how they *may* compete. "How did we get here?" and "What do we do now?" are separate questions. The former has been extensively studied as it pertains to great power rivalry. The latter has yet to be explored, and this book offers a first attempt to do so. As will be shown, the first part of evaluating what to do is knowing what the game is about—what one should seek, and how badly or well one is playing at any given time.

This study offers a way of understanding and evaluating how badly or well great powers are doing when they compete for control of a region of the globe—for regional hegemony. As

Introduction

such, it considers a host of variables and theoretical perspectives, ranging from international relations theory, to military and business strategy, to classical political philosophy. It is an attempt to offer a guide to those who seek to understand this game. While strategy is an art and not a science, and no framework can ever guarantee a competitive victory, knowing how the game works is the first step to winning it.

The concept of great power competition has become highly salient of late, as the United States enters an era of increasingly intense competition with its major geopolitical adversaries. A buzzword in defense and foreign policy circles, great power competition has become a thriving subdiscipline in international relations and geopolitical thought.[5] This study cannot claim to offer a definitive rundown of the nature of global great power competition, a complex subject that awaits a future treatment. It seeks, rather, to offer a guide to one particular highly relevant area of great power competition—as noted, those frequent occasions where great power competition specifically concerns a given region. No great feat of imagination is required to note the particular relevance of this subject matter to U.S. foreign policy decision makers, as competition for control of the first island chain, the South China Sea, the Indo-Pacific, various strategic chokepoints, and Europe have all been cited as top concerns; equally, many other regional examples involving other powers will be readily apparent to the reader as well. This book is not intended as a set of policy prescriptions, but readers may well find that policy-relevant ideas emerge from its subject matter. This work cannot tell U.S. policy makers or citizens what grand strategy to pursue—but it can offer guidance on what it will take to prevail in any of these regional contests and ultimately what they are about. In the process, it may offer insight into how the United States is currently doing in each of them.

As to that subject matter, the key concept at hand is that of a framework for inquiry. The great U.S. Department of Defense strategist Andrew W. Marshall, who for most of his life ran the Pentagon's long-term strategic think tank, the Office of Net

Introduction

Assessment, was famous for asking, relative to the national security problems his office analyzed, a simple question: "How do we think about this?"[6] The question was not just an attempt at cleverness, despite Marshall's nickname of "Yoda" for his perspicacity. It was a way of defining a problem about problems. Strategy cannot proceed without an analysis of the problem it is trying to solve, and this, in turn, requires knowing which questions are important to ask. When one speaks of a strategic framework, one is talking about a list of areas of inquiry that will explain how a great power seeking control of a region is doing in that competition and what it might do to achieve success.

This may seem backward, in that strategy is supposed to answer the question of what to do and not merely what is going on, but the one question answers the other. Just as a chess player gains insight into how to achieve checkmate by learning how to analyze games, in terms of observing various uses for different pieces; memorizing openings, gambits, and other repeatable situations; and assessing various positions, so too in any other competitive environment, the key to learning how to compete is knowing how to assess the game. In this case, if one can understand how great powers competing for regional hegemony are faring in that game, one can, as a policy maker for such a power, adjust strategy accordingly. Ultimately, then, it is the goal of this study to produce a framework for analyzing a state's prospects when engaged in competition for regional hegemony.

Skeptics are right to ask what can reasonably be expected from this line of inquiry. An objection that must be answered is that theory, to be theory, must explain and predict—and, as noted, strategy presumes the absence of strict determinism. Indeed, to be a strategist is in a sense to wrestle with fate, and, to some extent, to embrace a "great man theory," in that one's actions are not foreordained and, to some extent, neither is the outcome.[7] This is not to say that structure does not matter or that agency always trumps it—but it is to say that agency is important. To that end, it must be admitted that a framework such as this necessarily yields questions and not (or not only)

Introduction

answers: it builds a picture of a situation and allows those looking at it to understand it; it will not necessarily predict outcomes by itself (although certain answers to the questions it provides will be highly predictive). Such a framework is, however, highly necessary, since it is by knowing which questions to ask that one knows how to assess a situation and, therefore, how to thrive in a competitive environment.

To draw an analogy: one cannot predict the outcome of a football game in advance, but one can guess how a team will fare. A team with an inexperienced quarterback will fare badly against a team with one with most-valuable-player status. A team whose best linebacker has suffered an injury is worse off than it was. A team with a poor defensive line will likely fail against one with a fantastic offense. And so on. All is never equal, but assessment of variables builds a picture: once one knows the game and knows how it is played, one can ask relevant questions (and avoid irrelevant ones) that create a picture of how a player will fare.

The inherent limits of this are, in effect, the limits of strategic theory in its own right. Strategic theory is about winning games. A strategic theory that solved a game for all possible positions (as has been done, for example, with checkers and tic-tac-toe) would void the game—there would be no reason to play. What can be done is to search for best practices, allowing for the inevitability that adversaries will learn to counter them and that new ones will need to be found. What is being done here, in the creation of a framework for this specific game—the game of great power competition for control of a region—is to provide the questions that best practices are supposed to answer. More, of course, will remain to be done.

Such frameworks already exist, in various forms, in business strategy, and these may serve as a guide to what is possible. As an example, Michael E. Porter's groundbreaking study of firm competition, *Competitive Strategy*, offers just such a framework in a different context. Porter posits what has become known as the Five Forces model as a means of understanding the competitive environment a firm faces, offering a means of analyzing each of five contributing factors behind the level of competition a firm is likely to face if it enters a given industry: the respective bargaining powers of suppliers and buyers, the possibilities of new market entrants and product substitution, and the already existing competition within the industry. Each of these forces can be analyzed in turn to produce a holistic analysis of the prospects a firm would face upon entering the industry—effectively, an assessment of its chances in the competition.[8]

With some exceptions, these concepts in the main are not directly transferable to interstate competition for regional hegemony. They do, however, show what is possible in terms of strategic analysis in this area, and how much work remains to be done. This study proposes a framework that offers for interstate competition for regional hegemony what Porter's models offer in an economic context: identify the factors that drive the competition and suggest a means of analysis that can guide strategy. Such a framework must take into account the competitors' goals and also the diverse metrics for costs and benefits just discussed.

This study initially examines the concept of competitive strategy and synthesizes, from across the full, broad spectrum of strategic literature, essential principles of competitive strategy that are applicable to an international political context. It proposes a framework for the analysis of competitive strategy as practiced by states competing for hegemony over a given region, articulating a set of principles by which such competitive strategy may be scrutinized and judged. Through a case study involving the competition of Great Britain against Germany for dominance of the North Sea prior to World War I, and the international political consequences that entailed, this study tests this analytical framework against historical reality, employing the principles derived from the literature to understand and evaluate the competition between the actors involved. Having done so, it is thereby able to propose a framework with which to approach similar competitive scenarios between great powers seeking regional hegemony.

Introduction

The strategies of the states involved are asymmetric and must be analyzed as such: one state, the current hegemon, is attempting in effect to win a race with no finish line (as the above-cited quote holds), outlasting and outcompeting its challenger, while the challenger must in effect create a finish line by forcing a favorable decision. The goals of the states involved in such competition involve not only regional hegemony, but the specific political consequences of acquiring it. Such being the case, the framework employs principles drawn from business strategy to understand the degree to which states are able to focus resources on the competition and to make use of their inherent competitive advantages and core competencies. Regarding such resource allocation, the framework particularly articulates a set of metrics, based on existing international relations theory, for understanding the costs and benefits incurred by states engaged in such competition, and employs decision cycle analysis, drawn from the military strategic theories of U.S. Air Force colonel John R. Boyd, to understand the ways in which states may influence the outcome of such competition in real time.

This study, like strategy, sits at the crossroads of multiple disciplines and may have something to offer to many kinds of readers. It integrates its argument with existing international relations theory and offers a contribution to it. It of course offers a contribution to the literature of strategy in an integrated fashion. And, as a qualitative as well as quantitative analysis of an historical episode, it offers an historical narrative and some commentary on it. Like strategy, this study uses all the means at its disposal in pursuit of its end. If, at last, it offers readers a way of thinking about regional great power competition that offers both insight into how it is accomplished and how any given example of it may fare, it will have done its work.

Chapter 1 of this book lays out the theoretical underpinnings of this framework, defining important terms and locating the concepts under discussion within the academic discipline of international relations theory. In particular, it offers a definition of strategy to be employed throughout this work and a brief

Introduction

discussion of the idea of strategy as it can be applied to regional great power competition. It then illustrates the absence of a practitioner's perspective on great power competition, which this book seeks to begin to offer: whereas much international relations theory has sought to explain state behavior in competitive environments, very little seeks to suggest what states might do in such situations or how to analyze their performance.

Chapter 2 lays out the ideas behind the framework. Using the definition of strategy employed here as a guide, it reviews and synthesizes concepts drawn from a range of military and business strategy literature that can be used to analyze regional great power competition and which are to serve as the basis for the framework's essential components.

Chapter 3 lays out the framework and explains how it works. Drawing on the analysis in chapter 2, it puts together a way of investigating regional great power competition and discusses how to do so.

The book then proceeds with a case study to test and demonstrate the framework—namely, the Anglo-German naval arms race prior to World War I, colloquially known as the "Dreadnought Race," which had as its objective a preponderance of power in the North Sea and a political arrangement in that region that would support a German breakout from its British Royal Navy-controlled waters. Though Germany came close to eclipsing Great Britain as the dominant power in northern European waters, in the end a flawed strategy and a timely British response prevented it from doing so. It did not, however, prevent the Anglo-German enmity from becoming a contributing factor to the outbreak of war, which is ultimately given consideration in the analysis of the case.

Chapter 4 lays out the origins and aims of the Dreadnought Race, allowing for an understanding of its parameters for purposes of applying the framework. Chapter 5 offers a narrative of how the Dreadnought Race played out. Efforts are made in both chapters to locate various historical claims made within

Introduction

historiographical debates concerning them, though an effort is also made to avoid being bogged down on particular points.

Chapter 6 applies the framework. Drawing on the preceding narrative as well as quantitative analysis of the competitors' military and financial positions, it investigates the competition using the framework as a guide. It draws conclusions about the performance of the competitors—Great Britain and Germany—in light of this analysis, showing how the framework can serve as a guide to understanding their competition not only in retrospect but also, if it were to be replayed, in real time.

The conclusion to this study offers some insight into how the framework may be applied in the present day, as well as some general lessons that the foregoing analysis offers for strategy and statecraft. Fundamental to the idea of investigating a phenomenon—in this case, regional great power competition—using a framework focuses on the idea that important questions remain important, and asking and answering them with regard to one situation may yield lessons for others. It is hoped, here, that this framework will be the beginning of future inquiry into the dark but vital art of great power competition for supremacy in key areas of the globe.

Introduction

Endnotes

1. T. R. Fehrenbach, *This Kind of War: The Classic Military History of the Korean War* (New York: Open Road, Integrated Media, 2001), Kindle ed., loc. 10378–91.
2. Phillip Karber, conversation with author, January 2011; Phillip Karber, "Competitive Strategy: As an Approach to Business and Professional Life" (PowerPoint presentation at the Annual Fellows Conference of the Center for the Study of the Presidency, Washington, DC, 31 October 2003), slide 7; and Phillip Karber, "The 'Counter-Offensive' in Competitive Strategy: Lessons from the Reagan Era" (PowerPoint presentation presented at the Developing Competitive Strategies for the 21st Century Conference, U.S. Naval War College, Newport, RI, 23 August 2010), slide 11.
3. Michael Wood, *In Search of the Trojan War* (Berkley: University of California Press, 1996), 181–83, 186–88, 205–6.
4. Karber, "Competitive Strategy," slide 7; and Karber, "The 'Counter-Offensive' in Competitive Strategy," slide 11.
5. See, for example, Mahir J. Ibrahimov, ed., *Great Power Competition: The Changing Landscape of Global Geopolitics* (Fort Leavenworth, KS: U.S. Army Command and General Staff College Press, an imprint of Army University Press, 2020); and Thomas F. Lynch III, ed., *Strategic Assessment 2020: Into a New Era of Great Power Competition* (Washington, DC: National Defense University Press, 2020). The concept of great power competition entered U.S. foreign policy discourse in a central role with the publication of the 2017 U.S. *National Security Strategy*, authored primarily by Nadia Schadlow; see Donald J. Trump, *National Security Strategy of the United States of America* (Washington, DC: White House, 2017). It is now a well-established area of concern in U.S. national security circles. A major, and valid, criticism of the concept is its lack of salience to the American public, which is inadequately consulted about ultimate ends or kept in the loop about means employed and risks run; see, for example, Tanner Greer, "Introducing: Asabiyah," Scholar's Stage, 2 May 2015; and Tanner Greer, "You Do Not Have the People," Scholar's Stage, 3 March 2018. While this book cannot remedy all of these deficiencies, it is hoped that American readers in particular may derive some value and useful insight from it.
6. The anecdote regarding this question is widely circulated among alumni of the Office of Net Assessment. The mentality behind it is implicit in Andy W. Marshall, *Long-Term Competition with the Soviets: A Framework for Strategic Analysis* (Santa Monica, CA: Rand, 1972).

Introduction

7. The author owes this insight to Adam Elkus, who at one point noted that Italian diplomat Niccolò Machiavelli in particular saw politics as an exercise in bucking fate.
8. Michael E. Porter, *Competitive Strategy: Techniques for Analyzing Industries and Competitors* (New York: Free Press, 1998), 3–29.

Chapter One
On Competition and Strategy

Toward the end of the Warring States period... a ruler probably turned to a trusted minister and asked him to gather and record the wisdom circulating in Sun-tzu's name for use in formulating strategy and conducting campaigns. Perhaps the minister himself, despairing of his times, undertook the project on his own, to urge the ruler to better his performance. He would have queried scholars and military experts, sending scribes to interview them, if the budget of the court allowed. . . . Gathering the many sayings that had appeared over the centuries, he would have pared them down to those he thought genuine—and then discarded those he felt might lead his ruler down dangerous paths. . . . Perhaps the courtier was charged only to gather the general military wisdom of the past for a young ruler anxious to make his mark on the world. Perhaps, to that end, he "discovered" Sun-tzu, a wise ancient— perfect for a culture that revered old wisdom above contemporary insight. Whatever the details, a close reading of the text convinces that the chronicler was not content with what he had gathered: before placing the work in the hands of his ruler, he tempered the military prescriptions and proscriptions with . . . maxims that advanced his own view of the world.
~ Ralph Peters[1]

Any theory of strategy in international relations must wrestle with the concept of strategy writ large, as well as other definitional questions. To that end, some analysis is offered here of the important terms used in this study. First, competition as a concept must be defined and understood in its proper context. Following this, it is necessary to evaluate strategy as a concept and trace its conceptual development, with particular emphasis on its essential characteristics that cut across disciplines and fields. It will be shown here that strategy, although originating in the military realm, must be defined abstractly, and as such serves as both a heuristic for and a means of engaging in competition in multiple areas of human endeavor. Following this, it is necessary to address the nature of regional hegemony—the competitive goal at the end of the strategic decisions being discussed here.

This chapter will be a review for many. For students of the international relations discipline (IR, as it is occasionally called here), discussion of concepts such as regional hegemony is likely dilatory. Equally, for students of strategy, very little here is likely to prove new. It may be said that, inasmuch as this study integrates strategic concepts from several different strategic disciplines, business strategists may benefit from a review of the military side, military strategists may find something new in business strategy, and students of grand strategy may or may not find either of the above to be new. Since the goal is to integrate all these approaches and use them to build a framework for the analysis of a problem in IR, the reader's patience is begged for purposes of showing how these concepts do in fact intertwine. Once this has been accomplished, the next chapter can then show how this intertwining can be done. And, if it is sufficiently obvious that these concepts work together, readers are free to skip to the next chapter to see what is being attempted with them.

Competition as a Midpoint on a Continuum of Relationships

The concept of competition between human organizations would appear to be so deeply rooted in human nature as to need no introduction. It is necessary here, however, to offer a conceptual understanding of it as it pertains to states. Competition, as used here, should be understood as a midpoint between two extremes of human interaction: *cooperation*, in which actors pursue goals jointly and without the intent of hindering one another, and *conflict*, in which actors pursue goals antagonistically and may seek to harm one another. An alliance is cooperative; a war is a state of conflict. But a middle set of options in which actors neither work together nor directly damage each other is possible. For the purposes of this study, this state of interaction is defined as *competition*.[2] In a competitive relationship, the actors involved seek to obtain access to a scarce commodity or prize and often to hinder others' access, but they generally stop short of directly harming other players trying to do the same.[3] In effect, competitors are racing—albeit often, per the quotation referenced in the introduction to this book, without a finish line—toward some goal that they cannot both have. It is zero-sum, but not to the extent of imminently threatening the existence of both actors. It is not win-win, although it may be ended by negotiated settlement if some set of conditions prevails. It need not end, however, unless one side conclusively "wins" and the other side—per the classic answer to the question of how to win a race with no finish line—decides not to compete anymore.

Michael P. Colaresi, Karen Rasler, and William R. Thompson address international competition conceptually in their book, *Strategic Rivalries in World Politics: Position, Space and Conflict Escalation*, as "strategic rivalry." The authors understand rivalry in more purely security-related terms—namely, as a situation in which states are militarily hostile. Inasmuch as competition for regional hegemony typically involves at least this much, the concepts are comparable if not identical. In particular, they offer the following four-part conceptualization: that strategic rivalry involves incompatible goals; that the rivalry is ongoing as a "stream of conflict"; that the stream of conflict drives intersubjective perception and ultimately mutual hostility; and finally that, despite this, rivalry can differ considerably in intensity from one case to the next. The authors cite variances in capabilities as a major factor in the course that rivalries take. They also note that rivalries can take what they refer to as "positional" or "spatial" forms, depending on whether the rivalries in question are for actual positions within the international system (principally the concern of great powers) or merely control over certain areas (where regional powers may be more involved).[4] The subject has also been treated by Paul F. Diehl and Gary Goertz, who take a similar approach, referring to rivalries as "enduring" and "militarized" and proposing in particular that investigation of international conflict should refer to rivalry and not merely to war, which should properly be understood as its output.[5]

Colaresi, Rasler, and Thompson are principally concerned with how such rivalries develop and where they lead—Rasler and Thompson in particular, along with Sumit Ganguly, have offered a model for how such rivalries end (to be discussed in the next chapter). Diehl and Goertz are similarly concerned with what rivalries produce. It is enough to note here that, while this study is particularly framed around control of a given regional territory, there is considerable overlap in framing, and credit is due to these authors for foregrounding and examining the issue. Ultimately, competition for regional hegemony is, indeed, a form of such rivalry. As will be discussed, it is principally what Colaresi, Rasler, and Thompson would refer to as a spatial rivalry, but its positional implications form part of the analysis that is enabled by the framework offered in this study. This study, however, as will be discussed, takes a slightly different direction, being principally concerned with how to analyze and, if one represents a state participating, win such a competition, rather than merely what such rivalry entails.

Strategy: A Problematic Concept

The concept of strategy is a much more complex issue, however. Although the next chapter will address key concepts within strategic literature that pertain here, the basic definition must first be understood. In point of fact, the term applies to multiple areas of human endeavor, and while the connections between them are key to this study, it is important to understand exactly where those connections occur.

Strategy, for these purposes, is defined as "a plan of action and process of decision making for the allocation of resources in anticipation of a contingent event, orchestrating simultaneous and sequential engagement, to achieve an organization objective, in the context of a contest with other organizations."[6] As Colin S. Gray notes, strategy can be understood as a "bridge" on which "theory and practice meet": it is neither a goal (broadly understood as a policy), nor the actions taken to achieve it, but rather the plan that connects available means to sought ends.[7] Hew Strachan echoes the point, noting that it is easy in the present day to confuse *policy*, a statement of a state's objectives or desires, with *strategy*, which "lies at the interface between operational capabilities and political objectives: it is the glue which binds each to the other and gives both sense."[8] He uses public statements on the Global War on Terrorism made by the administration of U.S. president George W. Bush as a prime example. Strachan and Gray are concerned primarily with *military* strategy—the original usage of the term. To understand how the term evolved and to examine how its various meanings may be synthesized, it is necessary to look at its beginnings in the military realm before branching out to its later usages. It is quite obvious that there is no shortage of conceptual murkiness associated with the definition of strategy, and that on various points full agreement has been elusive. Nevertheless, a discussion of the various definitions and conceptual elements of strategy that have appeared over the years will at least show that certain points are agreed on, and that these points can serve as the rudiments of a definition. In brief, what appears to draw broad agreement is that strategy is about resource allocation when dealing with an adversary, first at the beginning of the adversarial relationship and then in the course of it.

As the term strategy originated in a military context, it is useful to first discuss its military application before moving on to its related applications. Perhaps the earliest author associated with the term is the historically ambiguous Chinese general and military writer Sun Tzu, whose short, 13-chapter work, with a title usually translated *The Art of War*, lays out a set of principles governing decision making at both the political and operational level of warfare, advising rulers and generals as to how to make best use of scarce resources.[9] Sun Tzu's ideas will be discussed in more detail in the next chapter, where they form a major part of the theoretical basis for this study. For the purposes of defining strategy as a concept, however, one must understand the term as it originated in Western thought, as the term strategy is of inherently Western coinage (and was obviously not used by Sun Tzu).

In brief, the term strategy is an outgrowth of a particular military school of thought that came to dominate Western—and then worldwide—warfare at the time of the Enlightenment era. As the military historian Martin van Creveld notes, the near-extinction in the West of the Greek language, from which the word strategy derives (via *stratos*, for army; *strategos*, for an army commander; and *stratagema*, for what a clever army commander might do), after the fall of Rome in the fifth century meant that the word was unknown to Western warfare until the early modern era. Medieval, Renaissance, and even early-Enlightenment military writers, including Niccolò Machiavelli and Frederick the Great, preferred *art of war* or similar terms. The Enlightenment-era article of faith that all human activity could be minutely analyzed led to the treatment of war in "scientific" terms, and by ultimate extension to the idea that commanders' plans and decisions—their strategies—could be objectively analyzed. Van Creveld cites an obscure eighteenth-century French military writer, Paul-Gédéon Joly de Maizeroy, as the first employer, if not the coiner, of the term *strategy* to describe the science of making military command decisions; he notes that it was in approximately the late eighteenth and early nineteenth centuries that a meaningful and now-familiar distinction between strategy (military planning) and tactics (what armies did on the battlefield) was made in European lexicons.[10] In this way, although the point is not always made, there is an intimate, and even confusing, connection between the plain-language usage of the term *strategy* to describe planning a campaign and the usage of the word to describe the discipline of scientifically studying military theory and affairs. As will be shown regarding business strategy, it is easy to lump together the study of, and quasinormative prescriptions about, actual decision making in a game on the one hand and objective analysis of the game on the other. This study, by analyzing the strategies of actual states as they engaged in competition, ultimately offers a route toward analyzing such competition from an outside perspective.

While defined slightly differently by each writer on the subject, the term strategy, in a military context, normally involves a few basic elements that are seen to be essential, and it is useful to examine some of the more eminent strategic literature to understand what those elements are. The term strategy is used by the most preeminent military theorist, Prussian general Carl von Clausewitz, as being differentiated from *tactics*, which in the aggregate amounts to the means strategy employs. In Clausewitz's formulation, "According to our classification, then, tactics teaches the use of armed forces in the engagement; strategy, the use of engagements for the object of the war."[11]

Per Clausewitz, "The original means of strategy is victory—that is, tactical success; its ends, in the final analysis, are those objects which will lead directly to peace."[12] Later, he reiterates:

> Strategy is the use of the engagement for the purpose of the war. The strategist must therefore define an aim for the entire operational side of the war that will be in accordance with its purpose. In other words, he will draft the plan of the war, and the aim will determine the series of actions intended to achieve it: he will, in fact, shape the individual campaigns and, within these, decide on the individual engagements. Since most of these

matters have to be based on assumptions that may not prove correct, while other, more detailed orders cannot be determined in advance at all, it follows that the strategist must go on campaign himself. Detailed orders can then be given on the spot, allowing the general plan to be adjusted to the modifications that are continuously required. The strategist, in short, must maintain control throughout.[13]

For Clausewitz, therefore, strategy is not simply the means of planning military success through a series of tactical engagements, but also the process of command that allows a leader to see the plan through to its successful conclusion. For Clausewitz, the strategist is also the commander, a point reiterated by other writers in other contexts (as will be shown later in this section) and of importance to this general discussion. In brief, then, Clausewitz's conception of strategy boils down to the following: that strategy in warfare (Clausewitz was unconcerned with the usage of the term in other contexts) involves planning the usage of battles and engagements (which Clausewitz viewed as essential to warfare) to achieve the end sought in war (the political objective to which military effort was to be subordinated); that strategy involves planning but also improvisation and on-the-spot decision making when planning is impossible or overtaken by events; and that strategy is differentiated from tactics in that the latter is solely the process of decision making necessary to win the battles that the strategist employs to win the war.

Other writers echo Clausewitz's basic intuition regarding the key elements of strategy, though with some variation. For B. H. Liddell Hart, strategy is " 'the art of distributing and applying military means to fulfill the ends of policy.' For strategy is concerned not merely with the movement of forces—as its role is often defined—but with the effect."[14]

Edward N. Luttwak discusses several varying definitions of strategy in a military context in the appendix to his book

Strategy: The Logic of War and Peace. He notes that several definitions involve not only the use of battle to win a war but also the preparation and creation of military forces.[15] Colin Gray, employing Clausewitz's principles while expanding their scope, views strategy as "the bridge that relates military power to political purpose; it is neither military power per se nor political purpose. By strategy I mean the use that is made of force and the threat of force for the ends of policy."[16]

Strategy in the military realm, properly understood, must therefore include the broader questions of means and ends that ascend to the political level of decision making, and by so doing incorporate the overall political objective of the war, not merely the immediate military goals. This is not inconsistent with Clausewitz, and may even correctly account for his understanding of the term, since Clausewitz argued that military operations must be subordinated to political goals.[17] What is clear, moreover, is that strategy—again, consistent with Clausewitz's understanding of the term—is best understood as an iterative process, a process of assimilating information and making decisions even after plans have been made.

The concept of strategy expands when it incorporates grand strategy, which is said to involve a state's military resources, but not exclusively. As noted by Peter Paret, strategy can be seen as something that not only generals but also politicians engage in, as it involves the planning and mobilization of a state's resources.[18] The term *grand strategy* has essentially two, not entirely related, meanings. It can refer to the mobilization of all a state's resources in the context of a major war or in preparation for one—the employment not only of the state's military, but of all other means at its disposal to achieve its war aims. It can also refer to the very broad decisions made when states are not at war regarding what policies to adopt, with whom to align oneself, what wars to fight, how strong one's armed forces should be, and so forth. In the first sense, the term was apparently popularized by British Army officer and military historian

Chapter One

J. F. C. Fuller, who discussed it in his 1923 book, *The Reformation of War*:

> The first duty of the grand strategist is, therefore, to appreciate the commercial and financial position of his country; to discover what its resources and liabilities are. Secondly, he must understand the moral characteristics of his countrymen, their history, peculiarities, social customs and systems of government, for all these quantities and qualities form the pillars of the military arch which it is his duty to construct.[19]

The concern for the overall arranging of national resources to meet a threat in peacetime and in war grew out of the emergence of total warfare in World War I and developed into a fine art in World War II. As the strategic scholar Hew Strachan notes, regarding Fuller's argument and the general sentiment that drove it,

> Strategy was now to be applied in peacetime, since how a nation fought a war would in large part be the product of the preparations, planning and procurement it had done in peacetime. Grand strategy was what Britain and its allies put into effect in the Second World War. It was the application of national policy in the war, and it involved the coordination of allies and of efforts in different theatres of war.[20]

Note, once again, the merger between preparation and actual military activity.[21] Although the military developments that drove this conception of grand strategy are outside the scope of this study, the foregoing discussion should serve to establish how the term grand strategy has been used in a wartime context.

The second sense—the allocation of national political priorities on the world stage—appears to now be the more common usage of the term grand strategy. For the historian Paul M. Kennedy, the study of grand strategy (in the context of the study of European states' policies) referred to "assessments of the success or failure with which various powers of Europe sought to integrate their overall political, economic, and military aims and thus to preserve their long-term interests."[22] Grand strategy, in U.S. public discourse, is seen by Kennedy to encompass "the proper balance of priorities . . . that should be carried out by the United States in the world today."[23] As will be discussed later in this chapter, it is precisely this understanding of grand strategy—as priority-allocation—that distinguishes it from what this study calls *competitive strategy*: political-level resource allocation once a decision to compete has been made.

For Luttwak, in comparison, "at the level of grand strategy, the interactions of the lower, military levels yield final results within the broad setting of international politics, in further interactions with the nonmilitary relations of states."[24] Grand strategy, therefore, in Luttwak's formulation, concerns resource allocation when dealing with potentially adversarial relationships within the international system. This second sense of the term grand strategy—the allocation of resources at the political level of conflict—is closer to the way in which competitive strategy as a concept is employed in this study, although Luttwak does not offer a detailed framework for its analysis. For now, it is adequate to point out that strategy, even in a purely military context, can refer to nonmilitary planning and decision making.

Since the term strategy came into usage two centuries ago, and also since the term grand strategy elevated strategy to the political level of decision making, strategy has acquired a wider meaning. It can be abused. As Strachan notes:

> The word strategy has acquired a universality which has robbed it of meaning, and left it only with banalities. Governments have strategies to tackle the problems of education, public health, pensions and inner-city housing. Advertising

companies have strategies to sell cosmetics or clothes. Strategic studies flourish more verdantly in schools of business studies than in departments of international relations.[25]

Despite Strachan's legitimate concerns regarding stretching the term beyond its bounds, strategy has come to be adopted as a descriptor for a similar process of planning and resource allocation outside a strictly military context. As Strachan notes, business strategy is a thriving discipline, and one with its own strategic literature, which, *mutatis mutandis*, uses the term in a similar fashion to its original military usage. How this came to be must now be addressed. In general, it may be noted that the advent of modern economics, combined with the advent of large firms that needed to make high-level competitive decisions, drove the adaptation of strategy from its original military context to that of economic competition, and in turn from a focus on destroying an adversary to a focus on outcompeting them.

The relevance of this for great power competition should be obvious enough. It may nevertheless be objected to here that business strategy and military strategy are distinct fields. This is assuredly true, but not only is there considerable overlap in conception (they are about allocating resources in a game), but business strategy is, as will be shown, actually more relevant than pure military strategy to competition between states. This is so because a state, like a business, is a going concern that, in normal circumstances, is not at war—though, as is frequently noted, some states are indeed more conflict-prone than others—whereas a competitive, vice strictly conflictual, environment involves the same concept of resource allocation as a firm in a competitive market, albeit, as will be discussed, with a more complicated understanding of what the bottom line is. As will be discussed in the next chapter, numerous concepts from business strategy are indeed applicable to statecraft, and in particular to great power competition for regional hegemony.

A short review of how the business discipline understands strategy is therefore provided here for greater understanding. As Walter Kiechel III recounts in his book, *The Lords of Strategy: The Secret Intellectual History of the New Corporate World*, regarding the intellectual and practical founders of the corporate strategic consulting field, the term strategy came to be used in a business context in approximately the 1960s. Two authors in particular, namely H. Igor Ansoff and Alfred D. Chandler Jr., are credited with introducing the concept to the study of management. As Kiechel notes, both authors offer definitions of strategy that are subject to opposing lines of criticism.[26] Ansoff avoids offering a single-sentence definition, but, citing the game theorists John von Neumann and Oskar Morgenstern, gives the following description:

> The concept of strategy is given two meanings. A pure strategy is a move or a specific series of moves by a firm, such as a product development programme in which successive products and markets are clearly delineated. A grand or mixed strategy is a statistical decision rule for deciding which particular pure strategy a firm should select in a particular situation.[27]

To Chandler, by contrast, strategy "can be defined as the determination of the basic long-term goals and objectives of an enterprise, and the adoption of courses of action and the allocation of resources necessary for carrying out these goals."[28]

As Kiechel notes, the two theorists are quite different in their initial approach. Chandler's concept is very similar, allowing for a difference in context, to Clausewitz's, essentially amounting to making actions follow from goals and objectives so that, when resources are handed out, the objectives may be achieved. It is broad and intuitive in scope, and so, in Kiechel's words, "did not offer much guidance to practitioners who might want to emulate his corporate examples."[29] The exact application of the concept of strategy, in any sphere, and the means of doing so, is the subject of the following chapter of this book; it is noteworthy here, however, that Chandler's definition echoes its military predecessor. By contrast, as Kiechel points out, Ansoff, who was concerned with the planning of operations and employed a hyper-detailed approach to analyzing decision making, defines strategy more in terms of setting "rules and guidelines."[30]

Kiechel notes that strategy, in a business context, originated in a practical sense as "strategic planning," concerned with what a military organization might call "staff work," rather than the decision making of the organization's senior leader.[31] Kiechel credits the business strategist Henry Mintzberg with rebutting the premise that strategic planning, properly understood, constitutes strategy per se. In Mintzberg's formulation, strategic planning (again, similar to military staff work) is organizational and analytic—breaking down problems into smaller pieces for information and decision purposes—whereas strategy, in its truest sense, is creative, intuitive, and sometimes synthetic, involving continuous decision making, judgments based on imperfect information, and reactions to circumstances, as well as the generation of new ideas for implementation.[32] One may note the similarity again with Clausewitz's assertion that the strategist must be the commander—given that situations are in flux and data is not always available, decisions must be made on the spot, with imperfect information, that nevertheless are expected to conform to the overall set of goals and account for available resources.

For another of the foundational theorists on business strategy, Kenneth R. Andrews, strategy does indeed constitute a planning function, although Andrews' definition appears to allow for modification in the course of competition. In his book, *The Concept of Corporate Strategy*, Andrews offers the following formulation:

> Corporate strategy is the pattern of decisions in a company that determines and reveals its objectives, purposes, or goals, produces the principal

> policies and plans for achieving those goals, and defines the range of business the company is to pursue, the kind of economic and human organization it is or intends to be, and the nature of the economic and noneconomic contribution it intends to make to its shareholders, employees, customers, and communities.[33]

Andrews does allow that, as with Clausewitz, it is the organization's leader (general manager, in Andrews' formulation), not its planning staff, who is the strategist in practice, and notes, again in a manner similar to Clausewitz, that "[c]hanging circumstances and competition produce emergencies upsetting well-laid plans. Resourcefulness in responding to a crisis is a skill which most successful executives develop early."[34]

Despite this, for Andrews, albeit to a lesser extent than for Ansoff, strategy is mostly about planning, not merely adapting to changing circumstances. In actuality, the relationship of planning to strategy is more nuanced than it may at first appear. Mintzberg notes in particular that strategy can be defined as planning, but it is equally correct to describe strategy as a pattern—that is, as a tendency in organizational decision making to favor particular objectives and choices.[35] Mintzberg actually favors a midpoint between these two understandings of strategy:

> For, after all, perfect realization implies brilliant foresight, not to mention inflexibility, while no realization implies mindlessness. The real world inevitably involves some thinking ahead of time as well as some adaptation en route.[36]

Mintzberg anecdotally cites an unpublished master of business administration (MBA) thesis by Claude Dube which notes that the Canadian armed forces from World War II onward discovered that they could plan or act, but not both: planning amounted to fighting battles on paper in peacetime when there were no real battles to be fought, and for this reason amounted to essentially giving an idle organization something to do while it waited.[37] Strategy, therefore, may be understood to involve planning as one goes rather than simply trying to predict and then control the future and future decisions with perfect accuracy. Moreover, in its organizational usage, strategy cannot be divorced from certain universal human traits—notably, the need to do something about the future even when one has little control and less to do.

Moreover, while Andrews' understanding of strategy accounts for "emergencies" that competition can produce, this earlier understanding of strategy did not incorporate the relationship of an organization to its competition and competitive environment to the same extent that later theorists of corporate strategy would. Kiechel recounts the generally accepted view that strategy, in a practical sense, emerged as a discipline and practice within the private sector as a result of its adoption by the newly formed Boston Consulting Group (BCG) as the firm's focus of operation and, in the form of consulting advice on the subject, its major product. Bruce D. Henderson, the founder of BCG, created the firm in 1963 as a strategic advisory business; a popularly recounted boardroom legend holds that Henderson, in the course of discussing the new firm's potential markets with his partners, was told that strategy, as a service to be sold to customers, lacked definite scope, whereupon he replied, "That's the beauty of it. We'll define it."[38] Perhaps true to form, the firm—and the larger strategic consulting industry—that Henderson founded did not produce a single definition of strategy. Henderson's initial emphasis, however, was on relating the decisions made by the firms that were BCG's clients to their competitors' actions (most notably in the area of cost analysis), something that had not heretofore been done, and that had in fact been largely ignored by Chandler and Ansoff.[39]

The involvement of a competitor's actions in the concept of strategy broadens the scope of the concept considerably. Leading theorists who have considered this aspect include

Michael E. Porter and Kenichi Ohmae. The application of Porter's work on competitive advantage will be discussed in the next chapter.[40] Ohmae, meanwhile, envisions strategy solely in terms of beating the competition, noting that

> What business strategy is all about—what distinguishes it from all other kinds of business planning—is, in a word, competitive advantage. Without competitors there would be no need for strategy, for the sole purpose of strategic planning is to enable the company to gain, as efficiently as possible, a sustainable edge over its competitors. Corporate strategy thus implies an attempt to alter a company's strength relative to that of its competitors in the most efficient way.[41]

It is noteworthy that Ohmae is not simply discussing the way to achieve a company's objectives, or a means of staying profitable or breaking even, but rather a means of actively gaining a victory over the competition in a zero-sum game for market share. One can debate whether such a focus is productive for a firm; certainly, a respectable argument can be made that sustaining a competitive edge is one means to a firm's ultimate end (which involves a bottom line) rather than an end in itself. That said, Ohmae is consistent with the theorists cited above, and with the general evolution of the concept, in identifying strategy as something that in its essence is created and employed in a competitive context.

Perhaps the grandest attempt at a synthesis of all previously existing theoretical understandings of strategy is put forward by Lawrence Freedman in his magisterial recent work, *Strategy: A History*. Freedman's emphasis on the role of adversarial relationships—he uses the word "conflict" in its broader sense—imperfect information, and reaction to surrounding situational variables offers a broader understanding of strategy than the individual authors cited above, without contradicting the overall picture formed by synthesizing their work. In Freedman's words,

> A productive approach to strategy requires recognizing its limits.... As strategy has become so ubiquitous, so that every forward-looking decision might be worthy of the term, it now risks meaninglessness, lacking any truly distinguishing feature.... [However], [i]t really only comes into play when elements of conflict are present.... [A]t moments of environmental instability, as latent conflict becomes actual, when real choices have to be made ... something resembling a true strategy become[s] necessary. So what turns something that is not quite strategy into strategy is a sense of actual or imminent instability, a changing context that induces a sense of conflict. Strategy therefore starts with an existing state of affairs and only gains meaning by an awareness of how, for better or worse, it could be different. This view is quite different from those that assume strategy must be about reaching some prior objective.... This is why as a practical matter strategy is best understood modestly, as moving to the "next stage" rather than to a definitive and permanent conclusion.[42]

Freedman's understanding of strategy is valid, and not entirely inconsistent with the elements just presented, insofar as it seems to require a fluid plan modified on the fly by a leader making decisions in pursuit of an objective while working against an adversary. The main point of disagreement appears to be that Freedman views strategy as a mental planning process that can exist with complete fluidity, with even ultimate ends changing as necessary. Contrary to Freedman's view, this study adheres to the idea of strategy as the allocation of resources towards a definite end, rather than mere reaction to circumstances—but it is clear that there is room for discussion on the merits of one conception or the other.

The above-cited definition of strategy that is to be employed in this study—a plan of action and process of decision making for the allocation of resources in anticipation of a contingent event, orchestrating simultaneous and sequential engagement, to achieve an organization objective, in the context of a contest with other organizations—is a synthesis of key elements of strategy across disciplines and contexts. Phillip Karber's "Competitive Strategy" presentation, from which the majority of the definition derived, explicitly noted the transferable nature of the principles of strategy in various contexts.[43] In effect, the definition attempts to account for the majority of the basic elements of strategy so far discussed. It recognizes that strategy, far from merely being about linking battles to a campaign objective, is ultimately a question of making resources do what they are supposed to do, first on paper as a plan and then in action as ongoing decision making. It also recognizes that strategy is most meaningful as a concept when there is an opponent who is making decisions of their own.

Where differences exist, however, they must be accounted for. Although the definition used here describes strategy as a plan of action, this must be understood, per Mintzberg and others, as being continuously subject to modification as circumstances demand, not a rigid decision path determined in advance—for this reason, it is also understood as a process of decision making.[44] Moreover, as will be shown, a strategy can be more or less coherent or consistent, especially when made by committee. To decide badly or inconsistently is still to decide. The larger point is that one thinks ahead and makes decisions based both on old plans and new information. Similarly, this definition does not adopt Ohmae's seemingly extreme position that the sole purpose of strategy is to gain an edge on one's competitor. Rather, it should be understood as accepting that the final goal of strategy is whatever that strategy's formulator has set out to achieve while in competition with a rival organization, not competitive advantage for its own sake.

It remains to say a little regarding locating this study's usage of the term strategy within the range of uses to which it is commonly put. As noted above, this study does not address grand strategy in the sense of determining one's place in the international system and allocating resources to account for the maintenance of that place. Rather, it concerns *competitive strategy*: strategy applied to a competition between states for regional hegemony once the decision to engage in competition has already been made.[45] Although grand strategy can have relevance for competitive strategy, a distinction between these concepts is maintained here and is discussed in detail below.

The Concept of Regional Hegemony

It is likewise necessary to discuss the concept of regional hegemony that is used in this study. As noted, it is the purpose of this study to analyze states' competitive strategies in the hopes of understanding how they may play the game of international competition badly or well. Because strategy is understood as the process of applying specific means to the pursuit of specific ends, and because such ends are idiosyncratic, any analysis of strategy must seek principles that transcend specific instances. The case study considered here—involving Great Britain's competition with Germany for dominance of the North Sea prior to World War I—attempts to analyze first the degree to which the principal players achieved their goals and then the broader lessons for competitive strategy and the analysis of it that may be drawn from such understanding. However, because it is argued here (as will be discussed in more detail in the next chapter) that states competing for regional hegemony, as Great Britain was with its strategic rivals in each of the cases cited, must adopt asymmetric strategies—the one playing a long game and the other forcing a short one—the concept of regional hegemony deserves some discussion here.

The concept of *hegemony* in the international realm is a contested one in international relations theory, with multiple definitions having been advanced. The most conservative type of definition is probably that offered by certain adherents of the realist school, with that of John J. Mearsheimer being the most prominent. As characterized by Mearsheimer,

> A hegemon is a state that is so powerful that it dominates all the other states in the system. No other state has the military wherewithal to put up a serious fight against it. In essence, a hegemon is the only great power in the system. . . . The best outcome a great power can hope for [Mearsheimer does not believe global hegemony is possible as defined] is to be a regional hegemon and possibly control another region that is nearby.[46]

However, other schools of thought (even within the realist sphere) note that hegemony has a social and moral, as well as physical, component to it: a *hegemon* is, as the word suggests, a leader, possessing not just a preponderance of capability but also the influence that comes with it. For the hegemonic stability realist Robert Gilpin, echoing E. H. Carr, although hegemony constitutes possession of a predominance of power, "prestige" is more important than power, since a state's reputation for enforcing its will preclude a constant need for it to do so.[47] Further down the spectrum, long-cycle theorists such as Karen Rasler and William Thompson understand "global leadership"—they prefer this term over its Greek equivalent—or "world power" status to combine concentrated hard power (they cite "global reach capability") with an element of social consent to the system's norms. In particular, the leader's hard power "then enable[s] the world power to develop policies and rules for the management of the world's economy."[48] Perhaps the best articulation of hegemony as a concept involving more than mere possession of disproportionate capability is that of Torbjørn L. Knutsen:

> And a hegemon is more than a great power. To be hegemonic means to possess the authority of command. It includes a notion of primacy based on a component of just and legitimate leadership. Preeminence in wealth and force is a necessary but not a sufficient precondition of hegemony. Hegemony involves preeminence which is sustained by a shared understanding among social actors of the values, norms, rules, and laws of political interaction, of the patterns of authority and the allocation of status and prestige, responsibilities and privileges.[49]

For the purposes of this study, *regional hegemony* is understood in the latter sense of hard power plus influence.[50] It is to be understood as the possession of a predominance of deployable military capability within a geographic region (and the economic capability to back it), along with the political influence that comes with that dominance. It also provides a conceptual understanding of the type of similar situations to which the framework offered here may be applied.

The term *power* is a similarly loaded word. It is defined in fairly precise terms for these purposes as productive and destructive capability (classically known as *hard power*). However, other definitions are possible, and they deserve some discussion here. Classically, the great divergence in the understanding of power in world affairs has been between those who regarded it as synonymous with influence (reducing to a near-tautology the argument that power is the determinant of a state's status in the global system) and those who regarded it as describing the ability to destroy and create (as, one might say, to buy things and break things).[51] The former is said to create epistemological problems, since influence on outcomes seems only to be demonstrable after it occurs, following which it remains unclear whether it can be exercised again.[52] The latter mostly eschews such problems but does not account for the full range of a state's capabilities. For convenience's sake, control over outcomes is understood in this study as *influence*, whereas material capability is understood as *power*. This will not solve all the problems associated with these definitions, but it will hopefully allow for some clarity of terminology.

It is also necessary to admit that the concept of regional hegemony as it is employed here is atypical in a sense: it need not necessarily apply only to regional powers. In the case studied here, the Dreadnought Race, while the region in question—the North Sea and its littoral environs—can be delineated explicitly, one of the two competitors for it, Great Britain, was very much a global power as well as the regionally dominant one. How regional dominance pertains to a global hierarchy of great powers is a complicated subject: a power can be a global hegemon or global leader—in the understanding of a theoretical framework such as Robert Gilpin's hegemonic stability theory, or the leadership long-cycle theory of George Modelski, Karen Rasler, and William Thompson—and yet remain weak in a given region, perhaps even to the point of having to defer to that region's hegemon. An obvious example of this that is conceded in these theoretical arguments is the Cold War, in which the globally dominant United States nevertheless was forced (however grudgingly) to acknowledge Soviet primacy in Eastern Europe most of the time. Long-cycle theory in particular acknowledges this through its emphasis on control of the global commons—and not land—as the determinant of global leadership (what others would call *global hegemony*).

Students of power transition (and the circumstances under which it can drive conflict), and based on the power transition theory of A. F. K. Organski, implicitly acknowledge this paradox, particularly when they study the likelihood of war under subordinate powers who may find themselves in conflict over primacy in a region.[53] Douglas Lemke, in particular, has applied power transition theory at the subordinate power level, which necessarily involves competition for regional primacy even if one is not the top state in the system overall, and therefore implies that such primacy has significance of its own independently of an overarching power's control.[54] Conversely, the refusal to incorporate this paradox into the concept of hegemony leads John Mearsheimer to the position that there has never been a global hegemon, only regional ones. This study finds the latter position oversimplistic and unsatisfactory, and prefers to acknowledge that global leaders (and others) can indeed nevertheless be required to compete for control of a given region (as Great Britain did in the case demonstrated here)—but even if one prefers this framing, the framework for strategic analysis that this study offers can be applied to regional competition among powers all the same.[55]

Of note, perhaps, is the related concept of a *sphere of influence*, which is virtually coterminous with the concept of regional hegemony, as it implies an area over which a great power has military and diplomatic preeminence. If only for reasons of brevity, the phrase *competition for regional hegemony* is preferred here to *competition for an exclusive sphere of influence* or something similar. But it should be noted that the term is valid as a means of describing what is sought in the competitions that this framework is meant to study.[56]

Notwithstanding the general notion that regional hegemony was the prize of competition in the case reviewed in this study, the competing powers in question had specific outcomes in mind in the course of the competition, and their success or failure in achieving these outcomes must be the final test of their strategies, and the same is true in similar cases. The possession or search for regional hegemony therefore defines the scope of the type of competition studied here, but should not be assumed to define its exact nature and structure. The latter requires a detailed analysis of the competition itself, which is provided here in the case under examination and which will compose a key aspect of the framework that this book develops.

The State of Theory: Absence of a Practitioner's Perspective

It is here that we can see the importance of these concepts—strategy in particular—for the international relations theoretical discipline. The general focus of international relations theory as it relates to adversarial relationships among states tends to be on the reasons for which states compete, or the reasons for which they take specific actions during competition. Realists concern themselves frequently with questions such as how competitive relationships among states can lead to certain outcomes, such as open warfare and balancing alliances.[57] Neoliberal institutionalists often discuss the means by which states may avoid competing for security through the establishment of institutions and the feasibility of doing so in general or specific contexts.[58] Constructivists, broadly, argue that competition is the result of interacting worldviews and seek to understand how this happens.[59]

To analyze the strategy of states competing for regional hegemony, however, it is necessary to assume a subjective, actor's-eye point of view rather than the objective point of view usually offered by theory. While an outside observer might wish to know how two states became adversaries, the participants might wish to know how to make the best of the situation—or even who, if either, is likely to prevail, based on what has happened so far and what resources and options exist. Whereas theory, in its many forms, usually attempts to answer the question of *why* states compete, to discuss the strategy of such competition, one must address the question of *how* a state *may* compete.[60]

Relatively little has been done regarding the ways in which states may fare better or worse once they have become locked in competition with another state for hegemony within a given area, or even on how to define success or failure while the game is in process. The great exception is the literature of grand strategy, which is discussed below, and which, though often lumped together with realism, is tied only tangentially to the broad disciplinary theories of international relations.

A key component of such analysis is being able to see costs and benefits in real time, so as to understand the impact of deliberations and decisions and, by such means, analyze strategy. In particular, realism, despite its focus on rational decision making and security competition, contains gaps regarding just exactly how cost-benefit analysis in such competition may be done. Although realists typically regard power as either the thing that states seek or the means by which they seek it—or both—it is incomplete to point out that states must seek power (or security, or any other singular goal) and yet offer no analysis of how they may succeed or fail in acquiring it.[61] This book intends to fill some of this gap.

In fact, the problem of assessing success or failure in competition has led to a rethinking of realist assumptions in recent decades, including some key observations that are relevant to this study. In effect, there has been a convergence of theories through synthesis. In a perceptive and confrontational article appearing in *International Security* in 1999, Jeffrey W. Legro and Andrew M. Moravcsik argued that realism had become defined in such a way as to encompass virtually all possible theories in the discipline, having shied away from its original deterministic tenets regarding the immutability of state preferences and the need for policies to address fixed needs. In particular, they noted that realists were increasingly willing to accept a wide range of goals as being the principal objectives of states—goals ranging from mere survival, to acquiring power, to dominance of the system.[62] While the proper definition of theoretical schools is not the topic of discussion here, the general point regarding the apparent confusion of theorists as to what it is that states are supposed to be seeking is highly relevant. The fact is that states have options regarding what they seek and how they seek it, and their choices among those options are not without consequences, nor are those consequences uniform at all times. Such choices, when combined with the allocation of resources in pursuit of them, constitute strategy, and they can be studied as such. In fact, as Peter D. Feaver et al. noted in response to

Jeffrey W. Legro and Andrew Moravcsik's arguments, a major lacuna in realist theory is understanding of the ways in which the international system "punishes" states that do not optimally use resources, implying both that states face strategic choices and that the implications of this are not fully understood.[63] This is particularly true, Feaver notes, when a state has not yet "lost"—when it has not suffered a catastrophic military defeat or similar calamity.[64] In this sense, the Reagan quotation from which this book takes its title is of particular relevance: when competition is ongoing and seemingly without end (a race with no finish line), it is difficult to tell (at least based on theory as it currently stands) if one is playing badly or well.

What applies to realism applies *a fortiori* to neoliberalism and constructivism. Neoliberal theorists have occasionally attempted to explain the strategic rationale for states' joining or forming international compacts. G. John Ikenberry in particular has argued that dominant states benefit from sacrificing short-term gains in order to form institutions that lock other actors into foreseeable and favorable behavior patterns.[65] These insights, while interesting and valuable, nevertheless do not address the question of how states can behave in a competitive, vice cooperative, environment, since the focus of neoliberal theory is on obviating at least part of the need for competition through the formation of institutions. They also skirt the question of how to quantify costs and benefits, and what a state with a specific set of competitive goals should do to achieve them in an institutional framework.[66]

As noted, constructivism has largely avoided the question of how to analyze competition by treating it as irrelevant, or at least beneath serious theoretical analysis. If what is important for study is the (essentially unpredictable) intersubjective construction of the international system, states' interests, and international norms, then detailed analysis of the options facing states in a competitive environment and the likely consequences of choosing any particular option is either impossible or undesirable. It is impossible if the system and its constituent actors are too fluid to be subjected to analysis on the basis of material costs and benefits, and it is undesirable because (particularly when constructivism crosses over into critical approaches) it can be seen to be powerful enough to create a self-fulfilling prophecy.[67]

It would be remiss at this point not to mention a major subdiscipline that would seem to offer a way forward on this issue, namely, foreign policy analysis. Foreign policy analysis as a subdiscipline is concerned with both the causes and effects of foreign policy—both the influences that drive it and what it achieves. Even this subdiscipline, however, is mostly concerned with causes of phenomena rather than options, sometimes over the lamentation of its participants. The primary focus of foreign policy analysis has been and remains the factors that drive states' foreign policies, rather than whether there are lessons to be drawn from them.[68]

To summarize, it is important to discuss exactly what is and is not being attempted here. International relations theory is widely cognizant of the importance of strategy to states in the international system—strategy being defined here merely as the choice of options for allocating state resources. However, the subject of competitive strategy as defined above—the allocation of resources in the context of an ongoing competitive relationship with another state—is less well-discussed. Most notably, it is useless to discuss strategy unless there exists some framework for assessing its results. Metrics for cost-benefit analysis as it is used in foreign policy are in fact rather difficult to come by. To date, there does not exist within the theoretical literature a detailed analysis of how such a state can assess its performance in a competitive relationship with other states or what a state's options are once it has become locked into said competition.

Indeed, the emphasis of existing theory appears to preclude such study. A theory that seeks to argue that states' actions are determined by a given set of factors—be they security imperatives, domestic politics or regime structure, subjective construction of the other, trading relationships, the demands of binding institution, or any similar driving force—or otherwise to explain why states act a certain way, will by its very nature seek to argue that a given set of choices by a given state is in some sense foreordained, without necessarily seeking to analyze what options states have available to them.[69]

Put differently, and to draw an analogy, the bulk of existing international relations theory may be considered (as noted) to be analogous to economic theory, with both seeking explanations within a complex social system for observed social phenomena. Economic theory has an engineering analog in business and management literature, which seeks to instruct practitioners within the economic realm—the leaders of firms, their investors, and similarly interested parties—in how to apply the insights gained from theory to solve their problems (the same can be said of finance). A similar situation prevails in the military realm: military history and war studies literature analyzes military phenomena from history to the present, while strategic literature exists to instruct military leaders in how to apply the lessons of pure theory to the serious business of warfare. By contrast, no analogous literature currently exists at the political level for the study of international relation—apart from classic works such as Niccolò Machiavelli's *The Prince*, very little has been done to suggest a systematic means of applying the lessons of theory to political decisions, far less to a specific type of decision making on a specific type of problem such as the strategy of international security competition.[70]

A Contrast: The Limitations of Grand Strategy

The major focus of what literature exists on strategy in international relations literature is on grand strategy. As previously noted, it is important to contrast what is being attempted here with the considerable literature on grand strategy currently extant. As discussed previously, the term *grand strategy* in fact has two nearly opposite meanings as it is generally used in this context. On occasion, as noted, it is a subgenre of military strategic literature, focused on the question of how to allocate military resources in order to win a war. In that sense, it has the same relevance as the military literature just discussed.

However, it is in its other incarnation that grand strategy is of greater relevance. The term can often be used to describe the overall allocation of resources by a state in its national security policy, in peace or war. Paul Kennedy's *The Rise and Fall of the Great Powers: Economic Change and Military Conflict from 1500 to 2000* is considered a classic of the genre. Kennedy, like Karen Rasler and William Thompson, who include and partially critique his insights in their discussion of the rise and fall of global leaders, is concerned principally with the ways in which great powers—highly influential states—can increase or decrease their position within the international hierarchy.[71] As interesting as these insights are, they do not address resource allocation when in direct competition for control of a specific region.

Edward Luttwak comes closer in his understanding of grand strategy to this study's usage of the term competitive strategy, in that he understands it applies to situations where states are in some kind of potentially military enmity. He writes,

> We may note incidentally that one way of evaluating the state of global politics on some normative index of progress is to examine how many of its relationships are significantly strategical. To be sure, grand strategy also exists outside international politics, for it includes the highest level of interaction between any parties capable of using force against each other.[72]

As promising as this analysis is, Luttwak does not take it in the direction offered here, toward a total appraisal of a state's resource allocation while competing for regional hegemony.

Robert J. Art, in his book *A Grand Strategy for America*, describes developing a grand strategy as first involving determining what the United States' interests are before proceeding through an analysis of challenges those interests face and possible strategies to safeguard them.[73] Other theorists have followed a similar pattern of analysis. The former U.S. national security advisor Zbigniew Brzezinski offers a review of U.S. global priorities in his book, *Strategic Vision: America and the Crisis of Global Power*, but he neither addresses the concrete steps that the United States (or anyone else) would take in the course of a full-fledged competition with another great power, nor speaks to a means of analyzing whether the competition was going well or badly.[74] Whatever the particulars, there is consequently a near-comprehensive tendency among geostrategists and grand strategists to view strategy as mere policy—the setting of goals—rather than as a plan that allocates resources and is implemented in real time. This has begun to change: recently the topic of grand strategy in this context—the allocation of resources and priorities by a state in the international realm—has been heavily analyzed, most notably in treatments by Nina Silove and Rebecca Lissner, with Lissner in particular describing it as a "conceptual minefield."[75] Hal Brands, in turn, despairs of a precise definition: "The fact that there are so many competing conceptions of grand strategy should probably tell us that the concept is subjective and ambiguous enough that it defies any singular definition. The best an analyst can do is offer a definition that is, in the strategic theorist Colin Gray's phrasing, 'right enough'."[76] But while these authors acknowledge interest in grand strategy as something more detailed and comprehensive than mere policy, it is rarely (if ever) discussed how this is done.

In any case, competitive strategy, as noted, must be understood as the strategy of a state engaged in an already determined competition, not the overall strategy of a state that (in theory) prior to that strategy's conception was agnostic about its place in the world and where, if anywhere, to compete for anything. It is nonetheless interesting to note that Art views grand strategy defensively, as a means of "protecting" interests rather than advancing them or accomplishing goals.[77] In the context of this study, competitive goals may be affirmative (needing to be accomplished) as well as negative (involving safeguarding the status quo). Indeed, as will be discussed in the next chapter, a key difference between two states engaged in the type of competition for regional hegemony under study here is the asymmetric nature of their goals, one having an affirmative goal to accomplish, the other needing only to frustrate its competitor.

To conclude, and to reiterate, grand strategy, except when used in a purely military context, normally concerns decisions regarding what role one will seek in the international system and how one will seek it. In contrast, this study merely offers a framework for evaluating the decisions and resource allocation of states that already have a defined role in the system and have already chosen to compete for hegemony within a given region, along with what that hegemony entails.[78]

Going Forward

This study, through an analysis of an historical case, will show that states engaged in competition for regional hegemony face strategic choices, that these choices have consequences, that such choices can be rigorously analyzed, and that general strategic lessons can be drawn for states facing similar competitive scenarios in the future. In short, the approach suggested here, while employing the insights offered by existing theory, seeks to augment that theory through a different approach. Assumptions implicit in this approach include the following:

- Notwithstanding the question of to what extent systemic structure rewards and punishes states' actions, states within the international system possess something like free will; they are capable of making decisions that are either to their detriment or benefit, according to some definition of each. They do not always "play" optimally and can make mistakes in pursuit of their goals; international competition does not occur in an "efficient market."
- It should be possible to assess whether states succeed or fail according to the goals they themselves choose to

- pursue, in the context of the international social system in which they operate.
- It should be possible, given a set of competitive goals, to understand how a state's actions lead to success or failure along these lines, and therefore what actions it should pursue or avoid in the course of competition to achieve its goals.
- It should be possible to model what states give up in both concrete and opportunity costs in pursuit of a given set of goals, and in so doing to assess the net impact of their decisions.

It is from these assumptions that this study will proceed. From here, now that terms have been explained and the differences between this study's approach and the existing theoretical literature have been examined, it becomes necessary to spell out this study's approach in detail. Synthesizing those insights of the theoretical literature on strategy that are of particular relevance to competition for regional hegemony, and laying out a structure for the examination of cases of such competition, are the subjects of the next chapter.

Chapter One

Endnotes

1. Ralph Peters, "The Seeker and the Sage," in *The Book of War: Sun Tzu, The Art of Warfare, and Karl von Clausewitz, On War*, ed. Caleb Carr (New York: Modern Library, 2000), xix–xx.
2. Phillip Karber, conversation with author, January 2011. A discussion of similar concepts occurs in Martin van Creveld, *The Transformation of War* (New York: Free Press, 1991), 117–18. Van Creveld notes that war is unique among all human relationships in that no external force compels compliance with a set of rules limiting what the participants may do to each other. He draws a distinction between competition, in which actors merely vie for the same goal, and conflict, in which they disrupt and antagonize each other directly. However, a significant part of van Creveld's thesis holds that participants in war do obey limiting rules that they impose on themselves and one another. Such a "no-holds-barred" game would be purely conflictual by Karber's formulation.
3. These three concepts partially overlap and "bleed into" one another: cooperation may be more intense between some actors than others; some competitors may also choose to cooperate on occasion; and some competition may begin to resemble open warfare. For purposes of elucidating the continuum, a charity is basically cooperative, a for-profit corporation is basically competitive, and an army is basically conflictual.
4. Michael P. Colaresi, Karen Rasler, and William R. Thompson, *Strategic Rivalries in World Politics: Position, Space and Conflict Escalation* (Cambridge, UK: Cambridge University Press, 2007), 4, 285–90, https://doi.org/10.1017/CBO9780511491283.
5. Paul F. Diehl and Gary Goertz, *War and Peace in International Rivalry* (Ann Arbor: University of Michigan Press, 2001), 19–26, 131–32, https://doi.org/10.3998/mpub.16693.
6. This definition is by Phillip Karber; it is modified to account for the other strategic literature discussed below. The definition in quotation marks appears in Phillip Karber, "Competitive Strategy: As an Approach to Business and Professional Life" (PowerPoint presentation at the Annual Fellows Conference of the Center for the Study of the Presidency, Washington, DC, 31 October 2003), slide 2. As it is adequate for purposes here, the definition is retained verbatim, with additions seen outside the quotation marks.
7. Colin S. Gray, *Modern Strategy* (Oxford, UK: Oxford University Press, 1999), 16–47.

8. Hew Strachan, *The Direction of War: Contemporary Strategy in Historical Perspective* (New York: Cambridge University Press, 2013), 11-12, https://doi.org/10.1017/CBO9781107256514.
9. It is not necessary here to discuss exactly who Sun Tzu was historically or whether one individual or several composed his works—except, perhaps, to repeat an old joke made about the *Iliad*: that the great work was possibly composed not by its legendary author Homer, but by another person with the same name.
10. Van Creveld, *The Transformation of War*, 94-96. Azar Gat also cites Maizeroy as the first user of the term, though he notes that "[strategy's] origins in modern military theory also [like those of the term "tactics"] seem to have been lost. Maizeroy . . . was the one who introduced the concept that derived from the Greek word for general and was used by Emperor Maurice as the title for his military treatise *Strategikon*." Azar Gat, *A History of Military Thought: From the Enlightenment to the Cold War* (Oxford, UK: Oxford University Press, 2001), 43-44. Hew Strachan also pays homage to Maizeroy's coinage of the term. Strachan, *The Direction of War*, 28-29. Edward N. Luttwak notes drolly that *strategy* is "a Greek word that no ancient Greek ever used." Edward N. Luttwak, *Strategy: The Logic of War and Peace*, rev. ed. (Cambridge, MA: Belknap Press, an imprint of Harvard University Press, 2001), 267.
11. Carl von Clausewitz, *On War*, tr. Michael Howard and Peter Paret (New York: Alfred A. Knopf, 1993), 146.
12. Clausewitz, *On War*, 165.
13. Clausewitz, *On War*, 207.
14. B. H. Liddell Hart, *Strategy: The Indirect Approach*, rev. ed. (London: Faber and Faber Limited, 1967), 321.
15. Luttwak, *Strategy*, 269. Van Creveld echoes this point, noting that the confluence of the two activities (preparation and usage of armed force) into the same description began approximately with World War I. Van Creveld, *The Transformation of War*, 97-98.
16. Gray, *Modern Strategy*, 17.
17. See Clausewitz, *On War*, 99-100. There, he makes the famous statement that "war is not merely an act of policy but a true political instrument, a continuation of political intercourse, carried on with other means."
18. Peter Paret, "Introduction," in *Makers of Modern Strategy: From Machiavelli to the Nuclear Age*, ed. Peter Paret (Princeton, NJ: Princeton University Press, 1986), 3.

Chapter One

19. Col J. F. C. Fuller, *The Reformation of War* (London: Hutchinson, 1923), 218; quoted and cited in Strachan, *The Direction of War*, 33.
20. Strachan, *The Direction of War*, 16.
21. This is also noted in van Creveld, *The Transformation of War*, 98–104. This was a break with Clausewitz, who, as Michael Howard notes, was concerned solely with the usage, vice the organization, of military force. Michael Howard, "The Forgotten Dimensions of Strategy," *Foreign Affairs* 57, no. 5 (Summer 1979): 975–76. In addition to reflecting the military preoccupations of the time, it can also serve to demonstrate not only that multiple understandings of strategy are possible but also the overall blurriness of the concept. Despite a lack of conceptual clarity, this study assumes that certain elements of a universal definition of strategy can be agreed on.
22. Paul M. Kennedy, "Grand Strategy in War and Peace: Toward a Broader Definition," in *Grand Strategies in War and Peace*, ed. Paul M. Kennedy (New Haven, CT: Yale University Press, 1991), xi.
23. Kennedy, "Grand Strategy in War and Peace," xi.
24. Luttwak, *Strategy*, 211–12.
25. Strachan, *The Direction of War*, 26–27.
26. Walter Kiechel III, *The Lords of Strategy: The Secret Intellectual History of the New Corporate World* (Boston, MA: Harvard Business Press, 2010), 24–25.
27. H. Igor Ansoff, *Corporate Strategy: An Analytic Approach to Business Policy for Growth and Expansion* (New York: McGraw-Hill, 1979), 105.
28. Alfred D. Chandler Jr., *Strategy and Structure: Chapters in the History of the Industrial Enterprise* (Cambridge, MA: MIT Press, 1962), 13; quoted in Kiechel, *The Lords of Strategy*, 25.
29. Kiechel, *The Lords of Strategy*, 25.
30. Kiechel, *The Lords of Strategy*, 24–26.
31. Kiechel, *The Lords of Strategy*, 23–25.
32. Kiechel, *The Lords of Strategy*, 24; and Henry Mintzberg, *The Rise and Fall of Strategic Planning: Reconceiving Roles for Planning, Plans, Planners* (New York: Free Press, 1994), particularly 23–29, 66–67, 329–30, 393–94.
33. Kenneth R. Andrews, *The Concept of Corporate Strategy*, rev. ed. (Homewood, IL: Richard D. Irwin, 1980), 18.
34. Andrews, *The Concept of Corporate Strategy*, 5.
35. Mintzberg, *The Rise and Fall of Strategic Planning*, 5–15, 23–29. Further raising the bar, Mintzberg discerns two other uses of the term strategy: as "position" and as "perspective"—the former a summary of the work of Michael E. Porter (discussed

later in this study) and the latter derived from the work of management theorist Peter F. Drucker. However, it is clear that Mintzberg is focused mostly on the synthesis of the first two meanings. The "plan" and "pattern" concepts occur verbatim in Nina Silove's discussion of conceptions of grand strategy, and the concept of "position" is closely related to her designation of "organizing principle" as an aspect of grand strategy. See Nina Silove, "Beyond the Buzzword: The Three Meanings of 'Grand Strategy'," *Security Studies* 27, no. 1 (2018): 27-57, https://doi.org/10.1080/09636412.2017.1360073. Something similar is offered by Rebecca Friedman Lissner, who conceptualizes grand strategy variously as "variable," "process," and "blueprint." See Rebecca Friedman Lissner, "What Is Grand Strategy?: Sweeping a Conceptual Minefield," *Texas National Security Review* 2, no. 1 (November 2018): 52-73, https://doi.org/10.26153/tsw/868. Suffice it to say that the subject has been thoroughly analyzed.
36. Mintzberg, *The Rise and Fall of Strategic Planning*, 23-24.
37. Claude Dube, "The Department of National Defence and the Defence Strategies from 1945 to 1970" (MBA thesis, McGill University, 1973), 71-72; quoted and cited in Mintzberg, *The Rise and Fall of Strategic Planning*, 114-15.
38. Kiechel, *The Lords of Strategy*, 24.
39. See Kiechel, *Lords of Strategy*, 32-37. The argument was that greater experience relative to competitors in delivering a good or service entailed lower costs of production and therefore what would later be known as "competitive advantage." The general idea of maintaining one's inherent strengths relative to one's competitors is discussed in detail below in chapter 2 of this book, and it forms a key part of the framework offered here for analyzing states' strategies when competing for regional hegemony.
40. Michael E. Porter, *Competitive Strategy: Techniques for Analyzing Industries and Competitors* (New York: Free Press, 1998), 127-28; Kiechel, *Lords of Strategy*, 131-32; and Kenichi Ohmae, *The Mind of the Strategist: The Art of Japanese Business* (New York: McGraw-Hill, 1982), 91-92.
41. Ohmae, *The Mind of the Strategist*, 36.
42. Lawrence Freedman, *Strategy: A History* (Oxford, UK: Oxford University Press, 2013), 610-11.
43. Karber, "Competitive Strategy," slides 3-5.
44. The understanding of the chief strategist of an organization as its leader, not its advisor—a point made by Karber, Andrews, and implicitly by Clausewitz—further confirms the essentially dual

nature of strategy: it is both the plan that the leader has in their head and the ad hoc decisions that the leader makes in furtherance of the plan's objective when the plan is not sufficient. See Karber, "Competitive Strategy," slide 17; Andrews, *The Concept of Corporate Strategy*, 5; and Clausewitz, *On War*, 207.

45. Phillip Karber, "The 'Counter-Offensive' in Competitive Strategy: Lessons from the Reagan Era" (PowerPoint presentation at the Developing Competitive Strategies for the 21st Century Conference, U.S. Naval War College, Newport, RI, 23 August 2010), slides 6, 8–9.

46. John J. Mearsheimer, *The Tragedy of Great Power Politics* (New York: W. W. Norton, 2001), 40–41.

47. Robert Gilpin, *War and Change in World Politics* (Cambridge, UK: Cambridge University Press, 1981), 29, 33, https://doi.org/10.1017/CBO9780511664267.

48. Karen Rasler and William R. Thompson, *The Great Powers and Global Struggle, 1490–1990* (Lexington: University Press of Kentucky, 1994), 16. Long-cycle theorists are concerned with global, vice regional, leadership, but their understanding of leadership as hard power plus influence is transferable to a regional context.

49. Torbjørn L. Knutsen, *The Rise and Fall of World Orders* (Manchester, UK: Manchester University Press, 1999), 11.

50. There are, of course, other definitions of *hegemony*, including ones that eschew a hard power component altogether. Because of the nature of the subject matter in this book, however, it is clear that hard power was a necessary component of what the competitors in each case sought, and it is therefore used as part of the definition employed here.

51. For a discussion of these competing understandings of power, see Brian C. Schmidt, "Realism and Facets of Power in International Relations," in *Power in World Politics*, ed. Felix Berenskoetter and M. J. Williams (London: Routledge, 2007), 43–63.

52. Mearsheimer, *The Tragedy of Great Power Politics*, 57–58. One might add that if "political capital" is thought of in finite terms, it becomes all the harder to say that, because a given state has exercised influence successfully in the recent past, it can do so again.

53. A. F. K. Organski and Jacek Kugler, *The War Ledger* (Chicago, IL: University of Chicago Press, 1980), Kindle ed., loc. 1009–1021, 3082–95. Organski and Kugler sum up their conclusion that differing rates of change in power drive the push for great power war.

54. Douglas Lemke, *Regions of War and Peace* (Cambridge, UK: Cambridge University Press, 2002), Kindle ed., loc. 2327–67, https://doi.org/10.1017/CBO9780511491511.
55. For Mearsheimer's rather austere understanding of hegemony, regional or otherwise, see Mearsheimer, *The Tragedy of Great Power Politics*, 40–41.
56. As Amitai Etzioni has discussed, there is actually a comparative lack of literature on precisely what is meant by the term *sphere of influence*, but it is sufficient to note here that it means approximately the same thing. See Amitai Etzioni, "Spheres of Influence: A Reconceptualization," *Fletcher Forum of World Affairs* 29, no. 2 (Summer 2015): 117–32.
57. One could choose innumerable examples. To take just a few, see Stephen M. Walt, *The Origins of Alliances* (Ithaca, NY: Cornell University Press, 1987), 147–80, which discusses states' decisions to form balancing alliances; Mearsheimer, *The Tragedy of Great Power Politics*, 138–67, which discusses states' options (including war) for survival in anarchy; and Kenneth N. Waltz, "The Origins of War in Neorealist Theory," *Journal of Interdisciplinary History* 18, no. 4 (Spring 1988): 615–28, https://doi.org/10.2307/204817, which discusses the effect of competition in anarchy on the likelihood of war. This tendency to look at causes of competition rather than examine the process by which it occurs predates neorealism. See also Hans J. Morgenthau, *Scientific Man vs. Power Politics* (Chicago, IL: University of Chicago Press, 1974); and Hans J. Morgenthau, *Politics among Nations: The Struggle for Power and Peace*, 7th ed. (New York: McGraw-Hill, 2006). These classic works on the nature of the international system focus on the fundamentals of international interactions or, in the case of the former, explaining the difficulty of explaining Hobbesian international realities to a modernist audience. Morgenthau uses the term strategy in numerous articles, but he does not examine the competitive strategies of states for lessons or articulate a framework for analyzing them. See also Edward Hallett Carr, *The Twenty-Years Crisis, 1919–1939: An Introduction to the Study of International Relations* (London: Macmillan, 1958). Though Carr's work is often thought of as an essential statement of classical realist thought on par with Morgenthau's, it too is primarily focused on philosophical questions rather than instrumental ones. Perhaps Carr's most timeless argument is as follows: "In both physical and political sciences, the point is soon reached where the initial stage of wishing must be succeeded by a stage of hard and ruthless analysis. . . . [R]ealism . . . places its emphasis

on the acceptance of facts and on the analysis of their causes and consequences. It tends to depreciate the role of purpose and to maintain . . . that the function of thinking is to study a sequence of events which it is powerless to influence or to alter." Carr, *The Twenty-Years Crisis*, 9–10. This mindset has characterized most realist thought and is a weak starting point for an analysis of how a state may make its situation better or worse by careful planning and judicious decision-making.

58. The classic example is G. John Ikenberry, *After Victory: Institutions, Strategic Restraint, and the Rebuilding of Order after Major Wars* (Princeton, NJ: Princeton University Press, 2001). Ikenberry discusses the long-term advantages of adopting institutions to "lock in" a victor's gains from a systemic war. Such analysis is helpful, though it concerns why states avoid competition, not how they can prevail in it.

59. See, for example, Alexander Wendt, *Social Theory of International Politics* (Cambridge, UK: Cambridge University Press, 1999), 251–312, https://doi.org/10.1017/CBO9780511612183. Wendt writes, "Five hundred British nuclear weapons are less threatening to the [United States] than five North Korean ones because of the shared understandings that underpin them. What gives meaning to the forces of destruction are the 'relations of destruction' in which they are embedded: the shared ideas, whether cooperative or conflictual, that structure violence between states." Wendt, *Social Theory of International Politics*, 255. Wendt does not use "competitive" as a midpoint between "cooperative" and "conflictual," as this study does, but one may see the point readily enough. Again, this work deals with why states compete, not how to win a competition.

60. For a discussion of the practitioner's perspective and a discussion of John Lewis Gaddis and A. Wess Mitchell's respective treatments of the subject, see Martin Skold, "Book Review: On Grand Strategy and The Grand Strategy of the Habsburg Empire," *Contemporary Voices: St Andrews Journal of International Relations* 1, no. 3 (August 2019): 117–26, https://doi.org/10.15664/jtr.1512. The two works under review are John Lewis Gaddis, *On Grand Strategy* (New York: Penguin, 2018); and A. Wess Mitchell, *The Grand Strategy of the Habsburg Empire* (Princeton, NJ: Princeton University Press, 2018). Both offer a groundbreaking perspective on strategy as a practice.

61. For a summary of the arguments concerning states' appetite for hard power versus the security it can provide, see Mearsheimer, *The Tragedy of Great Power Politics*, 18–22. The closest that realists

come to analyzing how states can achieve what they supposedly want is to cite their options or note their tendencies. Mearsheimer, *The Tragedy of Great Power Politics*, 138–67. This relates to Stephen M. Walt's argument that states are more likely to assuming a balancing stance toward a state when they specifically view it as hostile. Walt, *The Origins of Alliances*, 133–41.

62. Jeffrey W. Legro and Andrew Moravcsik, "Is Anybody Still a Realist?," *International Security* 24, no. 2 (Fall 1999): 5–55, https://doi.org/10.1162/016228899560130.
63. Peter D. Feaver et al., "Brother, Can You Spare a Paradigm? (Or Was Anybody Ever a Realist?)," *International Security* 25, no. 1 (Summer 2000): 165–93.
64. Feaver, "Brother, Can You Spare a Paradigm?," 166–67.
65. Ikenberry, *After Victory*, 53.
66. See Ikenberry, *After Victory*, 53. The author writes, "The state seeks to use its power as efficiently as possible. It wants to preserve and extend its power position into the future, and it is willing to give up some 'returns' on its power in the short run in favor of a greater longer-term return on its power if such a possibility exists." One may note that Ikenberry is quite vague as to how to measure whether a state is achieving this goal or not, even if one posits that this is in fact every state's goal. As an aside, it may be pointed out that the existence of institutions does not preclude competition for regional hegemony of the type discussed here; indeed, it may frame it if the institutions do have an independent influence on state interactions. The "finance and welfare" and "intangibles" metrics discussed below account for this. Ikenberry appears to be arguing that states are competitive even when they are not: in his argument, they form institutions to maximize their competitive gains over time. Even so, more can be said about how this may be done.
67. See, for example, Karin M. Fierke, *Changing Games, Changing Strategies: Critical Investigations in Security* (Manchester, UK: Manchester University Press, 1998), 111–29, 210–23. Fierke argues that how the two superpowers in the Cold War saw and spoke of one another affected the intensity of the competition between them.
68. For a representative sample of key foreign policy analysts who characterize their focus as primarily the factors that drive policy, rather than how to make policy do what a state wants it to do, see Ryan K. Beasley and Michael T. Snarr, "Domestic and International Influences on Foreign Policy: A Comparative Perspective," in *Foreign Policy in Comparative Perspective: Domestic*

and International Influences on State Behavior, ed. Ryan K. Beasley et al. (Washington, DC: CQ Press, 2002), 323-38; Christopher Hill, *The Changing Politics of Foreign Policy* (Basingstoke, Hampshire: Palgrave Macmillan, 2003), 5-11, 15-19; and Charles F. Hermann, Charles W. Kegley Jr., and James N. Rosenau, *New Directions in the Study of Foreign Policy* (Boston, MA: Allen and Unwin, 1987). See also Gregory A. Raymond, "Evaluation: A Neglected Task for the Comparative Study of Foreign Policy," in *New Directions in the Study of Foreign Policy*, 96-110. Raymond refers to foreign policy evaluation as an "important but neglected task," lamenting that enthusiasm for policy evaluation in domestic policy had not created a similar drive for foreign policy evaluation. He articulates a framework for evaluating foreign policy, although he does not specifically address the strategy of competition for regional hegemony, the subject of this study. Lamentably, at this point, that it has been more than 30 years since the publication of Raymond's article, and foreign policy analysis remains focused, as shown, on cause rather than evaluation.

69. This is not to say that all or even most theories purport to do this. Rather, it is to say that attempting to answer the question of why states behave in a certain way effectively forces a deterministic outcome—the notion of "why" implies causality, which in turn implies an absence of choice.

70. See Skold, "Book Review." "As the famous saying has it, those who speak do not know, and those who know do not speak." Intuitively, one might expect the leaders of states, not to mention their subordinates, to exercise considerable discretion regarding the reasoning behind their decisions. Similarly, given the exclusive nature of the group of decision makers involved, one would not expect there to be much of a formal discipline dedicated to training them (as is the case in less secretive areas such as private enterprise). Given Machiavelli's historical fate following the publication of his policy prescriptions and political analysis in *The Prince* (1532), one might say in the first place that such analysis can be done, and in the second place that there are excellent reasons why it is rarely done.

71. Paul M. Kennedy, *The Rise and Fall of the Great Powers: Economic Change and Military Conflict from 1500 to 2000* (New York: Vintage Books, 1989), 539-40. For the critique, see Rasler and Thompson, *The Great Powers and Global Struggle*, 143-46.

72. Luttwak, *Strategy*, 210.

73. Robert J. Art, *A Grand Strategy for America* (Ithaca, NY: Cornell University Press, 2003), 2, 12-81.

74. Zbigniew Brzezinski, *Strategic Vision* (New York: Basic Books, 2012).
75. Lissner, "What Is Grand Strategy?"; and Silove, "Beyond the Buzzword."
76. Hal Brands, *What Good Is Grand Strategy?: Power and Purpose in American Statecraft from Harry S. Truman to George W. Bush* (Ithaca, NY: Cornell University Press, 2014), 2.
77. Art, *A Grand Strategy for America*, 2.
78. This contrast between grand strategy and competitive strategy is drawn in Karber, "The 'Counter-Offensive' in Competitive Strategy," slides 6, 8–9.

Chapter Two
Some Principles of Competitive Strategy

There are those who say: "I am a farmer," or, "I am a student;" "I can discuss literature but not military arts." This is incorrect. There is no profound difference between the farmer and the soldier. You must have courage. You simply leave your farms and become soldiers. That you are farmers is of no difference, and if you have education, that is so much the better. When you take your arms in hand, you become soldiers; when you are organized, you become military units.
~ Mao Zedong[1]

How, then, should one think about the strategy of regional great power competition? What should one look at when trying to understand it? When one wants to know who is winning, what does one examine, and how does one know? It is to these questions this study can now turn.

To generate a framework for inquiry, this chapter now offers a discussion of the nature of strategy as it applies to regional great power competition and synthesizes the insights of a number of key strategic theorists that are of direct relevance to it. To that end, it will first clarify what is meant by strategy and strategic decision making when these terms apply to states. It will then synthesize the relevant strategic insights as they apply to the various components of the definition of strategy being used here.

Strategic Decision Making as Applied to States

When applying the lessons and concepts introduced in the last chapter, it is important to specify precisely what is meant by strategy in the context of competition for regional hegemony. Given the definition employed previously, a strategy would appear to be a product of human mental processes, implying both a cohesive set of decisions (such as plans of action) and someone to make them. In practice, as is obvious, it is more difficult to speak of such things in the context of the international realm, first because states are not individual human beings and therefore in one sense do not perceive and decide as unitary actors, and second because even to the extent that they do, they are unlikely to commit to a single set of decisions over time (unless, perhaps, they are led by dictators with unusual levels both of latitude in decision making and of consistency in thought and action). States also are unlikely to publicize their exact plans with any great specificity, notwithstanding the proliferation of published "national security strategies" in the democratic Western world.[2] It is worthwhile, then, to consider in what sense it is proper to speak of strategy in this context.

The problem of agency at the state level of international relations is much discussed in the existing literature and needs only a cursory recapitulation here. Briefly, although dominant schools of thought in international relations theory (realism and institutionalism, most notably) conceive of states' actions as being forced on them by the system, there is a considerable body of literature that treat states' actions as the product of specific internal processes. This study takes for granted that states have choices to make (whether this truly constitutes agency may be debated elsewhere); it remains to be specified what this means for the concept of strategy.

That states are rarely, if ever, internally unified in their foreign policy decision making is manifestly clear, and a vast literature exists attempting to parse to what extent the various components of a state influence policy, and how. In Robert Gilpin's formulation, the state is a "coalition of coalitions," and its interests are in effect the sum of the interests of its constituents—understanding states' objectives and interests is therefore a matter of knowing how priorities are apportioned among their members, although certain constants regarding security and control over the world economy may be observed across history.[3] Robert O. Keohane and Joseph S. Nye Jr. revolutionized the discipline of international relations by noting that because of the various interests that compose a state, there is not always a clear hierarchy of policy priorities.[4] Other theorists have shown, or attempted to show, the ways in which regime type plays a role in states' policy choices, as for example in Jack Snyder's work regarding war-prone nascent democracies and in the arguments underpinning the democratic peace thesis.[5]

But while it is clear that states' internal processes can be shown to influence policy in various ways, this complicates rather than simplifies the question of how a state may engage in strategy, since at any given time a multitude of actors within a state may be trying to push policy in one direction or another for one reason or another. Graham T. Allison Jr. famously showed that the level at which one analyzes a state's behavior impacts how one analyzes its decisions: in the case of the Cuban Missile Crisis, the actions of the key players at various times can be explained in one way by assuming that the state is a unitary rational actor, in another by analyzing the internal processes of its bureaucracy, and in another still by examining the conflicts of interests of key internal decision makers. Individuals' actions make sense in light of their psychology and can in some sense be understood; collective actions appear differently depending on what factors one considers. What is "rational" in one sense is just happenstance, a product of internally aligned interests and prejudices, in another.[6] Since strategy is understood as a set of decisions, it is pertinent to ask in the case of a state whose decisions they are, and how in fact they are made.

The matter appears not to be helped very much by the traditional conception, within military and business strategy, of a unified leadership. For military strategists since Sun Tzu, the general is the final authority—Sun Tzu famously argues that the field commander must be allowed discretion not only by subordinates but even by the sovereign he serves, since meddling from the top can only interfere with effective planning.[7] Carl von Clausewitz, with his emphasis on war as an instrument of policy, in principle disagrees, though this merely shifts the final decision a level up the chain of command. It may be noteworthy that Clausewitz's admirers in subsequent generations often forgot this aspect of his philosophy, though, as noted above, within the realm with which he is chiefly concerned, namely military strategy, Clausewitz is in agreement with the concept of the organization leader as strategist.[8] Business strategists repeatedly note that the individual with the final say on strategy—the only true strategist—is the organization leader.[9] The concept of a single executive making decisions is perhaps appropriate in simple and highly hierarchical organizations, but it appears too simplistic for an organization as large and complex as a state.

However, in the final analysis, it is indeed possible to reproduce this conception where states are concerned. States' decisions are, in the end, articulated by decision makers who, although beholden to innumerable domestic interests and possessed of their own priorities, nevertheless can be counted on to articulate, either publicly to their constituents or privately to their subordinates and colleagues, what they intend to do and what they claim to want to accomplish by doing it. Although in one sense it may be true that individual decision makers are motivated by many factors and may hide their true motivations, for purposes of understanding a state's strategy, decision makers' motivations and private intentions may be seen as irrelevant. If a decision maker outlines a course of action and then proceeds to execute it, explaining a rationale for it in the process, that set of plans and stated intentions may be analyzed as a strategy, whether or not the decision maker had something else in mind at the moment it was articulated.[10]

Similarly, the fact that a state, during a long period of time, may not articulate (even behind closed doors) a comprehensive plan of action, much less any of the other components of strategy as defined above, should not be seen as an argument that states do not have strategies or that these cannot be understood. It should be possible, based on the stated intentions of political leaders and the subsequent actions of their states, to piece together what their states were trying to do and when, even if, as the case may be, their strategies may seem quite incoherent as a result. The mere fact that a state may have a bad strategy—or hardly any at all—is no reason not to analyze its strategy; far from it, the lack of a coherent strategy may in a given case constitute a strategic lesson on its own. It remains to be seen, however, in what way a state's competitive strategy may be understood. For this, the concept of strategy must be analyzed.

The Elements of Strategy

The exact nature of competitive strategy, and its component elements, can now be addressed. What follows is a detailed summation of the applicability of existing strategic literature to the essential components of competitive strategy. The definition of competitive strategy given above is therefore broken into its individual parts for detailed analysis. Each piece of analysis serves as a portion of an analytical framework that is tested in the following chapters against historic cases of competition for regional hegemony.

A Plan of Action and Process of Decision Making

A strategy is, first and foremost, a plan, albeit a flexible and intuitive one. Although, as noted above, a strategy need not be thought of as a predetermined set of decisions to be executed rigorously, and may instead be understood as a consistent set of preferences in decision making over time, in general, whether organized in advance or improvised on the run, it must be understood as planning actions several steps ahead. As Colin

Some Principles of Competitive Strategy

S. Gray notes (and as mentioned in the preceding chapter), the "strategy bridge" involves neither the choice of a goal nor the presence of some means, but the "bridge" between them—the detailed plan of action that explains how one will use one's resources to move from point A (the starting point) to point B (the desired objective). The impetus behind such a plan, the choice of a specific goal to drive strategy, can be described in the words of business strategists Gary P. Hamel and C. K. Prahalad as "strategic intent."[11]

The idea of strategic intent is as old as Sun Tzu, and as primordial. As the ancient strategist would have it,

> The first of [the "five fundamental factors" of warfare] is moral influence [the *tao*] . . . the fifth, doctrine. . . . By moral influence I mean that which causes the people to be in harmony with their leaders, so that they will accompany them in life and unto death without fear of mortal peril. . . . By doctrine I mean organization, control, assignment of appropriate ranks to officers, regulation of supply routes, and the provision of principal items. . . . If you say which ruler possesses moral influence . . . in which regulations and instructions are better carried out. . . . I will be able to forecast which side will be victorious and which defeated.[12]

Later he would claim, "Thus a victorious army wins its victories before seeking battle; an army destined to defeat fights in the hope of winning."[13] In this instance, careful preparation obviates uncertainty and ensures victory. Stripped of the military specifics, Sun Tzu's understanding of strategy therefore boils down to the following: the successful commander must have an inspiring vision (the *tao*) that can mobilize their followers, and must then unify the organization around that vision, adapting their policies to these intentions.

The specific phrase *strategic intent* as a description of this concept was coined by Hamel and Prahalad in 1994 in their book, *Competing for the Future*, in which they advanced the thesis that firms' market performance ultimately depends on their leaders' vision rather than mere short-term efficiency. In effect, Hamel and Prahalad's thesis, expanded on in several articles, can be understood as arguing that the most dramatic strategic gains a firm may obtain are the ones that cannot be inferred or foreseen from its current assets and strategic position; they are instead an outgrowth of the implementation of a new concept or vision. Basing one's decision making solely on an analysis of already existing factors invites stagnation and, in the long run, surrenders necessary initiative.[14] "Competing for the future" effectively involves changing an existing game through a new set of goals and direction.[15] Strategic intent is therefore best understood as an intense focus on a goal deemed worthwhile—an implementable concept that can shape one's future so that the future does not shape oneself instead.

Along this line, Hamel and Prahalad highlight the concept of "stretch" in their 1993 *Harvard Business Review* article, "Strategy as Stretch and Leverage," in which they describe an interesting strategic paradox for corporate strategy: businesses that dramatically lose market share to challengers often have dramatically more resources at their disposal than their challengers. Stretch is literally the gap between one's aspirations and what is known to be possible given one's current resources. Counterintuitively, Hamel and Prahalad argue that stretch is actually necessary to avoid the stagnation and loss of initiative they see as limiting one's long-term prospects.[16]

Indeed, the competition involved between an organization that is pursuing stretch and one that is not or cannot is inherently asymmetric. Hamel and Prahalad posit, as archetypical examples, a hypothetical industry leader, "alpha," and an industry underdog, "beta." Alpha, defending the status quo, is concerned with using existing resources according to existing processes and views strategy as simply a matter of attrition,

Some Principles of Competitive Strategy

outlasting its competitor while doing what it has always been doing. " 'Where do you go,' alpha's managers ask themselves, 'when you're already number one?' "[17] The answer of course, is nowhere—one can only maintain one's position at the top.

Beta, by contrast, is creative. It cannot win the competition by adopting alpha's processes; indeed, by alpha's reckoning, with the resources that it has, it cannot win at all. Yet, beta engages in what Hamel and Prahalad refer to as stretch—dreaming big and then finding resources it previously did not have to achieve its intended goal of unseating alpha. In finding such resources, it cannot count on obtaining more of what alpha has; it must instead take what the British military theorist B. H. Liddell Hart referred to as the "indirect approach," avoiding confrontation in favor of finding a weak spot:

> Beta, on the other hand, is likely to adopt the tactics of guerrilla warfare in hopes of exploiting the orthodoxies of its more powerful enemy. It will search for undefended niches rather than confront its competitor in well-defended market segments. It will focus investments on a relatively small number of core competencies where management feels it has the potential to become a world leader.[18]

In other words, having decided to exceed conventional expectations of its capabilities in pursuit of a goal, beta, the weak challenger, must find efficiencies and force a decision—focusing its limited resources on only those options likely to produce the highest payoff and employing ingenuity to stretch the limits of its capabilities.

Although it does not appear to follow that alpha will necessarily waste resources, Hamel and Prahalad suggest that there is a powerful psychological incentive for alpha's managers to do so. Locked into a conventional mindset, suffering from organizational inertia, possessed of inhibited morale on account of its inability to progress further, and sustained by a belief in its own superiority, alpha's natural state is one of lethargy—doing the same thing today that it did yesterday, and not bothering to coordinate its operations except to some limited extent.[19] By definition, to unseat alpha, beta must use resources more efficiently and more creatively, finding opportunities that neither party previously knew existed. Therefore, by definition, beta needs a strategy to force a decision; alpha can make do without one for some time, until the threat from its rising competitor can no longer be ignored.

As will be shown, there is a direct comparison here to competition for regional hegemony of the type engaged in by Britain and its challengers during the Dreadnought Race. The sitting hegemon is in an identical position to the industry leader, concerned with maintaining its position since it cannot go anywhere but down. The challenger is aspirant, seeking to dislodge the hegemon by employing the limited resources at its disposal. *Mutatis mutandis*, the same analysis can apply: the challenger, with limited means and beginning from a weaker position, must mobilize effectively behind the goal of unseating its rival, while the sitting hegemon is in danger of strategic drift, with little to aspire to and powerful internal incentives to ignore the challenger's focused opposition.

The sitting hegemon must play a long game, setting a goal of maintaining its superior position and mobilizing its resources toward that goal, wearing down the challenger over time and, by so doing, seeking to win the "race with no finish line." The way to do so, as noted in the preceding chapters, is to persuade the challenger to quit the race, by convincing them through practice that its challenge is hopeless. By contrast, the challenger, behaving as "beta," must be proactive, focusing its limited resources on obtaining the capabilities it needs to unseat its rival. The point is echoed by the business strategist Kenichi Ohmae, whose concept of "foresighted decision making" requires that, "of the many strategic options open to the business, only a few may be chosen. . . . By concentrating more resources in support of fewer options, the company gains a bigger edge over its competitors."[20] Moreover, the challenger must force a decision, since neither the status quo nor a long game of attrition ultimately favors the incumbent's opponent. Even if the sitting hegemon eventually quits the race, the challenger is likely to have to force a crisis to obtain this result; it cannot assume that the sitting hegemon's retrenchment will occur automatically. Conversely, the sitting hegemon may, like an industry champion, experience a tendency toward avoiding strategic decision making. Until its position is threatened by a challenger, it may focus its resources on indulging other priorities than maintaining its position.

It is here that more international political analysis may be incorporated. As noted in the preceding chapter, with regard to international rivalry broadly, the question of how to terminate competition has been addressed recently by Karen Rasler, William R. Thompson, and Sumit Ganguly in their study of international rivalries, *How Rivalries End*. The authors argue that a combination of factors influences the end of such rivalry. Changes in "expectations"—specifically, the expectation of military attack—are necessary; the authors also argue that "shocks"—unexpected events that offer impetus for changes in policy—as well as "policy entrepreneurs"—leaders who seek a change in the status quo—and potential involvement by outside powers can all contribute. They contend, however, that the major contribution to the end of rivalry comes from "reciprocation"—the willingness of the adversary to signal a similar willingness to drop the rivalry.[21]

This does not have to be the only way in which a "finish line" can be created. One very viable alternative is simply warfare. Although this study is not about war per se, the framework offered here does not preclude the possibility of war as a final step after the competition has played out, provided it is accounted for by beta when seeking an outcome. Equally, collapse, bankruptcy, or other abandonment on alpha's part can constitute a finish line. Whatever the finish line consists of, it

Chapter Two

is incumbent on beta to produce one; equally, alpha will try to prevent this from happening (until beta is forced to quit).

The implications of a lack of a strategy are worth discussing. The need for strategic planning appears obvious in theory, but it is less so in practice. In fact, since planning amounts to a decrease in entropy, it requires some effort—and therefore some impetus to make that effort seem worthwhile—to begin it. The natural state of a reasonably secure competitive organization is, in fact, not to have a strategy. In the words of former management strategist (and management theory skeptic) Matthew Stewart,

> [I]t is important to note [that] the idea of strategy, like the owl of Minerva, typically arises just as the sun is setting on an organization. An old saw has it that strategy is when you're running out of ammo but you keep firing on all guns so the enemy won't know. As a rule, corporations turn to strategy when they can't justify their existence in any other way, and they start planning when they don't really know where they're going.[22]

Stripped of Stewart's cynicism, this assessment has merit: an organization that perceives itself to be doing well may not decide to expend resources on determining a strategy.

H. Igor Ansoff, one of the intellectual founders of the business strategy discipline, poses the question of why strategy is necessary as a matter for discussion in his book, *Corporate Strategy*. For him, the lack of a strategy entails maximum flexibility:

> [T]he alternative to strategy ... is to have no rules beyond the simple decision to look for profitable prospects. Under these conditions the firm does not select formal objectives, performs no appraisals, formulates no search and evaluation rules ... it would evaluate each new opportunity on the merits of its individual profitability.[23]

Allowing for the necessary change of perspective from firm competition to interstate competition, Ansoff's understanding of a lack of strategy implies a reactive approach: no plan or formal direction and no focused search for opportunities, but rather the mere seizing of useful opportunities when they present themselves. Ansoff acknowledges that there is some merit to this approach: no resources must be devoted to planning and coordinating actions, maximum flexibility of decision making is maintained, and decisions can be postponed until all necessary information is available. But he rebuts these advantages, noting that the firm cannot process information if it cannot focus its search for opportunities, that it has no yardstick with which to assess its own decisions, and that it cannot coordinate its actions. For Ansoff, having a strategy is superior to not having one for these reasons alone.[24]

However, for the sitting hegemon (the alpha in Hamel and Prahalad's formulation), with nowhere to go but down, one may hypothesize that passive reactivity may prevail over concerted strategy until a challenger emerges. A major challenge for the sitting hegemon is, at minimum, to recognize an emerging challenger early, when there is time to formulate a strategic intent of meeting the competitive challenge and winning the competition. For the challenger, on the other hand, even the possibly dubious advantages that Ansoff cites of lacking a strategic plan are false. The challenger is trying to undo the status quo, initially from a position of inferior resources; therefore, if it wants to unseat the hegemon, it must formulate strategic intent and focus its resources on implementing it.

The point is not, however, to adopt a plan so rigid as to lock out most options, or to seek to anticipate all possible contingencies.[25] On the contrary, strategic intent should set the tone for a competitor's strategy, and a plan should follow from it that coordinates the competitor's component parts, but not to the point in which that plan becomes a substitute for correct decision making in the face of new information. The military concept of a "commander's intent" is instructive here, being nearly synonymous with the idea of strategic intent: ideally, there should be a goal set by the organization's institutionally appointed decision makers that is promulgated throughout the organization and that serves as a guide to action whenever a decision must be made.[26] It is then possible to allocate resources according to the strategic intent's widely understood principles.

Allocating Resources

As noted above, a strategy, when properly understood, involves a choice among a set of options. Indeed, the business strategist Richard P. Rumelt identifies as a key component of "bad strategy" the failure to choose among alternatives and, likewise, the failure to identify a means by which the desired end is to be pursued—what he refers to as a process of "mistaking goals for strategy."[27] A strategy therefore entails a focus: one chooses to do this, not that; one allocates resources in pursuit of one option, not another. Although, as noted by Ansoff, the lack of a strategy can allow maximum flexibility in responding to one's environment, it does not allow for the coordination of an organization's resources in a proactive manner. The question, then, once a strategic intent has been formulated and communicated throughout the organization, is how the means available to the organization are to be allocated to achieve it. Although resources are generally thought of as anything that an organization possesses that may be useful in the pursuit of some goal, it is important to note that given the dynamic nature of competition in any field, what matters is not merely what resources one has but what resources and capabilities one acquires over the course of the competition.

A fundamental concept in business strategy is that of *competitive advantage*. The term is credited to Ansoff, who originally defined it as "characteristics of unique opportunities within the field defined by the product-market cope and the growth vector . . . [the] particular properties of individual product markets which will give the firm a strong competitive position."[28] More famously, competitive advantage was adopted and analyzed by

Michael E. Porter in his eponymously named book. Porter characterized competitive advantage as the possession of either an advantage in production costs or a way of distinguishing its product from competition.[29] The term, as originally conceived, was therefore understood as something external to the organization—a feature of the environment in which it chose to operate that conferred on it the ability to compete with relative ease, or else a happenstance position that involved choosing a segment of an existing market.

The idea of competitive advantage, however, can be understood in different terms as a feature of the organization that can be nurtured or neglected. On this point, Hamel and Prahalad are credited with coining the term *core competence* to describe a particular type of resource allocation conferring a competitive advantage.[30] In their view, once strategic intent has been formulated, it should be aligned with factors within the organization that constitute that organization's unique strengths. These strengths are referred to as core competencies, and in Hamel and Prahalad's formulation should have several key characteristics. They should be fundamental to the organization's work and applicable to a wide range of operations—in Hamel and Prahalad's words, where a business' core competencies are concerned, "a core competence provides potential access to a wide variety of markets."[31] Secondly, they should make major contributions to the organization's work—essentially offering a deep impact as well as a broad one. Additionally, "A core competence should be difficult for competitors to imitate."[32]

In the different context of competition for regional hegemony, a core competence can still be understood in a similar fashion, although the exact applicability of the concept remains to be explored. A state competing for regional hegemony should have available to it certain intrinsic capabilities that meet the above standard: broadly and deeply applicable to the competition and difficult for the rival state to copy. It is easy to see this concept applied to military matters, but it could as easily be applied to an economic advantage or a political one. As noted by Hamel and Prahalad above (and echoed by Ohmae), the challenger in such a situation will have to develop and nurture core competencies, eschewing courses of action and allocations of resources that do not advance this goal. Ultimately, to meet the challenge, the sitting hegemon will have to maintain and advance its own core competencies in the same manner. Although the concept of a competitive advantage may apply differently in interstate competition as opposed to in firm competition, it should nevertheless still apply. The strategic plan, therefore, must not only allocate resources to achieve the strategic intent, but it must also focus on developing capabilities that allow the state more resources with which to compete.

Orchestrating Simultaneous and Sequential Engagement: Anticipating a Contingent Event

A plan must have specifics. Where competitive strategy is concerned, a good strategy must bring together the actions and activities of the various parts of an organization—in this case, a state—over and through time, specifying which are to be done when and allowing for possible developments and the unknown.[33] It is necessary to chart a middle course between anticipating as many contingencies as possible and leaving flexibility to adapt to events in pursuit of overall strategic intent.[34] Nevertheless, in execution, a strategy is a series of decisions that relate to one another.

U.S. Air Force colonel John R. Boyd, in a famous series of lectures on strategy, identified the elements of decision in warfare as links in a cycle, which he referred to as the "OODA loop," the acronym standing for *observe, orient, decide, act* (Figure 1).[35] First, situational data is gathered (*observe*). Next, that data is processed in light of what decision makers already know or think they know (*orient*). Subsequently, a decision is made (*decide*) and then executed (*act*). The loop occurs when the decision maker reassesses the situation in light of their actions (returning to *observation*). The loop is also referred to in the same literature simply as a "decision cycle."[36]

Some Principles of Competitive Strategy

Figure 1. The OODA loop, as drawn by Col John R. Boyd, USAF

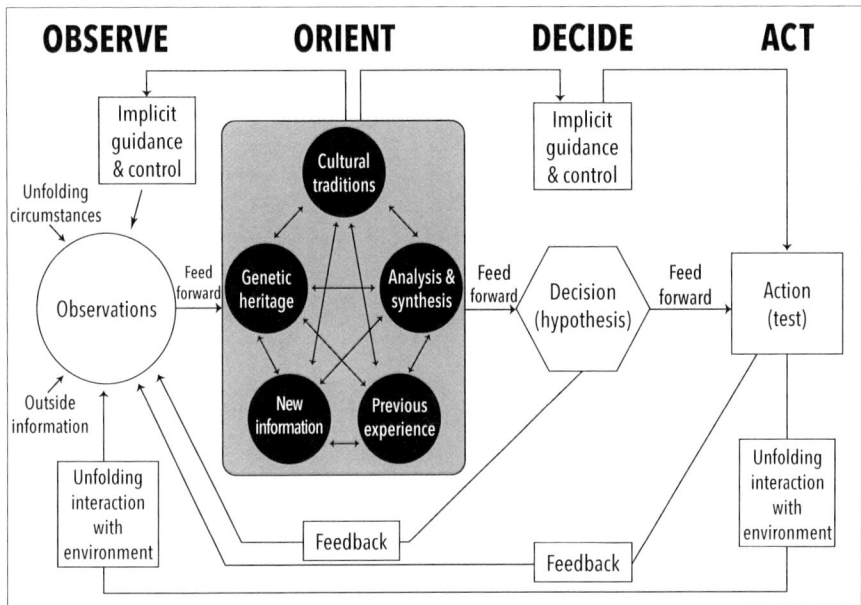

Source: Graphic attributed to Col John R. Boyd, USAF (Ret); found in Chet Richards, "Boyd's OODA loop" (PowerPoint presentation published online by the Defense and the National Interest, 2006), slide 7. Modified by MCUP.

Errors may occur at any step of the process, though in three of the four steps the conception of how they may be made is quite straightforward. In observation, data may be incorrect; in deciding, a decision maker may make a counterproductive choice; and there may also be a simple failure to properly execute plans (errors in action). The concept of orientation, however, is undoubtedly the most complex, and it is also the one in which perhaps the most significant errors can be made. Orientation, as used here, may be understood as being analogous to obtaining one's orientation in physical space: it involves processing what one knows about other things in relationship to oneself. Rather than process data about one's location relative to physical objects (as one does when one "orients" oneself in space or on a map), one orients oneself in this context by processing whatever data is deemed situationally relevant (this in itself involves a process of understanding relationships) to determine what they mean in light of the one's prior understanding, values, end goals, and plans, among other factors. In effect, orientation may be understood as the decision maker's construction of reality—the interpretation of relatively hard "facts" in light of subjective understanding, experience, intuitions, and assumptions.[37] Critical errors can be made here if the process of understanding the data is wrong for the decision maker's goals, or if it produces results at odds with objective reality, in cases where it is appropriate to speak of objective reality. Classically, what decision makers think they "know" about reality might be in some sense wrong, or a better model might exist, or old assumptions might not be applicable to a new case. Whatever the case, the programmer's maxim "garbage in, garbage out" applies: it is possible to orient oneself badly if there is a mismatch between one's conclusions about the data and what, at least for purposes of the decision maker's goals, is "really" going on.[38]

In his briefings on conflict in which he introduced the concept, Boyd saw the OODA loop as self-reflective as well as iterative: not only did one go through the loop before starting again, but one was engaged in multiple loops continuously. As figure 1 notes, one is engaged in all four steps of the loops simultaneously even as they reflect back on themselves—feedback from decisions and orientation leads to more observation even as decisions are driven into action, and so on. This understanding is critical, because the interaction of one's own loop with the adversary allows for multiple ways in which one competitor or the other may obtain an advantage or force the other side into a downward spiral.[39]

There are essentially two ways in which the OODA loop has relevance for a competition or conflict. The first involves the speed of the loop, and the second involves its interaction with the adversary.

As to the first: although it is obvious that the OODA loop can be executed well or poorly depending on what errors are made along the way and how significant they are for the overall strategy, Boyd further suggested that, *all else equal*, the faster decision maker in a game held an advantage.[40] An organization's OODA loop was therefore a source of competitive advantage or disadvantage, depending on how the various stages steps of the process were conducted within the organization. An organization that could dispense with nonessential processes in going through the loop, and therefore go through it faster and make more decisions in the same amount of time as its opponent, was more likely to achieve success at the opponent's expense. This was the case not only because, in a zero-sum contest, the actor with the faster decision cycle was more likely to gain control of whatever was being disputed before its opponent—in a battle, it might mean hitting the enemy before the enemy could strike at all—but also because the actor with the faster decision cycle could devote some of the surplus cycles to interfering with its their opponent's own cycle.[41] A faster decision cycle can also be a product of competitive advantage and core competence: the better one is at a particular action or process—the less a person or organization has to "think" to do something—the faster that action may be taken. One might hypothesize that an organization operating in familiar territory (physically or intellectually) would (again, all else equal) have a faster decision cycle, to its competitive benefit.

This, however, is only part of the concept, and not the most important. The second aspect of the OODA loop's relevance involves the ways in which adversaries interact. Boyd's briefings, as well as his (few) extant writings on concept generation, involve an interest in worldview construction—as noted above, orientation is the most complex and involved step of the OODA loop, and the one with the most potential for mischief. If one can impede an adversary's ability to conceptualize reality, either by deception involving what is really going on or by interfering with the adversary's ability to process it in some way, one can cause the adversary to, in Boyd's words, "become unglued." A faster decision cycle can do this—Boyd, drawing on his own experience as a fighter pilot, used the example of a pilot in a dogfight moving so quickly as to cause the opposing pilot to panic—but there is more potential than just this. If one has a superior ability not only to perceive reality but to manipulate it, one can render an adversary's decision cycle ineffective or even self-destructive. "[W]hat you want to do," Boyd opined in one briefing for which a transcript exists, "is generate a mismatch between that which he perceives and that which he must react or adapt to."[42]

To use the military context in which the OODA loop was first conceived, among two otherwise evenly matched opponent armies, the one with the faster decision-cycle not only can strike first, but in doing so it can potentially destroy its opponent's ability to strike at all by, for example, severing communications lines, destroying headquarters, killing officers, or putting needed units to rout. In this way, the slow enemy is harmed twice—first, by the damage inflicted by the initial move, and then because the faster side could act while the slow enemy was still trying to make its first moves, by the disruption of its ability to respond and its own psychological cohesion. A vicious cycle can then begin—for one side, success would beget success; for the slower side, failure would beget more failure. Boyd discussed this in particular with reference to aerial combat: "So any kind of maneuver where you're doing these very wild kind of maneuvers, and shifting from one direction to another, and you maneuver more rapid than me [sic], you're going to gain leverage on me."[43]

One can also, as per the second aspect of the OODA loop's relevance just discussed, mobilize other elements of one's organization to confuse or impede the enemy's decision cycle, again with the goal of paralyzing the enemy. This is particularly the case if one can interfere not only with the enemy's ability to decide and act but also with the more cerebral parts of the cycle, by preventing the enemy from getting needed information or causing the enemy to misorient. Boyd and his academic followers attributed the success of the *blitzkrieg* (and of similar combined-arms warfare) to this phenomenon.[44]

Boyd, in proposing this idea, was consciously borrowing from Sun Tzu, in particular from the latter's injunction that the best way to wage war (or perhaps to make national security policy) is to "attack the enemy's strategy."[45] Sun Tzu, in the same context, urged the head of state to "break up [the enemy's] alliances," again effectively citing the importance of interfering with resources an opponent might take for granted in order to disrupt his plans.[46]

Resources being finite, it would appear that there is liable to be a tension between quality and quantity where decision cycles are concerned, in that going through the OODA loop quickly but making poor decisions or situational judgments may not be helpful. The ideal is to be both faster and more accurate than one's opponent (or to minimize the relevance of inaccurate decisions), but if resources must be expended to get the decision cycle "right" as well as go through it quickly, then a competition between an opponent that emphasizes the former and one that emphasizes the latter may not have a competitor with a clear advantage. There is a risk, if one insists on being "right," that one's decision will be overtaken by events; there is also the obvious risk that a decision made in haste will be a costly blunder. Despite this, Boyd and his subsequent editors appear to have conceptualized the loop in an opposite manner—in their formulation, "improving" one's decision cycle meant getting both the orientation and the decision correct.[47] Exactly how this was to be accomplished, and how it differed conceptually from a simple simultaneous improvement in quality and quantity, is unclear, unless what is meant is simply that one should get better in the realms of both quantity and quality, reducing the concept to somewhere between a truism and a tautology. Be this as it may, an improvement either in orientation or decision making constitutes obtaining an advantage that did not previously exist.

Boyd was not entirely consistent in the application of this concept, but the fullest understanding of it is likely to be gleaned from the total corpus of his work, and specifically with reference to an essay he wrote titled "Destruction and Creation." This essay is concerned with the construction of mental models of reality and has wide application. It has been suggested, however, that Boyd's OODA loop is best understood as a corollary to the essay's thesis, which can be crudely summarized as suggesting that the refinement of mental models of reality detaches them from it by excluding new information. Boyd analogized this to Heisenberg's uncertainty principle, in that the refinement of a concept is similar to attempting to pinpoint the position of a particle. In effect, Boyd suggested that a master strategist confounds their opponent's conception of reality. This is not entirely clear in his initial presentation of the loop in which its principal application was to a dogfight between opposed pilots, in that in the original example the defining variables were speed and initiative, but it is not inconsistent with it: the ideal for a fighter pilot is to so confuse the opponent that they cannot maneuver.[48] To adopt the preceding chess analogy, the Boydian ideal might well be to somehow achieve the ability to make moves for one's opponent, or at least blind them to the arrangement and movement of pieces—or even rewrite the rules. Both understandings of the OODA loop are relevant here and are to be applied as part of the framework.

The applicability of these concepts to interstate competition in general terms should be easily understood. Indeed, when referring to the type of strategic disruption that Boyd later came to apply to purely military affairs, Sun Tzu was actually describing political decisions. It should be possible, then, to apply these concepts to competition for regional hegemony. A state that can cause an opponent to misread its intentions in a manner favorable to its overall competitive strategy (to misorient) or can interfere with its opponent's political process in a meaningful way (disrupting its decisions) could have an advantage, as could a state that simply acted quicker in a crisis.

Indeed, as U.S. defense strategist Andrew W. Marshall noted in a famous paper on strategic arms competition with the Soviet Union, at least in the abstract, one can take actions in an arms race that force one's adversary to compete in areas in which it is likely to have a disadvantage—effectively, a disruption of an adversary's decision cycle.[49] Ultimately, however, such decisions must be assessed in light of the goals they seek to achieve and the costs that they run. It is to such means of assessing progress that this study now turns.

Achieving an Organization Objective

A strategy can only be judged against a specific and observable set of results to be achieved. Given the employment of business strategy literature here in an interstate context, one might note that a major point of difference between a firm's business strategy and a state's political strategy is the relative clarity of the metrics involved. In the case of business strategy, a firm can easily gauge its success by measuring its performance against a quantitative metric, such as profits, shareholder value, market share, or something similar.[50] The metrics for the success or failure of a state are more elusive unless, as Peter D. Feaver et al. note, the state blunders so badly that it ceases to exist or suffers a catastrophic setback.[51] This is particularly true since, as Jeffrey W. Legro and Andrew Moravcsik note, theorists are not in agreement about what states seek, and since, in practice, there is no reason why they should necessarily compete for the same thing at all times.[52] Again, as noted above, competition may continue with no end in sight, making assessing "victory" difficult or meaningless. Yet, it would seem premature, not to say counterintuitive, to suggest that a state that has chosen (whether freely or for lack of a better option) to compete with other states cannot play the game relatively badly or well.

In the context of this study, the prize of competition is regional hegemony. Yet, it would be overly simplistic to impute a one-size-fits-all set of objectives to each competitor. Each such competition is unique in its scope and must be analyzed as such. Ultimately, the costs as well as the benefits of a strategy must be thrown into the scale: the success of a strategy in this context is contingent not only on whether the goal of protecting or attaining regional political and material supremacy has been met but also what that achievement cost. Such metrics are particularly important in the course of such competition: ultimately, the leaders of the states involved must know whether they are on the right track, and this involves knowing not only how close one appears to be on one's objective but also what one is expending and what opportunities one is giving up.

For the purposes of this study, the range of goals that a state may pursue through competition can be effectively categorized through the use of three metrics, roughly corresponding to the preferred emphases of the three major schools of thought in the international relations discipline. In the security realm—the emphasis of realists—a state may pursue goals ranging from mere security (which systemically may be a finite thing if the security of one state entails a threat to others) to the acquisition of military power (the degree to which power is acquired determining the intensity of the competition). Whatever the case, the state is measuring success in terms of how it can hurt other states and how other states can hurt it.[53] In the area of finance and citizens' individual welfare—the purview of various liberal and institutionalist schools—a state may seek prosperity either for the sake of its domestic constituents, as Robert Gilpin notes, or for the sake of other objectives, since wealth can be used to purchase other goods. Whatever the case, it counts its "winnings" in terms of the material goods it can distribute.[54] In the realm of ideas—the province of constructivists—a state at minimum may compete to ensure that its domestic constituents are able to preserve or enhance their mode of political and cultural life as they see fit, and at maximum may seek to defend or impose a cultural norm internationally or pursue a nonmaterial goal whose meaning is determined by the state's relationship to the system at large.[55] These three goals will be referred to hereafter as security, finance and welfare, and intangibles, and will be treated as metrics by which to gauge the results of states' actions in competitive scenarios.

These three metrics are to some extent fungible, in the sense that each is necessary for, and can be converted into, the other two; for this reason, they also constitute resources as well as policy aims. Money buys guns and security makes economic life possible; wealth can be traded for a preferred cultural practice (e.g., in the United States, guaranteeing rights to all citizens requires that the government spend money on courts and legal assistance that other governments might not choose to spend), which can in turn be representative of a way of life that must be defended against aggressors; religion, ideology, or nationalism can increase a population's commitment to its country's defense; and so on. As Gilpin notes in discussing the aforementioned guns-butter tradeoff, each state will somehow opt for a mixed basket of security and nonsecurity spending that satisfies its constituents' preferences. It is likely that there is a point of diminishing return in the case of each of the three metrics, at which point is ceases to be useful to acquire any more of one and relatively more useful to acquire more of the other two. However, this is not to say that all preferences are equal. A given allocation of resources among these three metrics, on the one hand, or a given preference for pursuing one over the others in a particular case, on the other, is likely to have consequences for the state in the course of competition. One can further conceptualize this set of tradeoffs in terms of investment and consumption: while security and ideology at the expense of wealth can be thought of as consumption, obtaining wealth at the expense of one of the others can be thought of as investment, as can the sacrifice of some amount of any of the three to obtain proportionately more of the same in the future.

The three metrics are not perfectly quantifiable. Moreover, the specific employment of these metrics is likely to be idiosyncratic and dependent on the nature of the competition involved. Certain constants may occur. For example, it is always possible to measure gross national product in some form, though the methods in which economic power are calculated are bound to change as economies change, and it is always possible to conduct budgetary analysis. But especially in the case of other, nonfinancial metrics, each must be assessed in light of the situation to which it pertains. Despite this, the fact remains that real human beings, in the course of participating in political life, do somehow manage to calculate them, however imperfectly, and as a result of this it should be possible to examine how states—and those who run them—calculate them and, where quantitative metrics can be found, to model the tradeoffs among them.

Each of these three metrics in turn encompasses any number of subgoals. Security can refer to protection from harm at home, to protection from harm in any given region (e.g., maintaining an advantageous military balance), to obtaining effective military and political control over a region. Finance and welfare may take the form of more goods to distribute to citizens, or a tax cut, or deficit reduction. Intangibles may take either an intersubjective or purely subjective form, and they may be constructed either by one state alone or by all of the states involved. The possible objectives may range from an ability to live freely in accordance with one's principles, to maintaining prestige among fellow states, to rewriting (or maintaining) principles of international law so that they fit with a domestic conception of how the world should behave.

What is crucial is not the exact form that goals take under these metrics, but that they form mutually exclusive strategic mindsets. Security involves the realist assumption that states may fight each other and that military power and the comparative absence of military threats are goods that are worth pursuing. Finance and welfare involve the liberal assumption that the goal of the state is the well-being of its citizens in material terms, to the exclusion of other objectives. Intangibles, in turn, involve a belief that the nonmaterial is worth sacrificing for, even at a cost of either military power or material wealth. One may note, anecdotally, that it is often difficult to persuade a person with goals in one of these areas to treat goals in another as worthwhile, a problem that encapsulates the difficulties of the states involved in the following case studies, as decision makers attempted to persuade others to given them what they needed to accomplish their goals.[56]

Competition may occur when a state seeks to pursue any of these three objectives at another state's expense. Once it does occur, a state's ability to allocate resources properly in pursuit of the objective should determine its relative success or failure. In the case of competition for regional hegemony under study here, one may assess not only whether the goal of the competition in its particular context is being met, but also the waxing and waning of these metrics in the process.

A competition for regional hegemony may emphasize any of these three metrics. Competing for an advantageous military balance in a region is not the same as competing for a controlling stake in a gold mine or to be a major financial hub, and this, in turn, is not the same as competing to be the one to write the rules for anything from international trade to human rights norms. Any of these could constitute an essential component of regional hegemony (for some value thereof) or the means to obtain that essential component. What is important is that any given competitive scenario entails competition along at least one metric, and, as shown in the coming chapters, a willingness to make sacrifices along one or both of the other metrics to obtain it. The ability to do so skillfully is the ability to create effective competitive strategy. The following case study serves demonstrate how this may be done.

Conclusions

From the above analysis, one can derive a few basic principles that should be applicable to the analysis of international competition for regional hegemony. They can be summed up as follows:

- Although states do not necessarily make decisions as unitary actors, it is possible to speak of them as having

strategies if one analyzes their actions and their leaders' intentions over time. This is not to say that their strategies will be coherent or correct, and much may be said in any given case as to why this is not so and how matters might be different.
- Strategy requires a strategic intent, an organization goal that is to be both a focus of effort and a guide to resource allocation.
- States contesting regional hegemony are not equal at the start of the competition. One is initially the reigning hegemon, while the other is challenging that state's supremacy. Whatever the means available or employed, their starting positions dictate asymmetrical strategic approaches.
- The sitting hegemon is less likely than the challenger to have a strategic intent, because until it is forced to recognize its challenger and compete it will not have a direct incentive to focus its efforts. If and when it does, its strategic intent is to remain supreme, and it must therefore ensure that in the "race with no finish line" that follows no finish line emerges, ultimately forcing the challenger to drop out.
- The challenger, starting from a weak position, must, if it is to compete at all, possess the strategic intent to unseat the hegemon. It must focus its resources to achieve a decision in its favor. It wins the "race with no finish line" by creating one and crossing it. Although war is usually to be seen as detrimental, it remains one option for a "finish line."
- A strategy must allocate resources. It must make use of its parent state's strengths and avoid playing to its weaknesses. As noted by Andrew Marshall, it is often possible to "move" the competition in one's favor by making one's strengths more applicable while neutralizing an opponent's strengths.

Some Principles of Competitive Strategy

- Strengths can be seen as the "core competence" or "competitive advantage" addressed by business strategy. Ideally, they should be unique or hard for the opposing competitor to imitate. The asymmetric nature of competition for regional hegemony should be a fertile ground for such core competencies, since one competitor's strength may not be relevant if applied from the opposite side. Competitors must nurture, grow, and protect their core competencies to succeed or, in the case of the sitting hegemon, prevent a competitor from succeeding.
- Strategies plan in time as well as space: they seek to "orchestrate" actions both together and seriatim. Following Boyd's OODA loop formulation, in implementing a strategy, an organization goes through a decision cycle in which it acquires situational information, processes it, makes a decision, and implements that decision. All else equal, an organization with a faster decision cycle will outperform one with a slower cycle, but accuracy of assessment and decision can quickly render that advantage moot. It is possible in competition to disrupt a competitor's decision cycle either by ensuring that the competitor has false information or by preventing the competitor from making a decision on their terms. Although these principles were originally designed for military conflict, there is no reason why they could not be applicable to interstate competition.
- A state's organizational objectives, against which strategy must ultimately be measured, are less well-defined than those of a corporation and less susceptible to quantitative analysis. Nevertheless, states work within a finite realm, in which material resources can be counted and even nonmaterial resources roughly assessed. In the final analysis, states have three responsibilities that they must fulfill: collective security, material welfare, and protection of cultural values. These three responsibilities constitute metrics for the success or failure of a state's

competitive strategy. Even in cases where precise quantification is impossible, it should normally be possible to model the tradeoffs between these three goals in order to assess a strategy's ultimate impact on a state's situation. Because the specific nature of these goals varies from state to state, attempts to measure them must be tailored to specific competitive scenarios.
- States within the international system often compete with one another, but, barring the direst necessity, the goals they seek in such competition—which may be categorized within the above three metrics—are also within their power to choose, and multiple equilibria are possible in any such competition. Their goals are the principal determinants of what is detrimental or beneficial within this context; in this case, without recourse to ethical judgement, a state's actions may be judged as beneficial to its "interests" if they advance those interests as the state itself defines them.
- It has been shown to be possible to devise frameworks for understanding the forces that drive competitive outcomes in an industrial context. It should be possible to devise a similar framework in the context of interstate competition for regional hegemony. This study offers such a framework.

Going Forward

Having outlined some principles by which strategy may be assessed, and having established as a goal the discovery of a more detailed framework by which international competitive strategy, as it applies to regional hegemonic competition, may be understood and analyzed, it remains to be seen how such principles may be applied in actual cases of competition.

It is important here to address an epistemological issue raised by the above-mentioned strategic principles: namely, the extent to which each principle—each conclusion—is susceptible of proof or demonstration on the one hand (implying that it is a hypothesis for testing) or, on the other hand, whether it is definitionally true, and deduced a priori, in which case its application to the realm of international competition only serves as a set of aphorisms for policy makers rather than a testable proposition that might or might not be true and that can (by its truth or falsehood) offer new insights into state behavior. In other words, it is necessary to determine whether state behavior is offering insight for strategy or whether strategic principles (taken to be true) are offering insight for states. Because of the deductive or intuitive nature of these principles, this study proceeds from a prima facie assumption of their truth and applicability. However, it does not treat them as being axiomatically true. Rather, each concept is applied to the analysis of the competitive scenario under study: the Dreadnought Race between Britain and Germany prior to World War I. To the extent that these principles are shown to require modification or qualification as applied to competition for regional hegemony, lessons are drawn as appropriate. What lessons are drawn from such application are, in the end, factored into the final analytical framework.

Chapter Two

Endnotes

1. Mao Tse-tung, *On Guerrilla Warfare*, tr. Samuel B. Griffith II (Urbana: University of Illinois Press, 2000), 73.
2. As Richard Rumelt notes, published "strategies" for national security in the United States are prime examples of "bad strategy"—various forms of vague or wishful thinking or organizational propaganda promulgated as true strategy. Richard Rumelt, *Good Strategy, Bad Strategy: The Difference and Why It Matters* (London: Profile Books, 2011), 33–37.
3. Robert Gilpin, *War and Change in World Politics* (Cambridge, UK: Cambridge University Press, 1981), 18–20, https://doi.org/10.1017/CBO9780511664267.
4. Robert O. Keohane and Joseph S. Nye Jr., *Power and Interdependence*, 3d ed. (New York: Longman, 2001), 22.
5. See, for example, Jack Snyder, *Myths of Empire: Domestic Politics and International Ambition* (Ithaca, NY: Cornell University Press, 1991), 31–60; Edward D. Mansfield and Jack Snyder, "Democratization and the Danger of War," *International Security* 20, no. 1 (Summer 1995): 5–38, https://doi.org/10.2307/2539213; and John M. Owen, "How Liberalism Produces Democratic Peace," *International Security* 19, no. 2 (Fall 1994): 87–125, https://doi.org/10.2307/2539197.
6. Graham T. Allison, *Essence of Decision: Explaining the Cuban Missile Crisis* (Boston, MA: Little, Brown, 1971), 246–63.
7. Sun Tzu, *The Art of War*, tr. Samuel B. Griffith (New York: Oxford University Press, 1963), 83.
8. Carl von Clausewitz, *On War*, tr. Michael Howard and Peter Paret (New York: Alfred A. Knopf, 1993), 99.
9. Kenneth R. Andrews, *The Concept of Corporate Strategy*, rev. ed. (Homewood, IL: Ricard D. Irwin, 1980), 2–16, 22; and Phillip Karber, "Competitive Strategy: As an Approach to Business and Professional Life" (PowerPoint presentation at the Annual Fellows Conference of the Center for the Study of the Presidency, Washington, DC, 31 October 2003), slide 17. For a criticism of this concept, see Matthew Stewart, *The Management Myth: Debunking Modern Business Philosophy* (New York: W. W. Norton, 2009), 183–85.
10. See Alexander Wendt, *Social Theory of International Politics* (Cambridge, UK: Cambridge University Press, 1999), 193–243, https://doi.org/10.1017/CBO9780511612183. Wendt's argument for corporate agency in the case of states can, in turn, bolster the argument that a state's strategy can be analyzed even if its components are the result of different individuals making

different decisions. The cases studied here do show frequent instances of decision makers being influenced by popular opinion, bureaucratic politics, parliamentary friction, and individual idiosyncrasies. Once again, however, the fact that a multitude of minds forges or influences a strategy should not be construed to mean that one does not exist.

11. The term was originated by Gary Hamel and C. K. Prahalad; see Hamel and Prahalad, *Competing for the Future* (Boston, MA: Harvard Business School Press, 1994), 129–33. It is used in Karber, "Competitive Strategy," slides 3, 5, 16, and 20.

12. Sun Tzu, *The Art of War*, 64. See also Sun Tzu, "The Art of Warfare," tr. Roger T. Ames, in *The Book of War: Sun Tzu, The Art of Warfare, and Karl von Clausewitz, On War*, ed. Caleb Carr (New York: Modern Library, 2000), 73–74.

13. Sun Tzu, *The Art of War*, 87.

14. Hamel and Prahalad write, "The important point is that the commitment to be a pioneer precedes an exact calculation of financial gain. A company that waits around for the numbers to 'add up' will be left flat-footed in the race to the future." Hamel and Prahalad, *Competing for the Future*, 37. They also write, "The dream that energizes a company is often something more sophisticated, and more positive, than a simple war cry. . . . *Strategic intent* is our term for such an animating dream. . . . Strategic intent implies a significant *stretch* for the organization. Current capabilities and resources are manifestly insufficient to the task. Whereas the traditional view of strategy focuses on the 'fit' between existing resources and emerging opportunities, strategic intent creates, by design, a substantial 'misfit' between resources and aspirations." Hamel and Prahalad, *Competing for the Future*, 141–42. Strategic intent involves a commitment to pursue a large goal and to mobilize all resources as well as ones yet unfound in pursuit of the effort—a commitment to avoid "business as usual."

15. Hamel and Prahalad write, "In short, strategy is as much about competing *for* tomorrow's industry structure as it is about competing *within* today's industry structure." Hamel and Prahalad, *Competing for the Future*, 42.

16. Gary Hamel and C. K. Prahalad, "Strategy as Stretch and Leverage," *Harvard Business Review* 71, no. 2 (March–April 1993): 75–84.

17. Hamel and Prahalad, "Strategy as Stretch and Leverage," 5.

18. Hamel and Prahalad, "Strategy as Stretch and Leverage," 6. See also B. H. Liddell Hart, *Strategy: The Indirect Approach*, rev. ed.

(London: Faber and Faber, 1967), 338–42; and B. H. Liddell Hart, "Foreword," in Sun Tzu, *The Art of War*, vii. In Liddell Hart's view, the aim of military strategy should be, contra Clausewitz, the avoidance of attacks on the enemy's main force in favor of attacks on weak points that allow one to surround and bypass the enemy, or even render a conclusion by combat foregone so as to obviate the need for fighting. This theory led to Liddell Hart's advocacy of mobile, combined-armed warfare as an alternative to the attrition warfare he experienced in World War I. One may debate the degree to which Clausewitz and Liddell Hart actually disagreed—Clausewitz did not necessarily advocate frontal assaults on fixed positions in all circumstances, and Liddell Hart was not ignoring the fact that one must ultimately do something to put the enemy out of action rather than avoid battle altogether—but Hamel and Prahalad clearly draw the same contrast as Liddell Hart.

19. Hamel and Prahalad do not quite say this explicitly, but it is implicit in their analysis. They write, "What bedevils Alpha is not a surfeit of resources but a scarcity of ambition." They do note that beta's use of resources will be more "focused," which in turn suggests a more refined strategic intent. Alpha, with an unfocused use of resources, does not necessarily behave as if it had a clearly identified goal. Hamel and Prahalad, "Strategy as Stretch and Leverage," 6.

20. Kenichi Ohmae, *The Mind of the Strategist: The Art of Japanese Business* (New York: McGraw-Hill, 1982), 242–43.

21. Karen Rasler, William R. Thompson, and Sumit Ganguly, *How Rivalries End* (Philadelphia: University of Pennsylvania Press, 2013), 16–20, 192.

22. Stewart, *The Management Myth*, 180–81.

23. H. Igor Ansoff, *Corporate Strategy: An Analytic Approach to Business Policy for Growth and Expansion* (New York: McGraw-Hill, 1979), 101.

24. Ansoff, *Corporate Strategy*, 102–3.

25. This is a major criticism of Ansoff's approach; see Stewart, *The Management Myth*, 178–89. The authoritative refutation can be found in Henry Mintzberg, *The Rise and Fall of Strategic Planning* (New York: Free Press, 1994), 39–40, 290–91, 334–35.

26. The commander's intent is found most famously in *Warfighting*, Fleet Marine Force Manual 1 (Washington, DC: Headquarters Marine Corps, 1989), 71.

27. Rumelt, *Good Strategy, Bad Strategy*, 31, 45–51. For Rumelt, "bad strategy" is pseudostrategy—it is a statement or plan that purports to be a strategy but lacks key elements of a strategy

as properly understood, and it is often articulated to mask the absence of a true strategy. See Rumelt, *Good Strategy, Bad Strategy*, 32. Of interest, one of the founders of the discipline of business strategy, Kenneth R. Andrews, does see strategy mainly in terms of setting goals and policies, though he does emphasize the importance of ensuring that a "strategy" is within a firm's capabilities. See Andrews, *The Concept of Corporate Strategy*, 18, 38.

28. Ansoff, *Corporate Strategy*, 99.
29. Michael E. Porter, *Competitive Advantage: Creating and Sustaining Superior Performance* (New York: Free Press, 1985), 11, 33.
30. Gary Hamel and C. K. Prahalad, "The Core Competence of the Corporation," *Harvard Business Review* 68, no. 3 (May–June 1990): 275–92; and Hamel and Prahalad, *Competing for the Future*, 226–36.
31. Hamel and Prahalad, "The Core Competence of the Corporation," 83.
32. Hamel and Prahalad, "The Core Competence of the Corporation," 84.
33. Karber, "Competitive Strategy"; and Ohmae, *The Mind of the Strategist*, 79.
34. Ohmae writes that "the true strategic thinker can respond flexibly to the inevitable changes in the situation that confronts the company.... Considering alternatives requires us to pose 'what if' questions. In other words: If the situation were such-and-such, what would be our best course of action." Ohmae, *The Mind of the Strategist*, 79.
35. What follows is a somewhat simplified elucidation of Boyd's conception of the OODA loop. Because Boyd preferred to present orally rather than in a written medium, most of his ideas must be traced either to presentation slides that Boyd prepared or to recordings of his briefings. Boyd's magnum opus, a seven-hour briefing created in 1986 titled "Patterns of Conflict," part of a folio titled *A Discourse on Winning and Losing*, has been transferred, as of 2005, from its original analog slide presentation to PowerPoint and made available online; a full-length video of Boyd's presentation of "Patterns of Conflict" is also available online. See Col John R. Boyd, USAF (Ret), "Patterns of Conflict" (U.S. Department of Defense briefing, December 1986), slides 6–8, 13, 178, 185, hereafter Boyd, "Patterns of Conflict" (slides); and Col John R. Boyd, USAF (Ret), "John Boyd Patterns of Conflict Part 1 of 7," YouTube video, online audiovisual material recorded by Jason M. Brown and posted by Steven Shack, 14 February 2015, 24:55–40:00. For a transcript of this briefing, see Col John R. Boyd, USAF (Ret), "Patterns of Conflict" (lecture, U.S. Marine Corps Command and

Staff College, Marine Corps University, Quantico, VA, 25 April and 2–3 May 1989; transcribed by Maj Ian T. Brown, USMC, 25 March 2015–11 January 2017), hereafter Boyd, "Patterns of Conflict" (transcript).

36. See, for example, Lawrence Freedman, *Strategy: A History* (Oxford, UK: Oxford University Press, 2013), 511–12.
37. Boyd, "John Boyd Patterns of Conflict Part 1 of 7," 26:00–26:20.
38. Boyd addresses this when he notes, for example, the paralysis with which existing regimes often face guerrilla operations: the guerrilla organization has caused the population to see the conflict from its point of view; the counterinsurgent does not know how to arrest this. Boyd, "Patterns of Conflict" (slides), slides 94–97. A particular emphasis of Boyd's thinking was that this did not happen always or even mostly by accident; drawing on his experience as a fighter pilot, he explains that pilots in aerial combat could be shocked into panic and poor judgment by an aggressor's superior maneuvering, which in turn could seal their own doom. Boyd, "John Boyd Patterns of Conflict Part 1 of 7," 32:00–37:30.
39. See Ian T. Brown, *A New Conception of War: John Boyd, the U.S. Marines, and Maneuver Warfare* (Quantico, VA: Marine Corps University Press, 2018), 117–19, https://doi.org/10.56686/9780997317497; and Frans P. B. Osinga, *Science, Strategy, and War: The Strategic Theory of John Boyd* (London: Routledge, 2007), 235–37.
40. Boyd was adamant, however, that it was relative tempo, not absolute tempo, of operations that mattered, and he even went so far as to suggest that varying the tempo of operation could be important—as with the second application discussed just below—in disrupting an enemy's decision cycle through unpredictability. See Osinga, *Science, Strategy, and War*, 235; and Boyd, "Patterns of Conflict" (transcript), 12.
41. Boyd, "Patterns of Conflict" (slides), slides 6–8; and Boyd, "John Boyd Patterns of Conflict Part 1 of 7," 38:15–40:00.
42. Boyd, "Patterns of Conflict" (transcript).
43. Boyd is said not to have originated the concept of the OODA loop with reference to air combat, but it is clear from his briefing that the latter was a significant contributor to his creation of it. See Boyd, "Patterns of Conflict" (transcript), 13; and Brown, *A New Conception of War*, 142.
44. Boyd, "Patterns of Conflict" (slides), slides 71–72, 80, 87, 100; Boyd, "John Boyd Patterns of Conflict Part 1 of 7," 28:10–28:25; and Col John R. Boyd, USAF (Ret), "John Boyd Patterns of Conflict

Part 4 of 7," YouTube video, online audiovisual material recorded by Jason M. Brown and posted by Steven Shack, 8 March 2015, 30:00–32:00. As a thought experiment, imagine a chess game in which a player makes normal moves but the opponent is allowed to make two moves with each turn. A player's demise would be not only practically guaranteed but also probably quite rapid, as losses would compound quickly. Alternatively, imagine a boxer who cannot strike a blow because they are being pummeled rapidly by a faster, harder-hitting opponent.

45. Boyd, "Patterns of Conflict" (slides), slide 14; Boyd, "John Boyd Patterns of Conflict Part 1 of 7," 49:35–55:03; Col John R. Boyd, USAF (Ret), "John Boyd Patterns of Conflict Part 2 of 7," YouTube video, online audiovisual material recorded by Jason M. Brown and posted by Steven Shack, 8 March 2015, 00:00–4:42; and Sun Tzu, *The Art of War*, 77.
46. Sun Tzu, *The Art of War*, 78.
47. Chet Richards, "Boyd's OODA Loop" (PowerPoint presentation published online by the *Defense and the National Interest*, 2006), slides 2, 6–7. Richards makes the key point that Boyd conceived of the OODA loop somewhat less than literally: it was not a repeated sequence, but a self-referential process that repeatedly drew on what had already happened. For this reason, the four components might be happening simultaneously as well as in sequence.
48. Col John R. Boyd, USAF (Ret), "Destruction and Creation," in *A Discourse on Winning and Losing*, 3 September 1976; and Adam Elkus, conversation with author, February 2021. An authoritative variant is found in Franklin C. Spinney, "Evolutionary Epistemology Talk at [the U.S. Marine Corps Expeditionary Warfare School, 15 January 2019]," YouTube video, posted by the Warfighting Society, 4 February 2019, which extensively links the OODA loop to Boyd's "Destruction and Creation" essay. For the more conventional (and simplistic) interpretation of the OODA loop adopted by the U.S. military, and the limitations thereof, see Stephen Robinson, *The Blind Strategist: John Boyd and the American Art of War* (Dunedin, NZ: Exisle Publishing, 2021), 340–41, 360–61. Robinson writes that "as Major Craig Tucker noted, there 'is considerable difference between maneuvering a fighter and maneuvering an army.'" This study attempts to apply the OODA loop concept in its broadest possible sense, allowing for variations in interpretation.
49. Andy W. Marshall, *Long-Term Competition with the Soviets: A Framework for Strategic Analysis* (Santa Monica, CA: Rand, 1972), 39.

50. In fairness, this has generated its own degree of confusion, in that business strategy, on the one hand, and the theory of the firm and business ethics, on the other, have not always moved in lockstep. Walter Kiechel III notes that the question of "What is strategy for?" (or "Strategy to do what?") has evolved along with the understanding of strategy. Initially, he notes, even the simple notion that strategy was supposed to serve shareholders was not intuitively obvious to corporate strategy theorists. Walter Kiechel III, *The Lords of Strategy: The Secret Intellectual History of the New Corporate World* (Boston, MA: Harvard Business Press, 2010), 202.
51. Peter D. Feaver et al., "Brother, Can You Spare a Paradigm?: (Or Was Anybody Ever a Realist?)," *International Security* 25, no. 1 (Summer 2000): 165–67, https://doi.org/10.1162/016228800560426.
52. See Jeffrey W. Legro and Andrew Moravcsik, "Is Anybody Still a Realist?," *International Security* 24, no. 2 (Fall 1999): 13–16. In this article, the authors discuss the problem of what they refer to as "fixed preferences," without which it becomes difficult to posit a default condition for states' foreign policies.
53. For the spectrum of "defensive" to "offensive" realism, see John J. Mearsheimer, *The Tragedy of Great Power Politics* (New York: W. W. Norton, 2001), 18–22.
54. Gilpin, *War and Change in World Politics*, 22–23.
55. A major limitation of most existing theories appears to be their insistence that a state will always pursue one of these goals, as well as a corresponding failure to admit that states in practice can and must pursue all three. Gilpin's candid acknowledgement that states pursue baskets of goods is instructive here. In fact, he acknowledges the impact of noneconomic ideology—specifically religion—as a motivator of state actions. Gilpin, *War and Change in World Politics*, 22. The recognition of this tripartite division of objectives is quite old. Thomas Hobbes writes, "So that, in the nature of man, we find three principal causes of quarrel. First competition, secondly diffidence, thirdly glory. The first maketh men invade for gain; the second, for safety; and the third for reputation. The first use violence, to make themselves masters of other men's persons, wives, children, and cattle; the second, to defend them; the third, for trifles, as a word, a smile, a different opinion, and any other sign of undervalue, either direct in their persons or by reflection in their kindred, their friends, their nation, their profession, or their name." Thomas Hobbes, *Leviathan: With Selected Variants from the Latin Edition of 1668*, ed. Edwin M. Curley (Indianapolis, IN: Hackett, 1994), 76. Thucy-

dides' famous concept of "fear, honor, and self interest," cited above, also accounts for the security, intangibles, and finance and welfare metrics, respectively.

56. A simple example may be found in the reaction to the terrorist attacks of 11 September 2001 in the United States. The decision to spend large sums of money on intelligence and homeland security, as well as on a military response, may (one hopes) have made the United States more secure from another terrorist attack than it otherwise would have been (a gain on the security metric). This came at an obvious cost to taxpayers, albeit one carefully hidden by deficit spending (therefore, a loss on the finance and welfare metric). It also entailed measures that damaged American civil liberties, hurt the United States' image abroad, and ruined many Americans' conception of their nation's unsullied moral goodness (an obvious loss on the intangibles metric). For a people accustomed not only to having their own way but also to believing that good things go together, perhaps the hardest part of these tradeoffs was the recognition that they had to be made at all, and would be made regardless of what combination of gains and losses Americans picked.

Chapter Three
A Framework for Strategic Analysis of Great Power Competition for Regional Hegemony

Now if the estimates . . . before hostilities indicate victory it is because calculations show one's strength to be superior to that of his enemy; if they indicate defeat, it is because calculations show that one is inferior. With many calculations, one can win; with few one cannot. How much less chance of victory has one who makes none at all!

~ Sun Tzu[1]

Therefore I say: "Know the enemy and know yourself; in a hundred battles you will never be in peril."

~ Sun Tzu[2]

It remains now to propose a framework for analysis of competition for regional hegemony, and then to test that framework through its application to an historical case. So far, this study has examined the nature of strategy as it might be applied to international competition, and specifically toward the competition between great powers for regional hegemony. Throughout history, great powers have had competitive adversaries—states that sought either to supplant their power and influence or to prevent them from doing the same. Such competition can go on for some time, and although it often precedes a military showdown, it need not necessarily do so. This study will employ a case study of a dyadic competitive relationship between great powers, namely the Dreadnought Race between Britain and Germany prior to World War I, as a means of illustrating the ways in which the strategies of the participants can be analyzed.

The counterpoint must be acknowledged: one overwhelming insight that is driven home by the case under study here (and intuitively by others) is the application of a time-honored truth readily accepted by realists: that states are often willing to fight for power and preeminence, and that once they compete for security, they have little reason to trust any settlement that might be reached. Moreover, as Sun Tzu famously claims, "there has never been a protracted war from which a state has benefited," to which one must also append Carl von Clausewitz's dictum that wars tend toward "the maximum use of force."[3] The literature of conflict resolution has provided a counterpoint to this general insight, but it must be said that exceptions to it rely on more optimism than history allows. Once competition is underway, it may lead to war; whether it does or does not, only so much precision in the use of a state's resources is possible.

That being said, although history argues for pessimism, it is clear even from the historical background to the case under study here that the notion that states must always fight to the death for power once they choose to compete for it will not bear overly close scrutiny. In fact, as noted above, studies of state rivalry have shown that deescalation is possible, and that a savvy competitor would seek to produce such a result on favorable terms. Otto von Bismarck, who served as chancellor of the German Empire in the late nineteenth century, was a master of producing advantageous political outcomes either without fighting or with short, sharp blows in limited wars that did not lead to total systemic collapse. Space prevents this study from addressing the "mother" of all great power competitions in recent memory—the Cold War—the end of which inspired the quotation which began this study (how to win a "race with no finish line"), but it is nevertheless clear from that example that it is possible for great power competition to end without a war that destroys the competitors. Although it is true that nuclear deterrence may have made the Cold War an exception, it is nevertheless true that the case studied here did not have to lead to open warfare, that other options existed for the competitors at the time, and that considerable debate exists as to the exact relationship, if any, of the competition under study here—the Dreadnought Race—to the war that followed it. In the case of the Dreadnought Race, it is possible to envision counterfactual scenarios in which, had they been sharper, the decision makers could have forced a political settlement in their favor. The states in question had agency, and their goals suggest that they at least considered such a settlement desirable.

On the other hand, the anticipation of conflict does not alleviate or obviate competition but clarifies it. A decision to compete for regional hegemony that entails a high risk of war must account for that possibility. As discussed, this means a willingness to sacrifice certain goods—namely material welfare—for either an advantageous military position at the outbreak of war or for the intangible gains that winning will bring, or both. Conversely, if a state does not wish to make this tradeoff, it must frame the competition in such a way as to make it less likely that it will have to make it.

Although the Dreadnought Race took place within a larger regional political context, it is possible to look at the competition from each competitor's point of view, and with regard to each competitor's adversary, in order to determine which decisions were more or less contributory toward each competitor's desired outcome. It is accepted here that during the Dreadnought Race, Britain and Germany were engaged in a major competition with one another, even if they did not always perceive it as such or pursued it unevenly. This is consistent with the general framework offered here: a competition for regional hegemony can occur even if other competitions are also underway or one—or even both—states are distracted. It is, in particular, possible to view the competition through the lens of strategic intent—of the marshaling of scarce resources around a competitive goal, to the exclusion of other goals deemed secondary. It is also possible, following the insights of the business strategists Gary P. Hamel and C. K. Prahalad, to note the difference in the formulation of strategic intent between a dominant, status-quo-favoring "alpha" and an insurgent, revisionist "beta." It is also possible to evaluate each competitor's goals, available resources, and strategic decisions through a threefold taxonomy of metrics derived from longstanding international relations theories: national security, welfare for one's people, and intangible goals, which correspond roughly to the assumed objectives of states in realism, neoliberalism, and constructivism, respectively.

    A general approach for analyzing such cases can now be suggested. Although existing factors may limit the utility of great power competition, one cannot assume that such will be the case forevermore, still less that there will never be lessons to learn from the analysis of great power competition in historic periods. In providing such an approach, this study will offer other analysts a possible way forward for understanding how such competition has played out in historical periods and how it may do so in future scenarios. It is this author's hope that it will also offer a chance to break out of the debate over state goals and motivating factors between adherents of the major theories by offering a chance to see how competitors in actual scenarios weigh and measure these goals and the means to achieve them—and, more crucially, to analyze present and future competitive scenarios in the same way.

Deriving a Framework

As a reminder, *strategy* is defined here as "a plan of action" and process of decision making "for the allocation of resources in anticipation of a contingent event, orchestrating simultaneous and sequential engagement, to achieve an organization objective," in the context of a contest with other organizations.[4] As was elaborated in chapter 2, each component of this definition is based on a corresponding tradition within the theoretical literature on strategy. Planning was discussed as the concept of strategic intent, based on the work of Hamel and Prahalad, Kenichi Ohmae, Sun Tzu, and others. *Strategic intent* is understood here as the formulation of a willingness to focus resources on the pursuit of an objective (the final part of the definition). As Hamel and Prahalad note, a reigning alpha and a revisionist beta will initially have differing levels of strategic intent; the more beta focuses its resources, the more alpha must concern itself with developing its own strategic intent. As was discussed in chapter 2, resource allocation requires throwing strength against weakness, utilizing core competencies to achieve a sustained competitive advantage over an adversary. Orchestrating simultaneous and sequential engagement and anticipating a contingent event requires focused decision making. As was discussed and will be recapitulated, this requires one party to ultimately to make better and faster decisions than its adversary, so as to render its adversary's strategy null. Each of these conceptual components is invoked here to assemble a framework for analyzing competitive strategy among states.

In chapter 1, *regional hegemony* was defined as a preponderance hard power (in military and economic terms) and political influence possessed by a given great power over a given region. The framework presented here governs competition that occurs within this conception of regional hegemony. Although regions can differ, a regional hegemon, if one exists, will have the above characteristics in some form. This regional hegemon is here referenced as "alpha," as an analogy to Hamel and Prahalad's similar formulation for economic competition between firms. Alpha will have been established as such for some time, and although it does not follow necessarily from alpha's position that such will be the case, there is good reason to believe that by the time it is challenged, although it will have some initial advantages, it will have become lax and complacent through inertia. Its challenger is referenced as "beta," again per the above analogy. Beta will start from a position of weakness. If it wishes to grow stronger and push alpha off its perch as hegemon, it will have to affirmatively decide to do it. It will need to allocate scarce and initially inferior resources accordingly, playing to its strengths—its *core competencies*, to once again use Hamel and Prahalad's terminology—to gain maximum advantage. It will have to focus its limited resources, avoiding distractions that do not contribute to its goal of unseating its competitor.

There are limitations to this formulation that must be acknowledged. The chief among them is that, in part for simplicity's sake, this framework is dyadic, focusing on only one hegemonic alpha and one challenger beta. In any real situation, it must be acknowledged, any given alpha may face several betas. Likewise, a beta more likely than not will already have competitors of its own, such was the case with Wilhelmine Germany (ca. 1890-1918) in the case study employed here. But although the focus on competitive dyads may be seen in part as an artificial formulation for purposes of simplicity, it is also valid: this study deals with competitive strategy, not grand strategy, and with the choices of particular actors who want particular things, not with everyone. As noted in chapter 1, while grand strategy deals with the question of with whom to compete, competitive strategy concerns how to compete. Given this, any potential distraction—including a separate, already ongoing competition—may constitute a drain on either alpha or beta's resources. As was the case with the Dreadnought Race, the need to focus resources on other competitive problems (as well as separate issues entirely) could limit both alpha and beta's competitive options and even cause them to lose the competition altogether. It is therefore legitimate and useful to look at competitive dyads in strategic terms, with grand strategy in the background. In effect, the competitor that is most able to make its grand strategy consistent with its competitive strategy is likely to have an advantage.

Although competition to maintain or gain regional hegemony is the defining subject of this study, each regional hegemonic competition will entail an idiosyncratic set of goals on the part of both alpha and beta. Such goals will contain a zero-sum component: for a competition to exist, both parties have to want something that both cannot have *in toto*. These goals can be assessed according to the three metrics discussed in the preceding chapters: national security, financial welfare, and intangible goals (the last of which encompasses either external or internal goals of a nonmaterial nature or import). In any given case, it may be possible to pursue these goals more or less intelligently, but efforts to do so would be expected to have to account for tradeoffs along the three metrics, absent an exogenous source of resources, for the simple reason that one cannot pursue everything.

Barring an exogenous increase in resources, a state in such a competitive situation must not try to pursue everything. It must think "long term" rather than merely of the immediate future, and it must recognize and indeed embrace tradeoffs along the metrics, never trying to pursue all three at once. In practice, this is difficult for most states; certainly the case presented here shows that such is true. In the real world, contrary to the model usually put forward by strategic theorists who posit that the single person in charge is the only true strategist, rarely is a single decision maker able to allocate resources and priorities according to a long-term plan; there are usually other constituencies to address. Even in the case of authoritarian Wilhelmine Germany, as will be shown, there were limits to the funds that Imperial German Navy grand admiral Alfred von Tirpitz and the emperor, Wilhelm II, could seek to appropriate for building ships, whereas in Britain democratic imperatives forced successive governments to limit both their policy objectives and the range of their strategic thinking. This may be so, but it does not change the essential nature of the game: tradeoffs exist whether decision makers acknowledge them or not, and not to decide is to decide, and often to decide poorly. This is particularly true if creating a "finish line" (for beta) or getting the other side to quit the race (for alpha) involves the risk or threat of war, which will entail an even greater need for focusing efforts in anticipation of that event.

As will be seen in the case of the Dreadnought Race, to compete effectively, a state must account for grand strategy when making decisions in a competition—in effect, making grand strategy and competitive strategy consistent with one another. This is once again consistent with the imperative of formulating strategic intent: to pursue a competitive goal, one must marshal resources to achieve it. Beta, in particular, starts off in such a scenario at a resource disadvantage and must therefore eschew other objectives to focus on the main one. In the case of the Dreadnought Race, this proved a problem for both competitors—a fatal one for the beta competitor. It is a truism that not only do policy makers' priorities vary, but so do states' security interests; it may well be too much to expect a state, having decided to compete with a rival, to focus its entire policy on that goal. It is certainly true, however, that failure to do so has demonstrable costs that can be understood in terms of the three metrics used here, as the following case study will demonstrate.

Furthermore, the concept of competitive advantage and core competence, as discussed by Hamel and Prahalad, as well as by Michael E. Porter, must be applied. Presumably, an alpha has some advantage that has so far allowed it to be the reigning hegemon. For beta to challenge it, starting with fewer resources, it needs to play to its strengths. This principle is illustrated well in the case of the Dreadnought Race, in which the key to Britain's Atlantic hegemony was a combination of geographic accident and longstanding institutional and organizational competence. Though the British Royal Navy was arguably quite decadent—and ripe for challenge—at the outset of the race, and though Britain had been politically and materially distracted, weakened, and delegitimized by the Second Boer War (1899-1902), it would require no small amount of technical, organizational, and political expertise to challenge it. In this, Germany was at a disadvantage: it was a land power without comparable experience in shipbuilding and naval administration, with a political process that was unused to treating its navy as an equal service to its army; politically, it was locked into a preexisting competition with France and Russia that limited its resources and its options. Overcoming each of these obstacles was possible, albeit to different degrees and with varying likelihood of success, but what is unmistakably true is that Germany, in challenging Britain, was not competing in an environment in which it could leverage strengths or nonmonetary resources that it already had. Its ultimate failure, as will be discussed, can be attributed in no small part to this fact.

The competition can then be understood through decision-cycle analysis. In the course of the competition to be studied here, each competitor had an opportunity to shape the competitive environment by making better and faster decisions than its opponent—in effect, to attack its opponent's strategy. As will be seen, merely making faster decisions (deciding and acting) is not enough: the decisions must be correct and relevant (observing and orienting) to inhibit the opponent's own decision cycle. In the case of the Dreadnought Race, Britain effectively achieved this over Germany, squeezing it politically and financially so as to make its naval program and its foreign policy irrelevant, although it ultimately failed to follow up this success.

Finally, there is the essential fact that competition begins in a lopsided state. As noted, beta must effect a change in the status quo—it must affirmatively *do something* or there is no point in competing. Alpha does not have to do this, except to the extent it is forced to react. As noted above, competitions such as this have a tendency to end in conflict unless they are carefully managed and skillful diplomacy is employed. The challenge for beta in any of these scenarios is to construct a finish line—force a crisis—that will allow for a *favorable* resolution of the competition at *acceptable* cost; it simply does no good to compete if one cannot win, and the risks and stakes are both quite high. Costs scale over time for the victor; less so for anything short of this. In the end, an exit strategy is needed, or beta will fail. It is also possible, however, for alpha, having risen to the challenge, to effect an exit strategy of its own.

The Framework

Recapitulating for the final time, it is possible to show that the method here can be applied to any dyadic competitive relationship between great powers in a regional hegemonic context to analyze it and possibly predict the outcome. The steps of such an analysis are as follows.

Any such competition will have a reigning hegemon, or alpha, which, perhaps for a complicated set of reasons, possesses military primacy and a preponderance of political influence over a geographic region. Its competitor, or beta, is at an initial disadvantage in terms of both hard and soft power, relative to it.

Once these competitors are identified, the nature of the competition can be assessed relative to their strengths and weaknesses along three metrics: national security, the material welfare of their citizens, and intangible goals, which may include aspirations that are both internal and external to a state. *National security* refers to the actual safety of the state. *Material welfare* refers specifically to those matters that improve an individual citizen's economic well-being; they are the "butter" to security's "guns." *Intangibles* can be anything that does not fall under these two categories. Most, if not all, states have them; they may refer to constitutional liberties, national or ethnoreligious autonomy, international prestige, imperialism for its own sake or in pursuit of some ideological goal, the shaping of a regional or even global political environment according to a state's desires or ideology, or even merely the respect of one's international peers. Such intangibles need not be deemed noble to be understood as important to those who seek them; they can, however, be understood as a resource that can be traded in pursuit of other goals, depending on how dearly they are priced.

These three metrics must be understood in context. National security is assessed in political and military terms according to its era. Material welfare is probably best understood in purely economic and financial terms, but it can vary from simple money left in citizens' pockets to the availability of resources for government programs or economic growth. Intangibles must be understood, qualitatively and quantitatively, in their own right, through the eyes of those who seek them and in light of the way in which they and their peers construct the world.

These three metrics constitute both resources available to competitors and competitive goals. They constitute a trilemma. Absent exogenous inputs, it is not possible to seek an increase in all three at the same time; something must be sacrificed to obtain something else. In a competition for regional hegemony, moreover, the situation is constructed such that the competitors cannot entirely resolve it without reference to a zero-sum equation: obtaining a competitive goal requires that a competitor lose something, even if that something is not exactly symmetrical. "Interests," in this framework, are replaced by "goals." At least for purposes of competition, the question is not what a state objectively *needs*, if such can even be determined, but what it *seeks*. The three metrics effectively render the three "-isms" of realism, neoliberalism, and constructivism redundant for these purposes: rather than posit what a state's interests are, one only has to examine what it actually wants, whether it makes sense (for some definition of the word) for it to want those things or not.

Once a competitive scenario is understood in these terms, the degree to which beta is able to supplant alpha will depend on each player's ability to leverage its national resources in pursuit of this goal. They may be sidetracked by their own internal processes, by the larger grand strategic context, or by other factors. In particular, whichever player is able to leverage a core competence will have a competitive advantage; this is particularly true of beta, which will have a difficult time winning the competition without doing so, as it begins from a weaker starting point. Beta may or may not optimally focus its resources in pursuit of its objectives, and alpha may or may not awaken to the challenge posed by beta in time.

A Framework for Strategic Analysis

Ultimately, for beta, the race must have a finish line of beta's making; for alpha, the goal is either to prevent beta from creating one or to win the race on alpha's terms. Whether this is done more or less elegantly can be the difference between peaceful resolution and major war, but the ability to create a finish line (for beta) or deny one (for alpha) is the final measure of each player's competitive skill.

Strategic intent is the capacity, on the part of a player, to focus resources on the successful pursuit of a goal, ultimately either creating a finish line and crossing it or, as the famous answer had it originally, getting the other side to quit. It can be inhibited by interference with a competitor's decision cycle by another competitor making decisions faster with the correct observation and orientation. Obstructing an opponent's strategy can buy time for alpha or gain an advantage for beta.

Any competition for regional hegemony can be assessed in these terms. Who are the players? What do they want, and where does this put them according to each metric? How do these relative positions shape what resources are available? Can they formulate strategic intent? Can they, through skillful decision making, obstruct their opponent from formulating strategic intent? And can they, in the end, win the race with no finish line?

The framework is therefore as follows:

1. Determine the competitive objective—what is at stake in the competition. Within the limits of this study, the competitive objective refers to regional hegemony, but exactly what that entails will vary in each case.
2. Determine which competitor is alpha, the reigning hegemon, and which is beta, the challenger.
3. Determine how that objective is manifested in terms of changes in the three metrics for each of the competitors. Determine what resources alpha and beta each have in terms of the three metrics. In so doing, also assess in particular which of the three metrics it is willing to trade for the others, and whether it has an abundance or a scarcity along

this metric, to determine whether it is utilizing a strength or a relatively weak area.
4. Assess more generally the ability of each competitor to formulate strategic intent—in particular its leadership, its ability to leverage resources and core competencies, and its overall understanding of the nature of the project it is facing. Assess in particular which competitor is more able to intelligently and ruthlessly make tradeoffs among the three metrics in pursuit of its competitive goals.
5. Assess whether either player has an endgame in mind—either to outrun the adversary (alpha) or to force a finish (beta), and whether that endgame is achievable within the context of the player's strategic intent—the decisions it knows to make and is able and willing to make.
6. Determine what core competencies each player has—what abilities it possesses that will enable it to obtain an advantage in the competition. Assess in particular each competitor's astuteness in leveraging these competencies behind strategic intent. Assess more specifically whether it has an advantage in performing the actions it will have to perform to win the competition, relative to its competitor.
7. Assess the ability of each competitor to read its adversary and make decisions that will correspondingly limit its competitor's options and situational understanding—the ability to operate "within" the adversary's decision cycle. Reassess as the competition unfolds.

A comprehensive assessment of a dyadic competition for regional hegemony in these terms will likely yield, even at the outset, considerable insight into the question of who, if either, is likely to prevail in the end. Once each factor has been assessed, a qualitative picture will emerge of each side's strengths and weaknesses and each side's relative likelihood of achieving its competitive goals. At the very least, however, after addressing such a competition, an analyst employing this framework will be equipped with a set of questions to ask to determine who is performing better and, perhaps, who is likely to prevail.

It remains now to demonstrate the usage of this framework through the analysis of an historical case, and it is to this analysis that this study will now turn.

Chapter Three

Endnotes

1. Sun Tzu, *The Art of War*, tr. Samuel B. Griffith (New York: Oxford University Press, 1963), 71.
2. Sun Tzu, *The Art of War*, 84.
3. Sun Tzu, *The Art of War*, 73; and Carl von Clausewitz, *On War*, tr. Michael Howard and Peter Paret (New York: Alfred A. Knopf, 1993), 83–84, 86.
4. Phillip Karber, "Competitive Strategy: As an Approach to Business and Professional Life" (PowerPoint presentation at the Annual Fellows Conference of the Center for the Study of the Presidency, Washington, DC, 31 October 2003), slide 2.

Chapter Four
The Origins and Aims of the Dreadnought Race

God of our fathers, known of old—
Lord of our far-flung battle-line—
Beneath whose awful hand we hold
Dominion over palm and pine—
Lord God of Hosts, be with us yet,
Lest we forget—lest we forget!

~ Rudyard Kipling[1]

It is easy today to forget that the English were Northmen just as much as the Vikings . . . sharing traditions of kingship and the ordering of society, sharing not least a common style of shipbuilding and a common maritime tradition; a world in which a warship was a natural present for a king.

~ N. A. M. Rodger[2]

The competition that emerged at the turn of the twentieth century between Great Britain and Germany amounted to a competition for North Sea hegemony—over the preponderance of naval power around the seaward approaches to northern Europe, and over the prestige, access to colonies, and freedom of action within the international system that that power afforded its possessor. As such, this competition had two major, related sets of aims. In the first place, it was, for Germany, a material, technical, and highly technologized competition to, at the very least, alter or maintain the naval balance in the North Sea and, at the outside, to maintain or supplant British naval predominance there, at acceptable or at least realistically possible cost. In the second place, the competitors sought the corollary political effects of such a change or maintenance of that naval balance, as well as (more imperfectly) certain political requisites. The two factors were intimately related in that each enabled the other: naval power was seen as leading to political respect, of which access to colonies was one tangible indicator, though not the sole one.

What is referred to here as the "Dreadnought Race" may be dubbed an "arms race plus," with the "plus" signifying the longer-term goal of success in the eponymous naval arms race—namely, prestige within the system and the (often intangible) gains that prestige represented.[3] The Dreadnought Race therefore contained elements that would satisfy the fundamental assumptions of either realism (concern for national security driving policy) or constructivism (policy driven by contingent interactions of states within the system, with states pursuing that which was important in the context of the other states, their often implicit values and priorities, and their resulting policies and the systemic norms derived from them). As has been noted, this study does not seek to resolve, or even take a side in, the longstanding philosophical argument as to whether tangible factors or less tangible social constructs are the fundamental determinants of the international system. It is enough to note that both were present in the case of the Dreadnought Race, and that it is actually futile to separate the two factors. Military power—specifically, naval power—in a key region and status within the international system were tightly intertwined and informed the motives and goals of the two major players.

It is critical here to understand how primary and secondary objectives within this competition affect not only each other but also the inquiry provided here. The framework offers lines of inquiry both for the primary objectives of a competition for regional hegemony, in terms of what is sought in-region, and for the competition's greater implications. First and foremost, the Dreadnought Race was a competition for naval supremacy, and the political environment necessary to sustain it, in the North Sea. As will be shown, the major component of the competition—a naval arms race to build and maintain battleship fleets—was only of use in this context for Germany (and of limited use otherwise for Britain) and was, in fact, a resource drain for other objectives. It was a regional competition with broader relevance for the grand strategies for the states involved, and the framework presented here offers means of inquiry into how these matters fit together.

To understand how this state of affairs came to be, it is necessary to understand not only the distribution of power within Europe at the time but also what that power meant, as well as how matters had come to this point. As is often remarked, the catalyst for Anglo-German rivalry prior to World War I was the way in which naval dominance in a particular area—namely the North Sea—was seen by Britain and the other states of Europe, both at the level of decision-making elites and by the general public.

This study will move from a consideration of these viewpoints to a narrative of the key decisions undertaken in the Dreadnought Race before proceeding to an analysis of those decisions in the context of competitive strategy. The narrative that follows may be summed up briefly. Beginning in 1897, Germany, under Kaiser Wilhelm II, sought to build a fleet of capital ships that would alter the naval balance in the North Sea, then dominated and controlled by the British Royal Navy, in Germany's favor. This would, per se, increase Germany's prestige, a goal sought by Wilhelm for personal reasons. It would also theoretically allow, at minimum, for a deterrent to a British blockade in the event of war, which would in turn allow Germany to pursue its imperialist ambitions—particularly in Africa, but also potentially elsewhere—without British approval. It would, at maximum, achieve sufficient dominance of Great Britain close to home as to force Britain into a subordinate position relative to Germany, again increasing Germany's international status and fulfilling Wilhelm's personal goals. Almost as a side note, it would also theoretically provide for a unification of the various factions opposed to socialism within German society, allowing for effective suppression of the German Social Democrats, who were feared to be gaining electoral ground. There were numerous flaws in both the conceptualization and execution of this plan, and in the end, both for reasons of its own and in response to German actions, Britain succeeded in outbuilding the Imperial German Navy's battleship program and in diplomatically isolating Germany. Although the Dreadnought Race did not come to a formal conclusion, it had effectively been resolved in Britain's favor by the time of the outbreak of World War I. Whether to count it among the contributing factors to that war is an open question, which, although this study need not take a definitive position on it, can be understood as having some impact on the assessment of the success or failure of each competitor's competitive strategy.[4]

A Note on Historiography

Before beginning, it is necessary to address a few points that may be of interest to specialists. Because this study is primarily concerned with the strategy of great power competitors as it pertains to their pursuit of regional hegemony and the significance of that goal to their broader aims, it is not primarily a work of history. Although readers will find a bit of original research here in the form of statistical analysis based on a consolidation of historical data and modern inferences about it (see the appendix), this work does not pretend to add any great detail to any of the existing historical analysis of the Dreadnought Race or its surrounding context. If it should provide such insight, such is the better, but that is not its objective, and this author will be successful to have merely kept up with, and addressed, existing scholarly debates and research. To that end, this work relies heavily on secondary sources, although where appropriate primary sources are referred to.

Although the main text of the narrative hopefully addresses all major points of contention regarding the historiography of the events understood as the Dreadnought Race, some additional commentary is provided here regarding the approach taken. Although, as noted, this study is not a work of history and is principally concerned only with a fairly cursory appraisal of the Anglo-German competition that is styled the Dreadnought Race (any longer treatment of the subject being prohibitive), and although space constraints necessarily (but unfortunately) prevent this work from considering every detail and aspect of British and German policy at the time, it is nevertheless important to consider some of the historical and historiographical disputes that underlie the major events under study, and to locate this study within some of these disputes where it is strictly necessary to take sides.

In the first place, this study proceeds, albeit cautiously, from the assumption that Germany's late-nineteenth-century expansionism from the accession of Kaiser Wilhelm II onward was driven at least in part by the kaiser's personality and personal motivations, and that these included envy of Britain's access to overseas colonies (particularly and primarily in Africa) as well as a fascination with naval primacy and the international prestige that that primacy conferred. It hews to the view, most notably promulgated by Paul M. Kennedy and Ivo Lambi, and echoed by Peter Padfield and Robert K. Massie, that the wishes of the kaiser, and the influence on him of a few key individuals (notably German statesman Bernhard von Bülow and Imperial German Navy grand admiral Alfred von Tirpitz) were key to understanding Germany's intentions at the state level before, during, and immediately after the Dreadnought Race ran its course.[5] It acknowledges that the drivers of German policy were complex, and that structural factors—more than can be discussed in the space available here—did indeed play a role in shaping and limiting German intentions, as Volker Berghahn and others have noted.[6] Where appropriate, it acknowledges some of these factors—particularly the influence of big business in the German Reichstag and the ideological and cultural concerns of both the Social Democrats and the Prussian nobility, all of which served as both limiting factors (to which, it will be noted, Wilhelm and his advisors paid insufficient attention) and drivers of policy. Without disputing or even minimizing these structural concerns, however, this study, because it is, in the end, a study of decision making and resource allocation, does nevertheless pay attention principally to the decisions taken at the higher levels. As is noted in chapter 6, there are ways in which it is possible to imagine Germany charting a different course, and many decisions that it is possible to imagine in the abstract that it is nevertheless impossible to envision Germany making in actuality. In that sense, the limitations of German society are acknowledged even as the focus is on the most influential decision makers themselves. As noted in the preceding chapters, states are treated in this study as having agency, even allowing for excellent arguments as to why they do not, and state decision making is treated holistically, even allowing for the problems of doing so. When this study speaks of a state—Germany or Britain—having a strategy, that strategy is understood through the thoughts and actions of its principal decision makers and those taking the initiative to make decisions.

This study also proceeds from the assumption that the individuals on which it focuses were the most influential decision makers on their respective sides. This is an essentially classical view at this point, in that a number of revisionist interpretations have now been put forward. On the German side, this study retains the focus on Alfred Tirpitz as the key driver of German naval policy in the Dreadnought Race that is employed by Kennedy, Lambi, and others (notably Jonathan Steinberg, Holger Herwig, Patrick J. Kelly, Padfield, and Massie), as well as on Wilhelm and Bülow. On the British side, it similarly treats Royal Navy admiral Sir John Arbuthnot Fisher as a center-stage character (per, especially, Arthur Marder, as well as Kennedy, Padfield, and Massie). It is necessary in doing so to acknowledge more recent revisionist interpretations that have attempted to qualify these figures' roles—notably those of Matthew S. Seligmann and Nicholas A. Lambert, who have argued variously that the foregrounding of Tirpitz's role has obscured equally important contrary efforts by other German naval officers and that too much has been made of Fisher's reforms in relation to Germany as opposed to other factors. These revisionist interpretations are not without foundation and, where appropriate, are accounted for in the narrative presented here. Nevertheless, this study works with the basic assumption that these figures' roles were, indeed, central to the Dreadnought Race.[7]

Perhaps most importantly, this study proceeds from the assumption that the Dreadnought Race was a major naval arms race that affected not only the defense budgets but also the foreign policies of the two competitors. In so doing, this study has unabashedly followed sources that treat the Dreadnought Race as an important and discrete series of historical events worthy of study in their own. These sources have included the more popularly accessible histories of Padfield and Massie, as well as a more recent narrative by Seligmann, Frank Nägler, and Michael Epkenhans, and the more exhaustive analysis of Lambi on the German side.[8] It is important to acknowledge key counterarguments, and this narrative attempts to do so as they arise—most notably, this study can readily acknowledge the objections of Lambert and others that Britain's naval decision making, particularly early on in the Dreadnought Race, involved other priorities besides Germany's battleship program.[9] It is enough to note here that, while one can say that both Britain and Germany were indeed attending to other matters as they competed against each other, and while neither side at any given time threw the full weight of its resources and resolution into the competition (a fact that will be noted in the strategic analysis in the final chapter of this book), a competition was underway and, in that competition, even not to have a strategy was in an important sense to have one. Likewise, this study (following Seligmann, Nägler, and Epkenhans) proceeds from the reasonable position that there was indeed a major naval arms race worthy of analysis from both perspectives.[10] In that sense, although this study can acknowledge objections that various decision makers had priorities besides the Dreadnought Race and factor them in, in the end, this study is about the competitive relationship between Britain and Germany in the military and political realm that existed from the last decade of the nineteenth century until World War I and seeks to understand the two competitors' decisions in light of that relationship.

Other matters can also be dealt with here. This study touches on the importance of imperialism and colonies for the international prestige that Wilhelm, and Germany under his direction, sought. It does not delve into the complexities of the history of nineteenth century European imperialism except as those complexities bear directly on the questions of strategic goals and strategic intent that are at hand. It can readily concede (though it does not necessarily have to) the thesis of Ronald Robinson, John Gallagher, and Alice Denny, which states that British imperialism was not directed by a well thought-out grand strategic plan or a singular ambition but rather evolved from specific British security concerns and political imperatives as they arose.[11] It can likewise concede D. K. Fieldhouse' arguments that colonialism was driven by as many historical factors as there were European colonies, while still acknowledging the subjective importance that colonies had to specific German decision makers—including Wilhelm and those in his inner circle—and the nature of that importance.[12] On these matters, Fieldhouse's insights actually serve as confirmation, in that he notes that colonies were often seen as having intrinsic importance that transcended material cost-benefit analysis, an insight that is relevant to the framework under study here.[13] This study is in broad agreement with the argument of P. J. Cain and A. G. Hopkins, who note that the German threat to British North Sea primacy was a major impetus behind the changes in British foreign and defense policy at the time, even as it also can concede their point that colonies were only a marginal prize in this rivalry.[14] Colonies did, however, serve as a lagging indicator of Germany's worldwide influence, which in turn Germany sought to enhance by its actions to overtake Britain's North Sea hegemony, as will be discussed.

Likewise, this study moves around the edges of a gargantuan debate into which it can only timidly dip—namely, the question of the causes of World War I. This study is prepared to accept—although it considers the alternative—that the Dreadnought Race did indeed contribute to the outbreak of World War I. It decisively accepts the conventional thesis promulgated by Kennedy and Lambi (as well as recently by Seligmann, Nägler, and Epkenhans, and in the narrative histories of Padfield and Massie) that the Dreadnought Race contributed directly to Germany's diplomatic isolation, which was inherently detrimental according to the framework discussed above, and also contributed to the enmity between Germany and other powers that made World War I, and British involvement in it, more likely.[15] It is not necessary here, however, to assume that World War I became inevitable because of the Dreadnought Race—only that the Dreadnought Race contributed to its likelihood (indeed, the final analysis considers other possibilities). Although some of this study's source materials do accept the thesis of German war guilt—most notably Padfield's narrative of the Dreadnought Race, which begins with this assertion—it is not strictly necessary for the narrative offered here to accept it.[16] In particular, this study does not accept, at least not in their entirety, the more provocative arguments of Fritz Fischer regarding German culpability for World War I. Fischer's crucial arguments—that Germany was incorrigibly expansionist prior to World War I as assuredly as before World War II, and that Germany made the decision to begin hostilities well before they broke out—do not have to be taken as valid for the analysis here to hold water.[17] As will be treated in more detail in chapter 6, all that is argued here is that the Dreadnought Race, at key points and in key ways, diminished Germany's position along all of the metrics employed in the framework offered above; that this ran directly counter to Germany's ultimate strategic aims; and that, to the extent that the Dreadnought Race had a role in increasing the likelihood of war, it did not serve the interests of Germany that it did so. British policy toward Germany is accounted for in a similar fashion: although British policy during the Dreadnought Race may have helped to set Britain on the road to war by shifting Britain's national security policy away from older adversaries and toward Germany, it is not necessary to take a strong position on the ultimate causes of the war for this study, in its final analysis, to comment on the results of British policies for Britain's competitive strategy.[18] In essence, the Dreadnought Race's effects on the international position and strategic goals of both sides can be understood whether or not World War I is ultimately included among them, but where it can be argued that it should be, this is discussed.

With historiographical matters addressed, this study now turns to the strategic theories and context that drove the competitors in this case.

Theoretical Underpinnings:
The Influence of Sea Power Upon History

To understand how the Dreadnought Race began, it is necessary to understand how the competitors understood seapower. The second half of the nineteenth century gave rise to dramatic changes in naval technology—specifically, the rise of steam-powered ships, armor plating (the first ironclads), and, eventually, swivel-turret gun mounts with breech-loading guns.[19] As befitted the leading seapower of its era, Britain largely took the lead in developing these technologies. As of the mid-1870s, Britain's four steam-driven, swivel-turreted, heavily armored ironclads constituted the most technologically advanced capital ships afloat, and they were seen as technologically unique, capable of defeating any ships then deployed by any power. These ships represented a qualitative break from the previous practice of either plating sailing vessels with steel armor (with the addition of a steam engine) or of producing ironclad hulls with guns in batteries.[20] These ships represented the beginning of the battleship revolution that defined naval competition in the late nineteenth and early twentieth centuries, and with it the overall challenge by Germany to Britain's Atlantic hegemony.

Conversely, a somewhat viable alternative was also in development in the form of small, fast-moving torpedo boats that could literally run circles around the big line-of-battle ironclads, moving faster than rotating turret guns could track. These boats ran the gamut from genuine torpedo boats to larger destroyers, which both fired torpedoes and hunted torpedo boats. They were limited in range: their unguided torpedoes (which traveled erratically) had effective ranges of only about a kilometer and were accurate at about one-half that distance. They likewise lacked combat radius, as they were ill-suited to long deployment at sea.[21] Big battleships were preferable, and no navy that aspired to great power status would seriously consider torpedo boats an adequate substitute (though they showed promise), but the latter offered a cheap, "good enough" alternative for a power that wished to content itself with coastal defense and perhaps short-range commerce raiding.[22] Such debates would become part and parcel of Germany's internal deliberations at the start of the Dreadnought Race.

The implications of these technological developments for naval procurement and doctrine were mirrored by the theoretical disputes of the time. For Alfred Thayer Mahan, a U.S. naval officer who became famous for his 1890 book *The Influence of Sea Power Upon History*, seapower came down to a few fairly simple concepts. Whoever controlled the ocean controlled sea trade, and the way to control oceanic trade was to be able to win naval battles. This view was neither intuitive in the minds of naval planners of the time nor without opposition. Commerce raiding (derisively called a *kleinkrieg*, or little war, in Germany) was a time-honored strategy of naval warfare and was in fact the dominant view of naval warfare at the time. In Mahan's view, it was an indecisive strategy. The way to command the ocean, in Mahan's analysis, was not to be able to make hit-and-run attacks on an adversary's commerce, but to be able to render that commerce inherently vulnerable by sweeping the seas clean of opposing ships. This, by definition, required possession of battleships—having enough of the most powerful and technologically advanced warships of one's era to be able to win sea battles and deny the sea to opposing navies. Once command of the sea had been achieved, a naval power could then blockade an adversary's coasts at will, arresting not only their trade but also military force projection, therefore inherently limiting the adversary's military options in the event of war and, thereby, their international power and influence.[23]

British naval historian Sir Julian S. Corbett critiqued some of Mahan's ideas in his 1911 book *Some Principles of Maritime Strategy*. Corbett noted that, if only because of the sheer size of the area involved, the default condition of the sea was not control by a particular state, but the lack of it: "command of the sea" had to be obtained; it could not be assumed simply because one state had a preponderance of seapower. For Corbett, one only assumed command of the sea when one had exclusive control of "maritime communications," a concept that linked naval force projection with protection of commerce; the key was to hold the right areas at the right time, not all areas at once, so as to be continually on the move and in play.[24] Corbett tied together the concepts of overseas force projection (and denial of the same to an adversary) and trade protection, since he viewed these matters as a question of interdicting seaborne logistics.[25] As a result, it was not overwhelming naval force but rather the right naval force in the right place at the right time—or even the risk to an adversary that one might exist—that was enough to protect what needed to be protected. Corbett referred to this concept as a "fleet in being."[26] Corbett wrote *Some Principles of Maritime Strategy* when the Dreadnought Race was at its height (he looked backward to "the golden age of our navy") and therefore possessed a spirit of caution and resource conservation that Mahan lacked.[27] For this reason, he was acutely attentive to financial reality: as he noted, the question of paying for naval warfare could not be dodged, and "the side with the longer purse wins"; accordingly, interference with trade only mattered if it affected the opposing state's fiscal options.[28] As will become clear, the Germans in the Dreadnought Race were motivated more by a Mahanian vision than a more modest Corbettian one.

Mahan's views, however, formed one endpoint of an ideological continuum in an increasingly intense debate about naval force structure and doctrine that took place across the European great powers at the end of the nineteenth century. Corbett's critique, moreover, lay at the midpoint of the continuum, in that it still partially concerned how to achieve Mahanian aims on a budget. It is important to understand this debate in some detail to understand (at least to the extent possible) what Germany ultimately tried to do in the Dreadnought Race.

At the other end of the continuum lay the French *Jeune École* (Young School), a naval strategic doctrine or concept that, in various forms and by different names, underlay the naval acquisition and doctrinal planning of some of Britain's rivals in the late nineteenth century. Its chief proponent, as the name suggests, was the French Navy, which devoted considerable time and effort to honing it. In essence, the *Jeune École* held that seapower could be used effectively by a weaker navy if it targeted an adversary's commerce instead of seeking to blockade their shores. In this, it was a direct rebuttal to Mahan: the smaller fleet could at least hold its own, and what mattered was not raw naval power but the ability to use smaller amounts of power in a highly targeted fashion.

The genesis of the *Jeune École* lay in an obvious fact of life: if France's navy was to have any use in a war against Britain, its historical adversary, it would have to overcome its essential weakness in ships of the line. Because of Britain's overwhelming naval strength, France would have to devise a strategy to remain relevant at sea without fighting a losing "big battle." French naval planners, beginning in the second half of the nineteenth century, therefore turned to the historical alternative that would be denounced as ineffective by Mahan: commerce raiding. What differentiated their analysis—what made "this time different," if one wills—was a somewhat optimistic assessment of the vulnerabilities created by the interaction of overseas colonialism and an industrial economy. French naval theorists believed Britain to be excessively sensitive to disruption of its overseas trade: because Britain's industries relied heavily on imports of colonial raw materials in a way that, for example, an agrarian economy in the days of the Spanish Armada might not have done, they could be easily disrupted and shut down by a coordinated attack on colonial shipping lanes.[29] In this, French naval theorists anticipated, in an odd fashion, later maximalist arguments concerning airpower: a few strikes against a few key economic pressure points would throw an entire economy into chaos.

These arguments, as overblown as they now seem, were not without foundation. In aggregate terms, Britain's economy was a world trader par excellence. As Arne Røksund notes in his history of the *Jeune École*'s influence on the French Navy, prior to World War I Britain alone accounted for about 40 percent of global trade by weight. It was heavily dependent on its colonies for industrial raw materials and for food, of which it was a net importer. Røksund even notes, amusingly, that Britain also could not function without its tea, coffee, and tobacco, all of which had to be brought in by sea. British war planners and political leaders were uniformly concerned about this vulnerability, even if they were not entirely sure what to do about it. It was a reasonable assumption that severe disruption of this trade would have profound economic effects. It was, perhaps, a slightly less reasonable assumption that this could be effected cheaply and easily in the event of war.[30] The basic idea, however, that commerce raiding could be an effective option in the event of a war with Britain not only permeated French naval thought from the late nineteenth century until its rivalry with Britain ended (during the Dreadnought Race, as will be seen), but also found its way into the disputes over naval force structure in other potential adversaries of Britain—including Germany.[31]

The Origins and Aims of the Dreadnought Race

As it happened, the *Jeune École* in its original conception was ultimately overtaken by events. France, in the end, concluded an entente with Britain and fought alongside it in World War I, thereby rendering moot its original, somewhat speculative plans for commerce raiding in the event of war. But, as Røksund notes, what France began in speculative terms, Germany ultimately accomplished in its submarine warfare strategy in World War I, although the actual effects were far less adequate than had been anticipated.[32] In any event, it is enough for now to note that the ideas of the *Jeune École* were widely discussed at the end of the nineteenth century, influenced naval planners in Germany as well as France and elsewhere, stood in stark opposition to Mahanian ideas about sea dominance, and had to be dealt with if one were to forge a naval acquisition plan in that era.

The basic tactics of the *Jeune École* were a modernized version of those used by weaker naval powers for centuries: the so-called "guerre de course" in which small, fast ships would attack enemy commercial vessels in piratical fashion, either relieving them of their cargo or sending them to the bottom of the ocean. The ships available for this task at the end of the nineteenth century varied. The oldest and most straightforward method was the use of small, fast cruisers that could keep up with merchant vessels. This, of course, had been practiced since time immemorial, either by navies or on a contract basis by privateers. A more outré version of this basic practice has been extensively documented recently by Matthew S. Seligmann, who notes that plans existed in Germany even during the Dreadnought Race to turn to commerce raiding in the event of war with Britain by conscripting civilian merchant vessels, which were required by regulation to be pre-equipped for the purpose, outfitting them with guns, placing them under naval command and control, and unleashing them—initially out of neutral ports—on British commerce. The normal counter to this method was to escort or screen merchant convoys with small friendly cruisers or similar light vessels, or to use these vessels to patrol shipping lanes for predators.[33]

Chapter Four

The end of the nineteenth century, with its increase in the rapidity of naval technological development, would bring new options for guerre de course. As noted, torpedo boats were in development from the 1870s onward, their development being led by the Imperial German Navy, which, despite organizational doubts about the viability of a *Jeune École*-driven doctrine, was concerned with economizing on costs. Torpedo boats created basing problems, since their short range limited their use to coastal waters unless inventive measures were taken. In a preview of what was to come, the German officer responsible for torpedo boat development at the end of the nineteenth century, Alfred Tirpitz, repeatedly sought to increase their range so that they could operate in the far reaches of the North Sea. The development of submarines at the beginning of the twentieth century would further refine the concept in predictable ways. For all this, however, the *Jeune École* primarily envisioned long-range cruisers rather than short-range small craft as its commerce raiders of choice, for the obvious reason that cruisers alone had the proven ability to strike at trade on the high seas far from home bases. The submarine, in its now-fabled World War I role as a commerce raider, was a comparative latecomer, long after France had abandoned the *Jeune École* and Germany had begun the Dreadnought Race.[34]

The two ends of the debate, therefore, could be summed up in somewhat alliterative terms: battleships would be used for battles and blockades; cruisers would be used for commerce raiding. Although a navy might seek a balanced mix of small craft and ships of the line, the weighting of this mix—constrained by state finances, a factor that would become crucial during the Dreadnought Race—effectively announced a decision, and an intellectual position, as to which side of the debate had more practical merit. In general, a basic choice faced the naval planners of a state such as Germany that wished to compete with Britain: such a state could compete directly, according to Mahanian principles, by building a main battle fleet capable of surviving combat with Britain's, or indirectly, by preparing for large-scale commerce raiding. Each end of this continuum contained risks, both doctrinal (the theoretical underpinnings had not been tested in a modern conflict) and strategic (committing to one frequently meant avoiding the other).[35]

From an international relations theory perspective, Karen Rasler and William R. Thompson (drawing on a line of analysis begun by Thompson and George Modelski), have presented a more modern version of Mahanian reasoning in the form of leadership long-cycle theory.[36] It is worthwhile to examine their contentions in greater detail here as a context for Britain's global position that drove its regional competitive thinking during the Dreadnought Race. Echoing Corbett's views on the fusion of state finance and naval power, Rasler and Thompson argue that waves of economic innovation, which is usually monopolized when it occurs by a single state's economy, drive temporary boosts in relative power by giving the states that initiate them temporary increases in economic production that can be mobilized to obtain greater naval power. This power, in turn, enables greater control of the seas and a corresponding increase in global political influence. Rasler and Thompson write:

> For a number of reasons related to winning global wars and supporting long-distance economic activities in war and peace, global leaders must develop sea power to be successful. Historically, naval leadership has been vital to both attaining and maintaining the ability to project force over long distances. That is why the global leader's [their term for a world hegemon] naval capability share, one method of indexing the level of military power concentration and deconcentration at the global level, is highlighted. . . . That is also the main reason why so much stress is placed on long-term fluctuations in naval capability within the leadership long cycle framework.[37]

Chapter Four

In Rasler and Thompson's model, global leadership by a given state rises and then falls (and may rise again) in cycles linked to cycles of economic innovation, which in turn drive cycles of naval power concentration. Rasler and Thompson use the term *global reach capability* to describe what seapower provides: the ability to project hard power on the one hand and push it away on the other. For almost all periods for which it is relevant, however, global reach capability is synonymous, for Rasler and Thompson's purposes, with seapower.[38]

The waves of economic innovation are styled "Kondratieff waves," named after Soviet economist Nikolay D. Kondratiev (a.k.a. Kondratieff), who first hypothesized their existence. Per the leadership long-cycle model, waves of global leadership driven by Kondratieff waves are bracketed by them and, in turn, by periods of global warfare. At the beginning of a cycle, a sitting global leader is knocked off its perch in a crisis—historically always a major global war—by a challenger state that has successfully monopolized a new wave of innovation that, in turn, allows its economy to grow faster and its share of global reach capability—seapower—to rival the sitting global leader. The new global leader then maintains its position for a while and legitimizes its position to some degree by offering the types of public goods that a monopoly on naval power can provide—notably safety of commerce and a degree of global governance. However, deconcentration of global reach capability—and therefore global hard power—occurs when another state is able to monopolize a new wave of innovation and use it to build up its own naval power. Typically, during this time, the global leader will also have fallen into a "territorial trap" in which it is seduced into maintaining its global reach by acquiring land-based territories and assets that must be defended by land forces, eroding its naval budget and therefore its ability to fend off challenges. Eventually, another crisis (effectively, another war) occurs, and unless the sitting global leader has caught hold of another wave of economic innovation, it is likely to lose this challenge to the upstart state. Rasler and Thompson posit and argue for five such cycles of global leadership: one by Portugal (and eventually Spain, after the union of their crowns) in the sixteenth century, one by the Netherlands after the Eighty Years' War (or Dutch Revolt) in the seventeenth century, one by Britain after the wars of Louis XIV that inaugurated the eighteenth century, a second cycle of British leadership in the eighteenth century after the failed attempt by France under Napoléon Bonaparte to supplant Britain in the Napoleonic Wars, and finally a wave of U.S. hegemony beginning after the collapse of Britain's world authority following World Wars I and II.[39]

At the end of the nineteenth century, Britain faced deconcentration. Although there was not yet a direct challenge to its naval dominance, all of the European powers now had rapidly growing industrial economies—on average, Germany's gross national product is estimated here to have been growing more rapidly than Britain's. The heavy industries that had been at the vanguard of the industrial revolution and are a good proxy for Rasler and Thompson's innovation wave—particularly coal and steel—were now possessed by all; Britain's share of global production was slacking dramatically, with Germany about to exceed Britain as the leading producer. As Paul M. Kennedy and others have noted, Britain's strategic position—a powerful military backed up by a new industrial economy—was not unique anymore, and by this time Britain was not exceptional or even well-off in this regard.[40] It is therefore perhaps far from surprising that British seapower should have come to be both a point of contention and a point of vulnerability for Britain at the end of the nineteenth century.

The British Royal Navy in the Nineteenth Century

It had begun well. To say that Britannia ruled the waves in the middle decades of the nineteenth century is to downplay the situation: there was no one else on the waves. Since the Napoleonic Wars, Britain had enjoyed a blue-water naval preeminence enjoyed by few states at any point in history. Following the defeat of Napoléon—which had most famously demonstrated

Chapter Four

British naval competence (as well as courage and sheer dramatic daring) at the Battle of Trafalgar (21 October 1805)—the Royal Navy was mostly demobilized, as there were simply no threats remaining. So unquestioned was British naval power that the Royal Navy was relegated to convoy and blockade duty during the Crimean War (October 1853-February 1856)—the Russian czar was more intelligent than to waste his navy in a futile effort to combat Britain far from port. With that much less to do, the Royal Navy's principal duties lay in a smaller imperial policing role, most famously in deterring piracy and mopping up the oceanic slave trade, which Britain banned. The fact that it was able to take the lead in banning slave-trafficking should provide an indication of how little Britain was opposed. Perhaps the most dramatic illustration of the Royal Navy's preeminence is the almost complete absence throughout the first half of the nineteenth century of anything like a defensive mentality: ships were widely dispersed globally instead of being concentrated in bases close to home, as there was simply no one in Europe left to fight. Although the Royal Navy's first priority remained guarding the seaward approaches to Britain and controlling access to the Mediterranean, it was not necessary to devote a large share of fiscal resources or ships to the accomplishment of this task.[41]

Still, Britain had come to view seapower in almost obsessive terms. This had somewhat to do with British anxiety about holding onto a good thing once they had control of it, but it also had to do with the exigencies of modern life. As noted, and as the proponents of the *Jeune École* had argued firmly, now that the Industrial Revolution had come to Britain, Britain could no longer feed itself—it was part of an interdependent global economy, and it was a net importer of food. As of the latter half of the nineteenth century, Britain had, at any given time, enough food within its borders to feed its population for three to six months, but probably not longer. Just as vital was protection for British commerce. Although trade is often seen as a barrier to hostilities, for Britain, overseas trade amounted to an Achilles heel. If a foreign adversary in wartime were to interdict it, Britain's domestic economy would essentially shut down, and with it Britain's war effort; food supply disruption, it was thought, did not even have to reach starvation levels to contribute to this theorized breakdown.[42]

National security was therefore a matter of protecting commerce and avoiding blockade through overwhelming strength.[43] Although this concern spoke most directly to the threat of *Jeune École*-style commerce raiding, it also argued in part—particularly where the North Sea was concerned—for the maintenance of Britain's essentially Mahanian naval policy: it was better not to be blockaded, and this argued for a preponderance of capital ships.[44]

For this reason, too, Britain also endeavored to keep the Dardanelles and Bosporus straits open to British trade (particularly as Britain's Crimean War nemesis Russia was seen as a threat to that trade), which meant maintaining as strong Mediterranean fleet on the one hand and keeping good relations with the Ottoman Empire on the other. For this reason, British prime minister William Ewart Gladstone's opposition to Turkey on what today would be called human rights grounds was a sharp break with more hardheaded British policy makers—including his predecessor, Benjamin Disraeli.[45] As will be shown, the need to safeguard Mediterranean waters ultimately clashed with even more vital priorities closer to home.

It was also a matter of avoiding invasion. Although at various points British politicians and defense planners discussed the possibility of enlisting the citizenry en masse into some sort of militia or home defense force (the idea was a hardy perennial among British governments hoping to shave a few pounds off the defense estimates), the reality was that the best way to defend Britain—a small, flat island—from invasion was to ensure that invasion never happened. This required ensuring that foreign fleets were weaker than the Royal Navy, and in particular those ships of the Royal Navy that were assigned to the British Channel Fleet (or Home Fleet, as such ships were occasionally designated).[46] As will be seen, Britain was capable (in a way that Germany was not) of linking its diplomatic and military policies in the service of a common set of goals. But in any event, keeping foreign ships away from British shores and keeping a favorable balance of naval power, both globally and crucially in home waters, were vital national security interests in the eyes of Britain's leaders. If other powers had an issue with this arrangement, it was their problem.

Naval Strengths and Estimates

To this end, Britain paid close attention to naval balances. After the mid-nineteenth century, these were in flux due to rapid technological change. The latter half of the century witnessed a massive revolution in naval technology that forced navalists in all the major powers to pay close attention to the latest developments. The viability and ultimate superiority of ironclad ships had been demonstrated by the experience of the American Civil War (particularly once it was shown that such ships could be made seaworthy for blue water deployments). Britain had led the development of ironclad naval technology. The first ironclad warships, of which Britain's HMS *Warrior* (1860) was a formidable and leading example, had copied the design of the hybrid steam- and sail-driven wooden vessels. After the development of rotating turrets (the first being that used on the fabled USS *Monitor* in 1862) and breech-loading cannon, however, ship design had undergone massive change, dispensing with sails altogether and beginning to mount guns on rotating, open-air turrets that would, in time, give way to true ones. About the same time, muzzle-loading cannon were replaced with breech-loaders.[47]

Conversely, Germany initially, before Tirpitz, had internalized the principles, if not the full doctrine, of the *Jeune École*, and was primarily concerned with protecting its commerce and, perhaps, threatening that of its adversaries.[48] It is perhaps to be expected, given all this, that Germany actually led the way in torpedo boat development, while Britain concentrated on battleships.[49]

The one constant in naval technology throughout this period was rapid change, which upset naval planners' best-laid plans. Ironically, this made it pointless and counterproductive to lock a navy into a long-term ship-building program, because the ships would likely be obsolete after launching if the plans were not continually updated.[50] This lesson, which British naval planners often took to heart, eluded Germany when it began the Dreadnought Race.

In the midst of all this, it was nevertheless possible to become complacent. Although for most of the nineteenth century there were no serious challengers to Britain, there was a natural tendency, in a country to which militarism was foreign and fiscal discipline was taken seriously, for governments to become lax. The most serious case of this occurred in the decade from about the mid-1870s to the mid-1880s, when Britain came close to falling behind its longstanding rival France. It was rectified by the 1889 Naval Defence Act, which led to agreement on the then-famous "Two-Power Standard": the Royal Navy would be kept at sufficient strength (technologically and numerically) to be superior to the next two most powerful navies in the world.[51] But even after this, France and eventually Russia would attempt to gain a more favorable ratio of naval forces to Britain's, a set of arms races that did not end until the three powers patched up their differences with their respective ententes (discussed below) in the decade preceding World War I, to meet the more urgent German challenge.[52]

The Dreadnought Race would strain the Two-Power Standard to the point of abandonment: a determined adversary would see to it that Britain could not maintain this level of capability without dramatically hiking its budget. But as the last decade of the nineteenth century began, Britain was committed to being untouchable at sea.

Chapter Four

Seapower and the Scramble for Africa

There was an intuitive, if not entirely logical, connection between seapower and the competition for colonies in which European powers were engaged. At minimum, it was plain that seapower had played a role in the colonization of the continent of Africa. This would not necessarily have been a major issue for German expansionists were it not for the closer and more deterministic relationship between German imperialism in Africa, British approval for that imperialism, and Britain's North Sea primacy, as well as the general mental picture created by the spectacle of European gunboat diplomacy in Africa. There was an implicit connection between naval power and access to colonies that was apparent to navalists and imperialists; there was, more importantly, a much more tangible and explicit connection between the ability to deter Britain's capital ship fleet and the ability to move freely in the colonial world.

Where European powers were concerned in the late nineteenth century, African colonization was a status symbol, even if its economic proceeds were an uncertain gamble.[53] The "scramble for Africa" (the phrase dates from the time and has entered the historical lexicon as a distinct set of events) was a remarkable process.[54] In the space of a generation—from the mid-1870s to the turn of the twentieth century—Africa transitioned from an almost uncolonized and unmapped continent to a veritable game board for European empires, with almost every major European power claiming a geographic stake. Beginning in 1876, the British explorer Henry Morton Stanley mapped the Congo River and what was to become the Belgian Congo, initiating British interest in the continent. In 1884, the Berlin Conference began the process of European partitioning of Africa, even as its host, the German chancellor Otto von Bismarck, was privately skeptical of what his own country could achieve by it.[55] Although colonies made up only a small fraction of Germany's policy goals when it ultimately challenged Britain's North Sea naval primacy, the ability to acquire them at will rather than by Britain's leave—along with an intuitive, if not strictly logical, understanding of the linkages between naval power and influence in this realm—formed a significant part of the impetus behind the Dreadnought Race and the thinking that informed it.

The Royal Navy had been a necessary component of early British colonization of Africa. As Daniel Owen Spence, writing for the National Museum of the Royal Navy, notes, a series of small naval actions and deployments had aided the establishment of the British foothold on South Africa in the first half of the nineteenth century, and a sizeable detachment of Royal Navy ironclads had played a critical role in the suppression of the Egyptian revolt of 1882, an action that set the stage for further British involvement on the African continent.[56] More generally, Britain had from the beginning treated its worldwide colonial commerce as an implicitly naval affair, appointing, from 1696 onward, "naval officers" in each colonial port whose job it was to enforce trade rules, even as it avoided regulating its oceanic commerce to the extent other powers had.[57] The relevance of naval power to the maintenance of an overseas empire probably goes without saying, but it is important to state nonetheless. It is even possible, as Spence notes, to see a mutually reinforcing relationship between the use of the Royal Navy for imperial expansion, on the one hand, and the economic development and availability of raw materials that drove naval technological development and fleet size, on the other.[58]

By comparison, although Germany possessed a small fleet that could be—and was—deployed to Africa and East Asia for the purpose of showing the flag, it lacked the resources to do so systematically, and its first attempt to do so in Africa from 1885 to 1886 was followed by an immediate drawdown of naval forces. The Imperial German Navy preferred to send to Africa small gunboats that were nearing the end of their service life to economize on costs.[59] In general, however, in the midst of the moves made by European powers to colonize a distant continent, German nationalism, imperialism, and navalism became intertwined. Those who wanted colonies came to also want ships, even if the means by which the one entailed or necessitated the other were not perfectly seen.[60]

But although the general mental connection between navalism and imperialism could be made by the impression that such small exercises of seapower made, the actual, tangible connection of seapower to colonialism—and to other worldwide influence—was made much nearer to home. Fundamentally, Britain's naval strength conferred not only a worldwide naval hegemony, but more specifically an Atlantic one, and with it the ability both to limit and even control what European powers could do in their home region and, accordingly, to obtain a measure of political deference. The method by which this was accomplished was as crude as it was effective: Britain, with its vast superiority in capital ships, could, by Mahanian logic, simply blockade an adversary that ran afoul of its imperial interests. The threat was so obvious as to be implicit and, in the inverse of a Rudyard Kipling line, was most certainly not "an hundred times made plain."[61] It did, however, occasionally have to be made. The most noteworthy instance of it may well have occurred in 1897, when, as Christopher M. Clark has noted, the British Foreign Office submitted a démarche to the German ambassador in London to the effect that German involvement in the Transvaal region of Africa could lead to war and specifically threatened a naval blockade alongside high seas commerce raiding (Britain then being in a position to easily do both) as an intermediary step in Britain's near-certain retaliation should such interference occur. The connection between the need for a naval breakout from the North Sea and the capability to interfere in matters farther afield was therefore quite apparent.[62]

For purposes of this study, it is enough to note that, at minimum, Germany came to covet Britain's status as a global power that was conferred by a local naval hegemony—namely, the ability to concentrate capital ships in the North Sea and the political influence that entailed. As discussed in the preceding chapters, the goals of a competitor in such situations are essentially idiosyncratic; the point of this study is not merely to look at regional hegemony as such, but to evaluate competition for control of some specific part of the globe.

German Strategic Doctrine

Germany played far less of a role in Africa than Britain or France did, and it had negligible imperial interests elsewhere.[63] It remained preoccupied with national security concerns closer to home: in the same way and to the same degree that Britain was a seapower, Germany was the quintessential land power. Although from its founding it possessed enormous geographic size, population, and industrial production, it was caught in a geographic straitjacket by its powerful neighbors. Unlike Britain, which before World War I never instituted a system of formal military conscription even during the darkest hours of the Napoleonic Wars (except for the haphazard, ad hoc, and much-hated naval press gangs), Germany had, post-Napoléon, adopted a mass conscription system.[64] Not only did Germany, as the successor to Prussia, require universal military service, but the overwhelming majority of its (sizeable) military budget was devoted to its army. Detailed analysis is offered in the following chapter, but for now it is sufficient to note that Germany's Army estimates between 1875 and 1898, which appear to have not always reflected actual expenditure but are a useful guide to the intentions of policy makers, averaged more than 90 percent of overall defense estimates.[65] Throughout this period, Germany could theoretically put 2.5 million troops into the field in the event of war, given some notice, although admittedly this would have come at enormous financial cost.[66] Perhaps the most telling, if anecdotal, indicator of how Germany treated naval power is that its two notable naval chiefs of staff in this era, Albrecht von Stosch and Leo von Caprivi, were in fact Prussian Army generals. Caprivi never even wore a naval uniform.[67]

As is so often mentioned in histories of the era, Germany's greatest national security problem—its policy makers' greatest fear—was the specter of a two-front war with France and Russia.[68] In theory, even a three-front war was possible, insofar as Austria, which Prussia had fought in 1866 to obtain some of the territory that was to be incorporated into Germany, shared a border with Germany and could theoretically also attack. Thankfully, political considerations—specifically the longstanding feud between Austria and Russia over the Balkans and the Black Sea coast—tended to prevent this from being a concern, and had allowed Germany to form an alliance with Austria instead.[69] Although in practical terms Germany was most concerned with France, which it had humiliated in the Franco-Prussian War of 1870–71 and with which Germany had the least cordial relations, Russia remained a serious concern because of its expansionist policies, quasireligious nationalist rhetoric, and slowly modernizing economy. While Germany's Army, then under the leadership of Field Marshal Helmut von Moltke the Elder, was prepared for the contingency of a two-front war, German diplomacy prior to Kaiser Wilhelm's accession was almost obsessively focused on preventing the conditions for a two-front war from arising. And although, as many have noted, there were inherent problems for German diplomacy in Germany's very existence as a powerful and rising state in the heart of Europe, it is nevertheless fair to say that, prior to the kaiser, Germany treated its security as a political problem rather than a purely military one and adjusted policy according to political conditions.[70]

Otto von Bismarck, the German chancellor of the era whose name has become a byword for shrewd diplomatic maneuvering, sought, during the course of a long tenure in office, to keep Germany's foreign relations on an even keel even as it became more and more difficult to do so. Despite his image, which he had cultivated, as a strongman willing to employ "iron and blood" to accomplish his goals, Bismarck preferred political maneuvering to bloodshed wherever feasible.[71] The essence of Bismarck's foreign policy lay in preventing foreign powers from forming a coalition against Germany without committing Germany to the defense of any foreign power, which might, in its turn, cause a coalition to form. Bismarck's grand strategy lay in forming overlapping agreements that mediated conflict between Austria and Russia (in which Germany might have to get involved) while tying Austria and, eventually, Italy to a German-led alliance. More specifically, through the Reinsurance Treaty of 1887, he managed to obtain an agreement on the part of Austria and Russia alongside Germany to avoid coming to one another's aid except if attacked (implicitly guaranteeing that Austria would not attack Russia) while simultaneously maintaining an alliance with Austria (guaranteeing that Russia would not attack Austria). In time, as noted, he added Italy to the German-Austrian alliance, ensuring by this that Italy would not be a problem for Germany either. Bismarck's particular insight lay in his assertion that, in a world of five major powers, success and safety lay in being in the coalition of three against the remaining two.[72] He particularly avoided alienating Britain in Europe for this reason, although he had no such compunction regarding imperial competition farther afield.[73]

Germany's African Diplomacy

With Germany, under Bismarck, concerned with matters much closer to home, at least in comparative terms, it mostly stayed out of the scramble for Africa, notwithstanding the fact that the pivotal conference that led to Africa's colonial partition was held in Berlin. Rhetorically, Bismarck was quite modest in his African ambitions. When asked about African colonies, Bismarck famously pointed to a map of Europe, indicated Germany's position sandwiched between France and Russia, and said, "that is my map of Africa."[74] But in practice, Bismarck was as deft a colonial maneuverist as he was in European politics, although it appears that he approached colonial matters with even greater caution and shrewdness. Precisely because image was so important to Bismarck both officially and privately, it is difficult to take any of his statements at face value. Much of his rhetoric may indeed have been designed to reassure while he plotted his next move. However, his actions with regard to African colonization were almost schizophrenic, and they bespeak a rather characteristic pragmatism. "In January 1883," writes the historian Thomas

Pakenham, "no one dreamt that Bismarck would try to grab a slice of Britain's unofficial empire."[75] In fact, Bismarck quite aggressively did exactly this, audaciously claiming modern Cameroon (*Kamerun*, as it was known to Germany), Togo (*Togoland*), and Namibia (*German South-West Africa*) in 1884 and arranging a British guarantee for a slice of East Africa in 1885 (roughly modern Tanzania), while tying German diplomatic assistance to Britain on disputes about Egypt to British acquiescence in the matter. In the case of German South-West Africa, as of 1889 Bismarck was already attempting to sell the colony back to the British on the grounds that it was not worth the expense required to maintain it.[76] As of the first decade of the twentieth century, the colony, still in German hands and about to undergo the atrocious Herero conflict, was costing Germany 9 million marks (or about 440,000 British pounds in that era's currency) annually, with no obvious return on the investment.[77] Bismarck made no further African land grabs. His successor, Caprivi, would obtain some land along the Zambezi River that bears his name in 1890 and annex modern Tanzania in 1891, bringing Germany's colonial project to its greatest extent.[78] Bismarck's colonialism, then, was of a rather pragmatic nature, and it was subordinated to more pressing concerns closer to home.

When it came to Germany's actual aims, however, more complex questions of national interest and intent were raised. German imperialism has been described by historian D. K. Fieldhouse as having initially approached colonialism pragmatically, even if its profitability came into question from quite early on.[79]

It is the profitability question that more directly impacts the understanding of Germany's imperial aims, which would ultimately impact the Dreadnought Race. In the main, Germany's African colonies lost money. Togoland and Kamerun were so unattractive that (unusually) no companies could be formed to administer them, so that they had to be administered by the German state directly. All of Germany's African colonies were losing money by the outbreak of World War I: they required massive infusions of government funds (estimated at between

50 and 100 million pounds sterling in the era's currency), which they did not repay in the volume of added trade they brought in, let alone in any other terms. German imperialists tended—through 1906, when the Reichstag effectively put the brakes on additional increases in colonial expenditure, and even afterward—to couch their arguments for colonies in noneconomic terms, ranging from national prestige to the need to secure raw materials for Germany's defense industrial base in the event of war.[80] Colonies did not need to provide Germany with any material benefits to be worthwhile; their value was psychosocial, providing Germany with the accouterments of a great power even when those trappings came at a cost.[81] As Fieldhouse notes, it is counterproductive to understand nineteenth century imperialism, on any great power's part, in strict penny-and-pound terms: a colony was a societal appendage that had intrinsic value to the great power that owned it, even if that value could not be expressed in terms of what money it brought in. Empires, Fieldhouse notes, were not businesses, and this basic understanding applied as assuredly to German colonial priorities as to British ones.[82] This will become important in assessing Germany's competitive goals, since the colonial ambitions that were at least nominally tied to the Dreadnought Race had more psychological than material roots.

Germany—though it was not alone in this—did not understand the extremely simple guiding purpose behind Britain's colonial strategy. Britain may not have had a coherent Africa policy, but when push came to shove it was guided neither by profit nor sentiment but rather security and geopolitical logic. Having secured control of the Suez Canal by alliance with a puppet regime—the Khedivate—in Egypt, whose hold on power necessitated greater and greater intervention, and seeking to protect access by it to India and its eastern possessions, Britain picked up African real estate that was not so much profitable as strategic, holding onto the Cape Colony while picking up possessions that allowed it to protect the Khedivate and related choke points. German strategy, seeing colonies as either profitable ventures—which they turned out not to be—or prestige vehicles, never approached even the relatively rudimentary level of sophistication that British imperial policy in Africa demonstrated.[83]

The bottom line, then, was not the bottom line: Germany pursued colonies for reasons besides their economic attractiveness (and while seemingly oblivious to their power projection value), on the one hand, and did so from Bismarck's time through the early years of Wilhelm II's reign with a degree of coolness, on the other. Germany prioritized politics and diplomacy over hard power alone, cultivated land power at the expense of seapower, participated only very tentatively (in relative terms) in the scramble for Africa, and in fact consciously avoided confronting Britain. It was not until the accession of Kaiser Wilhelm II that Germany looked to exert influence in a manner that might require a confrontation with Britain closer to home.

Weltpolitik, Sammlungspolitik, Renewed Colonial Interest, and the Coming of the Dreadnought Race

The accession of Kaiser Wilhelm II to the German imperial throne in 1888, and the subsequent eclipse not only of Bismarck but of his political vision, would have profound effects that would set the stage for the Dreadnought Race, turning the Anglo-German relationship entirely on its head and beginning the phase of direct competition. It is often remarked that the alignment of the major powers on the eve of World War I was historically anomalous to the point of absurdity. By the time of the outbreak of war, Britain was aligned with its almost atavistically established enemy—France—and its historic rival over the Crimea and Black Sea—Russia—against the country that held both at bay physically and that had historically been Britain's reliable ally when needed—Germany.[84] The turnaround occurred in a long sequence of events that began with Wilhelm's *Weltpolitik*, to the details of which this historical narrative now turns.

The Origins and Aims of the Dreadnought Race

It is easy to suggest that Wilhelm's change of policy toward Britain grew out of his fragile psyche.[85] Crippled from birth with a malformed arm (the result of a botched delivery), and educated in an unsympathetic manner by an unevenly caring mother; a martinet of a tutor; and, in young adulthood, a Prussian officer corps whose machismo was legendary, Wilhelm had acquired a noteworthy temper, a hostility to deep thought or contrary advice, and a vainglorious sense of his own destiny.[86] Concerned for their own professional self-preservation, officers who dealt with him repeatedly advised each other to flatter the kaiser as much as was feasible, and at any rate to keep him grounded and well-handled, lest he make an ill-considered move.[87] Wilhelm was also vain, almost to the point of pathos; loved (more than most of his European royal peers) to strut in military uniforms; and was given to gaudy bravado, longwinded monologues and social faux pas, which frequently bemused British officials who had the misfortune to spend too much time with him.[88]

To this emotional state must be added familial rivalry. Wilhelm's mother was Queen Victoria's daughter (Victoria, or more formally Empress Frederick), and Britain's King Edward VII (a man who, incidentally, possessed more than his own share of psychological baggage) was therefore his uncle.[89] Wilhelm, even in his prime, was therefore literally looked down on by a patronizing British (and incidentally Hanoverian) royal family, one of whose princesses was also his estranged mother. The fact that Wilhelm's accession into office had been necessitated by the botched diagnosis and failed treatment of his moribund father, Frederick II, by British doctors recommended by his mother's family, could scarcely have helped matters.[90] Nor could his reception by British high officialdom, which treated him with a mixture of amusement and scorn.[91] Although it is possible to make too much of such matters, it has seldom gone unremarked-on that Wilhelm vacillated emotionally between seeking his British relatives' respect and, when it was insufficiently or too slowly given, rejecting them in turn. He claimed, moreover, to have grown up admiring the ships of the Royal Navy at anchor.[92]

German decision making under Wilhelm is generally understood through two distinct lenses, one emphasizing individual agency and the other structural factors. As will be seen, these two viewpoints are more complementary than contradictory. Briefly, at the actual point of any given decision, the kaiser's personal influence, and in turn his psychological motivations, were of paramount importance. The kaiser jealously guarded his royal prerogative as the head of Germany's armed forces and, in effect, its warrior-king, with the result that all military decisions within a policy framework classically noted for its militarism ran through him. As will be discussed further below, the kaiser went so far as to divide authority within his own military still further—particularly within the Imperial German Navy—so as to emphasize his role as commander in chief and involve himself in the details of military organization and command, in which he took no shortage of pride. Alternately micromanaging when it suited him and delegating when he trusted his subordinates, Wilhelm would, along with a mere handful of trusted advisors and policy makers, bear a great deal of responsibility for Germany's military planning during the Dreadnought Race, even as his management of the details was predictably chaotic.[93] For this reason, the naval officer who Wilhelm entrusted with naval procurement and whose warped vision drove the Dreadnought Race, Alfred von Tirpitz, is normally brought to center stage in narratives of these events, and is so treated here, although his bureaucratic opposition will be given its due where appropriate. On foreign policy questions, however, Wilhelm was inclined to delegate to his chancellors, and for this reason they, Bernhard von Bülow and Theobald von Bethmann Hollweg, will feature prominently in this narrative. Wilhelm's essentially personal model of governance has justly been called "polycratic chaos," and the seams in Germany's grand and competitive strategies will emerge in this narrative.[94] But because of his firm, yet erratic, hold on German policy, his personal motivations and those to whom he directly entrusted his government are important guides to Germany's competitive goals.

On the other hand, a more structural interpretation of Germany's decision making hinges on the various factions within the German state at the time, the social changes Germany was undergoing in the course of rapid industrialization, and the social pressures that subsequently emerged.[95] This study cannot presume to account for all of the historical and historiographical debates that emerge from the study of Germany's industrial revolution. It is sufficient here to note, as Hew Strachan has done, that Germany industrialized both later and more quickly than many European powers, and that the effect on German society was dramatic. The upshot was a fluid political situation in which, classically, a moribund landed nobility with tight links to the army clashed with an industrial middle class that increasingly formed the backbone of the Imperial German Navy's officer corps, even as both opposed an increasingly insistent industrial proletariat whose rising political influence took the form of an increasingly aggressive Social Democratic Party. Into this social fire may also be thrown Germany's Protestant-Catholic divide, which periodically injected itself into political negotiations.[96] To this in turn, as Gordon A. Craig has noted, must be added the German Reichstag's essential political immaturity: with German democracy in its infancy, members of parliament were simply not used to making serious decisions and did not always know how to make the necessary compromises.[97]

Because this study concerns competitive strategy, and because strategy is understood to be found first and foremost within the minds of an organization's leaders, even as they may face pressure from below, this study will frequently focus on the motivations of Germany's key policy makers, oftentimes necessitating understanding the mindset of Wilhelm, but accounting where appropriate for their structural motivations. The same can apply to Britain. But because of the erratic nature of not only the kaiser's government but the society it governed, the structure of German society will occasionally play a role here as a driver of German strategic intentions, insofar as the need to bring some order to Germany's political maelstrom was therefore as much a factor in determining German aims in the Dreadnought Race as the psychology of Germany's leadership.[98]

Wilhelm possessed more grandiose (and nebulous) geopolitical ambitions than Bismarck. In 1890, scarcely two years after ascending the German throne, Wilhelm removed Bismarck from the chancellorship to prevent the latter from challenging his authority.[99] Throughout the following decade, Wilhelm and his advisors claimed to be pursuing *Weltpolitik*, a term that they never defined but that meant an attempt to make Germany's political influence felt across a wider area than merely central Europe and that implicitly required seapower.[100]

To understand what Germany's key decision makers—and particularly Wilhelm, around whom they revolved—wanted when they referred to *Weltpolitik*, it is necessary to trace the origins and evolution of the term. At its most basic, the term, coined by the kaiser's senior lieutenant Bernhard von Bülow, who would rise from the rank of foreign minister in 1897 to chancellor in 1900, meant an attempt to break out of the straitjacket of landward power and exert worldwide influence.[101]

Weltpolitik would ultimately serve as what the framework suggested here would categorize as a third-metric goal: concerned neither with relative security nor relative wealth but with intangible or visionary goals. As such, it had at least two major drivers and at least three major aims. Its driving forces lay in the kaiser's personal motivations, on the one hand, and the domestic needs and pressures of the new German state, on the other. These, in turn, determined its goals, of which increased access to colonies was perhaps the least important one. The first aim was domestic harmony and, in particular, maintaining Germany as a broadly nonsocialist state, a goal essentially agreed on by all German decision makers except the Social Democratic parliamentarians themselves, expressed in the curious neologism *Sammlungspolitik*, which referred to a national unity policy. The next aim was reordering Germany's position in the international system—and particularly within Europe—so that it was not merely a continental power but rather a state capable of dominating European waters and thereby reducing the dominant maritime power—Britain—to the status of a subordinate ally. The last aim would provide for the other two: it would unite Germany's various political factions behind a common nationalist goal, and it would pave the way for further imperial gains, in which the major international playground was Africa but which might occur elsewhere as easily.[102]

Each of these may be addressed in turn. Where Wilhelm was concerned, his personal motivations have been covered. These manifested in a generalized ambition for German expansion, which could take many forms but gravitated toward generalized imperial influence, on the one hand, and navalism for its own sake, on the other (again, driven as much by the kaiser's own personal obsessions as anything else).[103] Likewise, Wilhelm's calculus was in part driven by his own preoccupation with presenting a united front against the German Social Democrats, who were gaining electoral ground.[104]

The genesis of Germany's renewed interest under Wilhelm in acquiring colonies is traced by Gordon Craig to Paul Kayser, then the chief colonial affairs officer of Germany's Foreign Ministry and, ironically, a Bismarck protégé. Kayser advocated that the government take a greater interest in overseas colonization and—presaging Wilhelm's later interest in involvement in the politics of the Boer War—interfering in other European empires' colonial conflicts. As Craig notes, this entailed getting into disputes with Britain, which was the logical choice of an adversary where colonial disagreements were concerned. In what would be a repeated pattern throughout the Dreadnought Race, Kayser had only to win the enthusiastic support of Wilhelm (such support being an essentially all-or-nothing proposition) to make his department's influence felt. Once Kayser had accomplished this, the seeds of Wilhelm's interest in overseas power projection were sown. For all the interest he took in principle in acquiring colonies, however, Wilhelm left the specifics vague—in effect, Germany merely needed to be in a position to take advantage of opportunities as they arose. This Germany would do, repeatedly antagonizing Britain in the colonial realm, although without tangible results to show for the effort.[105]

The psychological, *vice* material, importance of colonies for Germany has just been discussed. What is equally important to note, however, is that from the mindset of the key German decision makers, colonies were the least important aspect of the problem and the goal most amenable to diplomatic accommodation. As Paul Kennedy has noted, at various points during the course of what would become the Dreadnought Race incentives existed, and attempts were made, to resolve Anglo-German tensions over colonization; indeed, as the British policy maker Eyre Crowe would ultimately suggest in a famous memorandum, if all that were in play during this period between Britain and Germany were a few colonies, absent gross aggression, Britain would have (and had) remained capable of reaching a reasonable agreement. And indeed, as Kennedy has famously noted, Britain was quite willing to engage in what would come to be known as "appeasement" toward Germany and others if it would resolve international disputes at acceptable cost.[106] Wilhelm, however, wanted more than this, and hoped to force Britain into the position of a subordinate ally who would compliantly acquiesce to further German overseas initiatives.

From the German perspective, as D. K. Fieldhouse has noted, Wilhelm's desire for colonies was driven not by their practical benefit, but by the intangible status they conferred as markings of world power status.[107] From the point of view of the framework offered here, colonies were first and foremost part of the intangibles metric. Whatever the case, while the connection between Mahanian seapower—predominance in capital ships—and overseas colonies was tenuous, it nonetheless existed. Germany in its current position was allowed overseas colonial adventures only by British permission. Otherwise, Britain's predominance in capital ships could make the North Sea a very unfriendly area of home waters for Germany. On this basis alone, pursuing seapower beyond the capabilities required by the *Jeune École*, and specifically in the North Sea, was a necessary intermediate goal for a Germany seeking to make its influence felt on a broader stage.

Opposing Britain also furthered the goal of *Sammlungspolitik*. In essence, this policy idea referred to fusing much of Germany's foreign and domestic policy together so as to unify the feuding factions that opposed the Social Democrats—the common political opponent of all of Germany's decision makers from Wilhelm on down—causing them to work on a common political project against a common foreign adversary and a common domestic foe. As will be seen, the fleet-building program that inaugurated the Dreadnought Race was *supposed* to accomplish this to the extent possible: although it would ultimately stoke tensions between the aristocrats and the business class (particularly regarding funding and direct taxation), initially it served to enlist multiple factions in pursuit of a hawkish foreign policy against Britain.[108]

As has been noted, there was an intuitive connection between imperialism and navalism, which Wilhelm, a naval enthusiast, had grasped but not fully developed. For purposes of making German influence felt globally, Wilhelm initially (as of the latter half of the 1890s) favored a policy of naval procurement, but of an entirely different kind from the policy that would ultimately drive the Dreadnought Race. Thinking solely in terms of power projection rather than great power relations, Wilhelm favored procuring cruisers rather than battleships, on the assumption that these relatively cheap ships would allow for gunboat diplomacy and overseas troop deployments. In this, Wilhelm was driven more by images than detailed logic, including everything from his perception that a strong cruiser fleet would allow Germany to intervene in the Mediterranean; to a newfound fear for the safety of overseas colonial possessions; to a grand leap of logic, from Mahan to the *Jeune École*, that held that the purpose of new ships would be to protect Germany's burgeoning overseas trade.[109] It was not until 1898, as will be discussed, that Alfred von Tirpitz would persuade Wilhelm that in order to make German influence felt outside European waters, Germany would first have to acquire the capital ships necessary to dominate those waters.

Where Britain was concerned, Wilhelm, having appointed the likeable, if not exactly brilliant, Leo von Caprivi to fill Bismarck's office, initially sought closer ties to Britain.[110] Practically Caprivi's first act, sanctioned implicitly if not explicitly by Wilhelm, was to refuse to renew the Reinsurance Treaty with Russia, whose expiration was pending, to curry favor with Britain, a rival to Russia since the Crimean War. In so doing, Caprivi, and presumably Wilhelm, sought to pull Britain closer to Germany and ultimately facilitate the establishment of a formal alliance between them. In this, they almost comically miscalculated. As Henry Kissinger notes, Britain simply did not enter binding alliances—in fact, it did not enter alliances at all except as a matter of convenience. Conversely, as Bismarck had understood and Wilhelm and Caprivi did not, what Germany needed, from the standpoint of its own security, was not Britain's allegiance but its acquiescence: a simple guarantee of British neutrality in any continental war involving Germany (which might have been possible given certain caveats) would have been quite sufficient to guarantee Germany's safety from almost any reasonably imaginable set of aggressors. It was only if Germany sought to subordinate Britain for its own sake that an alliance was necessary, and it was in precisely such circumstances that Britain was least likely to acquiesce in one. Britain rebuffed Wilhelm and Caprivi's requests for a formal alliance. At least in some part, the naval arms race that followed can be seen as a rather ham-fisted attempt by Wilhelm to convince Britain that Germany was better as a friend than an enemy. However, as Kissinger dryly observes, there was little in recent history to suggest that this was a good idea.[111]

The Origins and Aims of the Dreadnought Race

Germany was being slowly encircled politically and militarily well before the Dreadnought Race began. Germany's refusal under Wilhelm to renew the expiring Reinsurance Treaty, which had kept the peace with Russia under Bismarck, and its subsequent courting of Britain as a coalition partner pushed Russia abruptly into France's camp. The Franco-Russian Military Alliance Convention was formally concluded, after two years of informal agreement on its basic terms, in 1894. The agreement, which was kept secret between the two parties, merely stipulated that France or Russia, as relevant, would declare war on Germany if Germany (or one of its allies, Austria and Italy) attacked the other party. The agreement did not constitute a unified front against Germany in all circumstances.[112] For all that, it brought together Germany's two principal potential adversaries in a political coalition against it. In the midst of this, Germany would nonetheless embark on a collision course with Great Britain as well, partly in an attempt to win it over as an ally.

Exactly when Germany, and specifically Wilhelm, decided to stop treating Britain as a potential coalition partner and started treating it as an adversary is disputable, insofar as early on, the one was thought to entail the other. German might would force Britain to treat it with respect and instill a desire to join a German alliance as a junior partner.[113] In Kissinger's formulation, there was potential for an Anglo-German entente throughout the Dreadnought Race, but the problem was that Germany never ceased its demands for a formal alliance that Britain would not give.[114] This, however, is a form of retroactive wishful thinking regarding both sides. As Kissinger stresses, Britain during the Dreadnought Race quickly and uncharacteristically cast its lot with France and Russia; despite the absence of formal security guarantees by Britain to its entente partners during the decade prior to World War I, what had been done could not be easily undone.[115] As for Germany, while it may have continued publicly to hold out an olive branch to Britain on certain conditions, the Dreadnought Race effectively began when Germany decided to knock Britain off its perch for good. In its most genuine form, Wilhelm's early thinking on the matter clearly did not admit to even a short-term partnership with Britain.

By the middle of the 1890s, the kaiser's desire to make German influence felt abroad was already on a collision course with Britain's empire. During the New Year between 1895 and 1896, the British colonial agent Sir Leander Starr Jameson made an abortive attempt to start an insurgency of British settlers against the Boer government in the Transvaal Republic in modern South Africa. Although Britain disavowed Jameson, Kaiser Wilhelm took the opportunity to publicly congratulate the Transvaal government on repelling the invasion and in so doing obliquely humiliate Britain. Kissinger puts the gaffe (if it was) down to a "public-relations ploy" to burnish the kaiser's nationalist credentials (which scarcely needed help) and mobilize public support for his policies. However, given Wilhelm's subsequent interest in the Boer War, it seems more likely that he was making a rather characteristically frank statement of his views and had already decided that British interests were fundamentally hostile to Germany's.[116]

To sum up, therefore, Germany's key decisionmaker, an unstable and erratic individual with numerous personal eccentricities, was motivated variously by a desire to unite his country's competing domestic factions behind a common enemy, an acute sensitivity to social status relative to British society and its monarchy, an understanding of colonies as a status symbol and a desire to acquire them, a desire to subordinate Britain diplomatically, and a nebulous but distinct impression that building a navy would solve all of these problems. Wilhelm lacked the precise understanding of what form that fleet would take or how it would perform its task. In fact, as noted, his initial intuition ran to cruiser procurement, both because he was initially concerned with trade protection and because gunboat diplomacy—of which he cited the Mediterranean as a specific example—seemed to suggest the need for cruisers. The decision to challenge Britain's capital ship fleet in North Sea waters, and thereby to crystallize a strategy for accomplishing Wilhelm's nebulous goals, would ultimately lie in the hands of the man Wilhelm eventually put in charge of naval procurement, Alfred Tirpitz (later von Tirpitz).

Tirpitz, in his role as head of the Imperial Naval Office (*Reichsmarineamt*), would become the crucial mastermind of the Dreadnought Race. Tirpitz was a self-made man of middle-class origins (the nobiliary "von" was added to his name only during the Dreadnought Race in recognition of his services) who had worked his way up the German naval hierarchy in the era when the German Navy was a strategic afterthought. A brilliant, erudite man with impeccable manners and considerable (and un-Prussian) charm that made up for an intimidating appearance (a bulky frame, bald head, scowling face, and forked beard) and an arrogant temperament, Tirpitz had learned to work the internal politics of the German naval bureaucracy to his advantage, making more than a few enemies along the way. Although his major burst of creativity to date had involved overseeing torpedo-boat procurement, Tirpitz was a disciple of Alfred Thayer Mahan who internalized the latter's rejection of commerce raiding as a naval strategy and promotion of the big fleet and big battle as the route to power. As an ambitious officer and intellectual in a landlubber navy that attracted the Prussian Army's personnel rejects, Tirpitz was a big fish in a small pond.[117] In any event, he would come to sell the idea of a massive German capital ship fleet, which coincidentally would bring him renown and secure his place in history, as a solution to a heretofore questionable problem: the threat posed by British North Sea naval hegemony.

Wilhelm and Tirpitz shared certain character traits and a congenial relationship that made for easy cooperation, even if it took time for Tirpitz to convince Wilhelm to authorize his capital ship program. Wilhelm, who as noted had been enamored of ships from boyhood, first discovered Tirpitz and made him a protégé shortly after ascending the throne and accelerating the modernization of the German Navy;

Tirpitz reciprocated by lobbying carefully but aggressively for fast-track postings. Wilhelm saw to it that Tirpitz was given ever-increasing responsibilities as a strategy advisor, including him in discussions with more senior officers and accelerating his career, eventually making him chief of staff of the Imperial German Navy's Naval High Command.[118] The two men had similar predilections. Perhaps significantly, both were not only navalists but also imitators of British culture: Wilhelm by birth, and Tirpitz by education and affectation (the latter sent his children to English schools)—this notwithstanding the fact that both came to regard Britain as a major adversary and, in Tirpitz' case, an ideological one.[119] And while Wilhelm had provided the initial impetus toward naval modernization, proclaiming that "our future is in the water" and pushing for the construction of the four earlier battleships, his ambitions in this area were, as noted, quite vague and unfocused until Tirpitz took over naval procurement in 1897.[120]

But despite his Anglo-Saxon mannerisms and admiration for British maritime power, Tirpitz was a German nationalist of a radical bent. The congenial exterior concealed a darker set of influences. To understand these, it is necessary to trace the rapid ascendancy of navalism within Germany at the time.

German Navalism: Origins

The rise of German navalism tracked closely with the rise of both German industry and German nationalism. As the historian Lawrence Sondhaus has documented, from Prussia's earliest years intellectuals and policy makers had made the connection between the possession of at least some seapower and international recognition of a state's sovereignty, although Prussia had been slow to build a navy of any meaningful size. Frederick the Great's Prussia had assembled and maintained ad hoc fleets of small ships, but Prussia had, out of expediency, avoided building any substantive naval force even at the height of the Napoleonic Wars.[121] The industrial era changed this. One of the earliest proponents of seapower in nineteenth-century

Germany had been the neomercantilist economist and German nationalist Friedrich List, who in addition to his arguments for an active state role in Germany's industrialization also argued for the combination of German resources at sea. List in particular argued for flying the same national flag by merchant ships of the (then disunited) German states and, more ambitiously, a prototypical German fleet for littoral defense. In line with his nationalist credo, List believed that only by making a strong show of naval force could Germany hope to obtain the respect of Britain or other states. In this, he was well ahead of his time, and it was not until more than a half-century after he committed suicide in 1846 that the Dreadnought Race would test his intuition about the connection between German seapower and respect for Germany's international position. List probably never in his wildest imaginations envisioned that Germany would seek seapower on the scale to which it ultimately did. His argument as to the reasons for seapower's importance, however, was, if anything, too well taken.[122]

It was not until the 1880s, however, that the now-united Germany took any meaningful steps to maintain more than a modicum of naval strength. As noted above, as of 1875, Britain possessed four large, state-of-the-art battleships that were considered superior to any ship deployed by any state of the era, including Germany.[123] Germany only belatedly followed suit, beginning in the 1870s with a massive naval modernization program that was accelerated post-1888 by the newly coronated Kaiser Wilhelm.[124] Although this study will treat the Dreadnought Race as having begun with the passage of the German naval law nine years later, it is important to note that German naval modernization for ambiguous purposes was already underway well before this time. It reflected Germany's newfound national self-confidence; its nascent, if small, colonial policy; its new kaiser's commitment to *Weltpolitik*; and, above all, its statist national philosophy.

The impetus for shipbuilding went hand in hand with a trend, accelerated after Wilhelm's coronation, toward a more overt and incautious militarism that elevated military service to the level of society's highest calling and extolled obedience to a warfighting state.[125] As historian Barbara W. Tuchman observes in a small but telling vignette, in an otherwise patronizingly chivalrous era, when high-born German ladies encountered military officers on city streets, it was the women who were expected to step aside and make way.[126] This was an era in which, as the military historian John Keegan has remarked, professors and military officers vied for the top of Germany's social hierarchy, and it is easy to see why: the relationship was symbiotic.[127] Although a discussion of the intellectual origins of German militarism could fill a separate book, it is sufficient here to note that such origins existed and had many illustrious proponents. One of them is of particular relevance here. The spokesman for the next generation of German navalism—and the chief spokesman for German nationalism generally—was the German nationalist historian and philosopher Heinrich von Treitschke. Treitschke was a charismatic public speaker whose lectures extolled militarism and approved of an authoritarian state that would unite the German people against their enemies, end the divisiveness of democratic politics, and propel them to international greatness. Treitschke did not delve into the details of seapower and its requirements, but he thought he recognized its purpose. For Treitschke, it was Germany's destiny to dominate other nations, and the ultimate prize was Britain. Although Treitschke was careful to argue for naval parity with Britain rather than completely supplanting Britain's power, it was quite clear what he wanted.[128] Treitschke's navalism paired well with Social Darwinist intellectual fads of the time, and a number of Social Darwinist thinkers voiced similar sentiments and would lend their weight to the naval project that Tirpitz would ultimately undertake.[129]

Treitschke was an intellectual mentor to Tirpitz. Tirpitz had taken Treitschke's classes while at university and had spent long hours discussing politics with Treitschke over meals and drinks, almost literally taking notes on cocktail napkins.[130]

Tirpitz had also absorbed a series of almost protofascist philosophical tenets, including, as Paul Kennedy has noted, a fair bit of Social Darwinism, combined with extreme nationalism and a hostility to its more internationalist antitheses, a dose of anti-Semitism, and an affection for statism, all of which only increased in intensity as Tirpitz aged. As Kennedy notes, Tirpitz's respect for Britain did not ultimately entail a genuine liking for it—in fact, Tirpitz contrasted his own fervent nationalism with Britain's more internationalist pretensions, to the point of seeing Britain as the emblem of all that he had come to hate.[131] The union of nationalist ideas, Wilhelm's global ambitions, and Tirpitz' drive amounted to a perfect storm.

Tirpitz' own views existed at the far end of a spectrum of opinion on force structure and policy within the Imperial German Navy. As noted above, Germany's force structure to date had largely been designed with the *Jeune École* in mind: torpedo boats would defend the coasts in wartime while cruisers and small ships would raid enemy commerce; expectations of the navy in wartime (heretofore expected to involve conflict with France and Russia) were modest. Tirpitz's views ran to the opposite end of the range of opinion: he was a Mahan disciple who favored a capital ship fleet on ideological and practical grounds. He particularly drew a connection, supposedly founded in Mahan's thought, between the possession of a fleet of ships of the line and the ability to sustain cruiser warfare, in that the latter could supposedly only achieve success in the presence of the former. As an ideologue rather than a technocrat, however, Tirpitz had also internalized a Mahanian conception of the linkage of capital ships and national power, which dovetailed in turn with his Social Darwinist views and his increasingly personal jealousy and hostility toward Britain. In this he was not alone: a number of prominent defense intellectuals within Germany had adopted and refined a basic Mahanian ideology, as had the kaiser, although he had only digested it in outline form. In time Tirpitz would acquire a unique capacity to implement these views.[132]

Chapter Four

Germany's naval program before the Dreadnought Race was unfocused. It did not seek to challenge, much less supplant, Britain as a naval power and had initially the fairly modest goal of building only eight ironclad frigates, most of them of the old "casemate" design that aligned guns in batteries rather than rotating turrets. As of 1889 (the year Britain adopted the Two-Power Standard in response to concerns about naval modernization worldwide), the goal had been moved up to building a few large battleships; four were planned in the German naval estimates of that year.[133] The growth of the German Imperial Navy after 1888 was quite frenzied. Sondhaus argues that, although Tirpitz, the subsequent architect of the naval race with Britain, saw the years between 1888 and 1898 as a "wasted ten years," in reality the Imperial German Navy underwent extensive reorganization during this time, and only a few years were without significant naval developments.[134] The Dreadnought Race, in the end, built on a fairly solid foundation of German naval preparation that had been laboriously assembled during the course of two and a half decades. What was different about the Dreadnought Race was both the scale and scope of naval construction, on the one hand, and its intent, on the other.

Perhaps significantly, under Bismarck and in the early years of Wilhelm's reign, Germany outsourced much of its naval shipbuilding to British contractors. German ships were being built at the Bath Ironworks, Thames Ironworks, Poplar, and other locations in Britain, with no objection from either party.[135] Britain was in fact rather indulgent toward the Imperial German Navy, most notably when it offered a Portsmouth dockyard to repair the German ironclad SMS *König Wilhelm* (1868) after it inadvertently collided with (and sank) another vessel, the SMS *Grosser Kurfürst* (1878), in the English Channel on a clear day in 1878.[136] The complicated relationship between Britain and Germany in naval matters mirrored the complicated relationship between their monarchs. It was, however, about to be simplified by Tirpitz's plan to make the Imperial German Navy serve Wilhelm's goals.

Alfred von Tirpitz and the Coming of the Dreadnought Race

In 1897, Wilhelm appointed Tirpitz, now elevated to the rank of admiral, to the official position of state secretary for the Imperial Naval Office (*Reichsmarineamt*), the office responsible for naval procurement and force structure.[137] Although in theory the state secretary for the Naval Office reported to the German chancellor, Tirpitz's relationship to Wilhelm was such that he could, when it suited his purposes, go straight to the top of the hierarchy for approval.

To understand the strategic problems that followed, a word is necessary regarding the naval organization—or lack thereof—to which Tirpitz reported. Britain's famously conservative and sclerotic naval bureaucracy, with a uniformed First Sea Lord leading an administrative Board of Admiralty and a set of fleet commands, presided over by a single civilian First Lord of the Admiralty, was efficient compared to its German counterpart. There was no titular uniformed head of the Imperial German Navy. Rather, under reforms personally instituted by Wilhelm to consolidate and strengthen his authority, the kaiser was nominally that leader, while administration was divided between the *Oberkommando* (Naval High Command, responsible for operations, which dissolved when the kaiser assumed this role in 1899, with the remainder of its offices transferred to a newly formed *Admiralstab*, Imperial Admiralty Staff), the *Reichsmarineamt* (Imperial Naval Office, responsible for procurement and force structure—famously headed by Tirpitz throughout the Dreadnought Race), and the *Marinekabinett* (Imperial Naval Cabinet, which nominally handled personnel matters and doctrine but in some respects duplicated the other two). Tirpitz, as state secretary of the Naval Office (head of the *Reichsmarineamt*) was therefore the head of only one of the three bureaus and answered nominally to the chancellor, even as the kaiser was the nominal commander in chief. An intense bureaucratic squabble existed from the outset between the High Command and the Naval Office, as the division between fleet structure and tactical doctrine was artificial at best—it was an organizational conflict in which Tirpitz was eventually to have fought on both sides, first in his old capacity as chief of staff to the High Command and then as state secretary of the Naval Office. To make matters more confusing, Wilhelm, while mostly ignoring his role as commander in chief, would interject himself as he saw fit between the chancellor and Tirpitz. Suffice it to say that, in Weberian terms, Britain naval administration was institutionalized, while in Germany it was heavily charismatic.[138]

Tirpitz replaced Admiral Friedrich von Hollmann, whose ambitions for the German Navy were modest. In the years preceding Tirpitz's appointment as his successor, Hollmann had not seen the need for a powerful battle fleet, preferring to acquire midsize cruisers for coastal defense and for waging a *Jeune École*-style guerre de course. He in particular did not envision (ironically, as it turned out) that the Reichstag, which held veto over the military's budget, would ever vote the necessary expenditures for a large, blue water fleet. Pressured by Wilhelm to do more and unwilling to carry on, Hollmann resigned in March 1897, paving the way for Tirpitz's promotion and appointment.[139]

The Dreadnought Race Begins: Tirpitz and the Naval Laws

Tirpitz, who regarded the years preceding his appointment as a "lost decade," wasted little time.[140] In June 1897, mere months after Hollmann's resignation and only a week after arriving home from an extended leave abroad to take up his new office, Tirpitz penned a memo for Kaiser Wilhelm outlining his arguments for a powerful fleet that could challenge Britain. "For Germany," Tirpitz wrote, "the most dangerous enemy at the present time is England. She is also the enemy against whom we must have a certain measure of Fleet Power as a political power factor."[141]

This marked a change of attitude. It was the first time a serious policy document argued for treating Britain, as opposed to the more immediately threatening states of France and Russia, as the focus of effort for Germany's national security policy. It was also unwarranted, in that it was clear from context that deciding to compete with Britain's naval primacy was a strategic choice—the memo was largely silent on the question of why Britain had suddenly become a problem. Tirpitz elaborated that the purpose of the fleet should be to confront Britain in its home waters, "between Heligoland and the Thames," and that therefore large battleships would be required.[142] Tirpitz modestly requested that the fleet acquire 19 such warships: 2 squadrons of 8, which he had previously calculated to be the ideal squadron size, plus a flagship, with 2 more kept back to replace losses in either squadron. The construction of these ships would be accompanied by the construction of new cruisers and torpedo boat fleets.[143] Tirpitz expected that the fleet, which would come to be known as the High Seas Fleet, could be complete by 1905 and built within approximately the same budget that his predecessor had requested for a coastal defense force. The difference would be not cost but capability and intended use.[144]

The specifics of Tirpitz's logic regarding the construction of a fleet are as vital to the question of understanding Germany's competitive goals as they are disputable. Certain aspects of Tirpitz's plan can be taken for granted, certain ones are probable, and certain ones can be suspected but not proven. In the first place, Tirpitz was solving a problem for his sovereign: Wilhelm wanted a fleet, and although he was not specific or even consistent about what form that fleet was to take or what purpose it was to serve, his desires tracked well with the navalism of Treitschke and other intellectuals of the era, and Tirpitz, in creating a battleship fleet, was giving it both a concrete form and a purpose.[145] As noted, moreover, the procurement of battleships for opposition to Britain served the goals of *Sammlungspolitik* by giving various domestic factions an ideological and sometimes financial stake in the procurement program.[146]

Chapter Four

Where the latter was concerned, at minimum, Tirpitz intended to deter a British naval blockade of German shores, which had been threatened in the above-mentioned demarche of 1897 regarding the Transvaal, but which had always been a theoretical possibility, as noted. His extant documents demonstrate at the very least that he hoped to accomplish this by achieving a 2:3 ratio of capital ships between Germany and Britain, respectively, in the North Sea. At this point, the Royal Navy would have difficulty blockading German shores in wartime without incurring unacceptable losses. Given certain assumptions about superior seamanship and technology, it was even possible to imagine the weaker German fleet winning such a fight or at least achieving a draw. This was particularly true if, as Tirpitz appears to have envisioned, the Royal Navy was unable to bring all of its capital ship fleet into home waters, as this would require it to denude its farther-flung bases and leave more distant parts of the British Empire vulnerable. All that was required to achieve this was to build a fleet of sufficient size to achieve the appropriate force ratio in North Sea waters. Tirpitz referred to the fleet as a "Risk Fleet," emphasizing its primary purpose as a deterrent.[147]

There were a number of problems with the usage of such a fleet for deterrence, some more obvious than others, but all of them centering around the willingness of Britain to play into Germany's hands in wartime. As Paul Kennedy has noted, the "Risk Fleet" concept assumed a British operational doctrine of employing a close blockade in wartime—if Britain chose an option other than a direct confrontation in German coastal waters, including the highly probable one of a blockade farther out, the options for confronting it would diminish considerably.

The problem in particular was one of range: coal-fired battleships sortieing from German North Sea ports would not be able to operate for long periods at the distances necessary to confront a distant blockade, and although various solutions existed—all of them involving the seizure of foreign ports or acquiring agreements to allow their use—few of them were seriously contemplated. Kennedy, echoing the historian Wolfgang Wegener, suggests an audacious means of doing so that Germany megalomaniacally attempted in World War II: the seizure of Norway as a base for further power projection into the North Sea and North Atlantic. As much as this makes clear the degree to which the Dreadnought Race was holistically a regional supremacy gambit, there were obvious reasons why this was not attempted. But not gaining access to northern ports, by means fair or foul, made for a sizeable hole in German plans. Kennedy also notes that while there was considerable synergy in the future German strategy of attacking through Belgium on war's outbreak and Tirpitz's plans for naval dominion of the North Sea, Germany's various organs of government, perhaps mercifully, did not put this together, either.[148]

As for the ultimate size of the fleet and the uses to which it might be put, whether Tirpitz hoped for more than a deterrent from a position of weakness—or at least dreamed of this—is a more open question. At various points in the Dreadnought Race—particularly toward the end—Tirpitz not only spoke of achieving parity with the Royal Navy in the North Sea but also stymied arms control negotiations that would have potentially codified the supposedly acceptable outcome of a 2:3 force ratio in the form of a treaty. He also admitted privately that he could not be specific about his aims in public, and it appears that he regarded the 2:3 ratio as a minimum rather than a stopping point.[149] This study is in broad agreement with Paul Kennedy's assertion that although there is little documentation suggesting that Tirpitz sought naval parity or superiority, it made sense for him to do so if possible, and such was probably at least at the back of his mind—though, admittedly, other interpretations are possible.[150]

The purpose of the fleet, whatever size it ultimately achieved and whatever the extent to which Britain was able to build and move ships to oppose it, also is subject to a range of interpretations. At minimum, in its function as a deterrent, it would enable *Weltpolitik* by eliminating the ability of Britain to thwart German interference outside Europe by threatening a blockade. At the outside, if the fleet achieved something like qualitative or quantitative parity—or superiority—vis à vis the Royal Navy in North Sea waters, it would force Britain into an inferior bargaining position, perhaps even giving Germany the upper hand in wartime and reversing the then-current position of Britain as the potential blockader and Germany as the potential blockaded. This, in turn, would serve the aforementioned diplomatic purpose of forcing Britain to become a subordinate ally; failing this, it would at least allow Germany to engage in what Volker Berghahn has called "diplomatic bullying" with regard to colonies or similar matters. Obviously, the degree of leverage that Germany would acquire would depend on the force ratio it could achieve; on these points there were few specifics.[151]

German policy did not immediately crystallize around even this relatively vague goal set. Wilhelm was enthusiastic about the High Seas Fleet, but lacked detailed knowledge of its scope and purpose. Where enthusiasm was concerned, Wilhelm peppered his public appearances with navalism. In a speech given in 1898, he announced at the opening of a new harbor at Stettin (now Szczecin, Poland) and stated, "Our future lies upon the water."[152] The link to *Weltpolitik* appeared clearly in a speech Wilhelm gave at Kiel in 1900 on the launching of the battleship SMS *Wittelsbach* (1899), in which he extolled how "powerfully the wave beat of the ocean knocks at the door of our people and forces it to demand its place in the world as a great nation."[153] It appeared more notably in Wilhelm's now infamous speech in 1901, in which he lamented that "we have no such fleet as we should have" but affirmed that nonetheless "we have conquered for ourselves a place in the sun." The famous phrase, which referenced the successful resolution of the Boxer Rebellion in China, to the crushing of which Germany had contributed, echoed a statement by Bernhard von Bülow before the Reichstag in 1897 in which he had announced intervention in China and affirmed, "In short, we do not want to put anyone in our shadow, but we also demand our place in the sun." Wilhelm then repeated the fateful line: "[O]ur future lies upon the water."[154]

Knowledge and attention to detail, however, were another story. When first presented with Tirpitz's plan for battleships, and even after it was in process, in 1902, Wilhelm was interested in cruisers—normally understood as a *Jeune École* implement for commerce raiding and protection—at least as much as, and as an alternative to, a battleship fleet. Tirpitz had to convince Wilhelm of the importance of battleships (which was not difficult) and keep him convinced long enough to execute his plan (which was).[155] Indeed, despite the naval modernization program already underway, and despite a partial shift away from the *Jeune École*, there was not yet an impetus within the German Imperial Navy for an ambitious battleship program.[156] Tirpitz's plan, therefore, involved a twofold shift. First, the basis of German naval strategy would now be the construction and maintenance of capital ships, not the *Jeune École* and its corollary small ship and cruiser-based guerre de course (although the protection of trade was retained as a justification). And second, the fleet would be built, for the first time, with the intention of directly competing with British naval might.

The problem that Tirpitz did foresee—one that he grossly underestimated—was the need to cross what he referred to as the "danger zone" in military preparedness.[157] Somewhat like a modern state seeking an illicit nuclear arsenal, Germany faced the possibility of attracting attention to which it could not respond to in kind. Belying the notion that Britain was a direct and immediate threat, Tirpitz's conception of the project held that the most dangerous period for Germany would be, not that exact moment when Britain's attention was elsewhere, or the time when the fleet was fully operational, but the period in between, when Germany's nascent fleet was large enough to be a threat to Britain but not large enough to hold its own in a fight. This transitional phase Tirpitz labeled the "danger zone." In his professional and technical judgment, the danger zone could be crossed by 1905. As it happened, the danger zone was never crossed: in the years immediately preceding World War I, Tirpitz thought that the navy would be out of trouble by 1915. This did not turn out to be correct, either. The story of Germany's decision to locate itself indefinitely in Tirpitz's danger zone is, in brief, the story of the Dreadnought Race. But at the time he conceived of the idea for the fleet, Tirpitz thought the danger zone could be crossed.[158]

Doing so required policy coordination. As his biographer Patrick J. Kelly has remarked, it is no small irony that Tirpitz, in the service of *Weltpolitik*, actually lobbied Bülow to reduce Germany's foreign entanglements to allow the navy the opportunity to focus on the development of its fleet. Tirpitz in particular requested that Bülow adopt a conciliatory policy toward Britain so as to avoid a confrontation that might involve his nascent fleet when it was too weak to fight, and he also requested that Germany stay out of colonial confrontations, which paradoxically caused him to advocate both conciliating Russia and avoiding alliance talks with it. As Kelly notes, however, and as Rolf Hobson has intimated, Tirpitz appears to have been sufficiently dedicated to his building program as to pursue it regardless of foreign developments. As will be shown, these quickly got out of hand.[159]

The distance between conception and execution was a short one. Wilhelm approved the memo, and Tirpitz set to work drafting what would become known as the first of five Naval Laws. The law authorized ship construction and, within limits, allowed the Navy to spend money freely without having to go through the normal budget process, which the Reichstag ordinarily controlled.[160] Tirpitz personally supervised the political negotiations, which were no easy task. In principle, the bill would set limits to the size of the new fleet, but these limits actually worked in the Tirpitz's favor, insofar as they mandated a minimum fleet size that was well above what Tirpitz's opponents, not least among them the German Social Democrats, wanted to pay for. It would also give Wilhelm legal leeway to spend what was necessary to build the ships. Theoretically, this was to avoid the legislative pork barrel, on the one hand, and give formal authorization for the naval funding, on the other. In practice, it eliminated the necessity for any future consultation with the Reichstag (unless more ships were required, as they would be), while giving Wilhelm a blank check for naval construction.[161]

Buried in the political maneuvers, however, lay layers of domestic intrigue, bias, and machination. Tirpitz, it will be remembered, was not a member of the hereditary nobility, and as a naval officer he possessed no great affinity for the nobility's preferred service or for its privileges. The German naval plan, at minimum, had profound social implications.

Tirpitz has been lauded for his ability to manipulate and mobilize German industrial interests to not only support the project politically, but to deliver the ships at cut rates—sometimes even at or below cost, for the sake of both industrial pride and good advertising. But he also appears, from the start, to have anticipated—and also underestimated—a major fiscal concern that his program possessed: namely, the absence of sufficient funds that could be generated through normal means. The typical peacetime method of military funding—and of government funding in general—at the time was the indirect taxation of goods and services. It was apparent early on that this taxation would not produce enough tax money to fund a naval arms race, particularly once it turned into a race as opposed to a mere building program. The normal way of augmenting funding for such projects, understood in hypothetical terms because of its political gravity, was direct taxation, specifically of property and specifically of inheritances. At this, the German nobility could be expected to do no less than scream. Tirpitz appears to have understood this, and by dodging questions about the anticipated worst-case costs of his program, he appears to have sought to logroll the program to completion. In this way, Tirpitz could make the naval buildup a fait accompli before anyone could object, with the ultimate hopes of making its funding, in turn, a fait accompli, no matter what the nobles might have to say about it. In this, as will be shown, Tirpitz was both terribly duplicitous and naïve: indirect tax increases would float the naval buildup along for a while, but in the end the requisite inheritance taxes would not be forthcoming. On this basis alone, *Sammlungspolitik* was destined to fail.[162]

To sell this project, Tirpitz and Wilhelm (at the former's urging, and with difficulty, given Wilhelm's legendary penchant for braggadocio and filibuster) avoided any mention of the fleet being aimed offensively at Britain, portraying it as a means of commerce protection instead. They also downplayed the obvious implication that the navy would now have a legal minimum size. In pushing the Naval Law, Tirpitz exercised creativity. He was also not above resorting to more practical measures, ranging from legitimate political horse-trading (concessions to Bavarian Catholics in exchange for votes) to soft bribery (payments made for future expenses). The wide-ranging and varied opposition that Tirpitz encountered—which included Social Democrats, landowning traditionalists who favored the army, and even merely less ambitious politicians with more parochial concerns; the degree of prevarication and dissembling required to sell the project as an innocuous defensive measure; and even the difficulty Tirpitz had in keeping Wilhelm from indiscreetly discussing the project during this time—should have served as a warning. In Germany, the navalists were in the minority, and there were reasons for that. That it did not is a testament to Tirpitz's famous singlemindedness, a classic example of a vice made out of a virtue. In the end, thanks to Tirpitz's capable management and crafty political maneuvering, the First Fleet Act passed in March 1898, less than a year after he assumed his office and barely a year since his predecessor Hollmann resigned.[163]

Going forward, Tirpitz sought to mobilize intellectual opinion as well. In particular, he enlisted the intellectual support of the navalists in the German professoriate—the *flottenprofessoren*, as they became known. These individuals were perhaps most notably represented by the *Deutscher Flottenverein* (Naval

League), which was formed within a few weeks of the passage of the first Naval Law to promote German seapower, and which occasionally proved even more zealous than Tirpitz. Tirpitz also made good use of his access to propagandists within the Imperial German Navy, in particular to silence dissent from within the officer corps.[164]

Political Forces Align: The First Steps in the Race

Three events in 1898 and 1899 seemed only to further illustrate the value of seapower, and they served as clarifying motivations for Wilhelm and Tirpitz. The first was the Spanish-American War, in which modern battleships clashed for effectively the first time, and in which decisive U.S. sea victories brought a swift end to the conflict. Bülow, as Wilhelm's new chancellor, employed the war as a talking point in drumming up support for the new navy in the newspapers, claiming that seapower—or lack of it—was the reason "Spain lie[s] on the floor."[165]

The second was the Fashoda Incident. In the autumn of 1898, an unfortunate event in Africa between rival French and British colonial expeditions nearly got out of hand. France was attempting to establish a string of colonies along an east-west axis across the heart of Africa, which would interdict the British north-south colonial territory being established along the eastern coast of the continent. Germany would later attempt to duplicate this project, taking its first tentative steps during the Agadir Incident of 1911, and would meet with even less success. The French Jean-Baptiste Marchand Expedition made a dramatic 14-month trek across central Africa to establish a fort at Fashoda in modern Sudan, which the British were already in the process of colonizing (the war with the Mahdi was concluded in the bloodbath at Omdurman that year). A British force led by none other than Horatio H. Kitchener, the victorious general of the recent war in the Sudan and later the brains behind Britain's mobilization in World War I, met them there. Exercising a high level of military professionalism, the two commanders managed to avoid bloodshed and agreed to wait for instructions from their respective governments. In the end, in October, France was forced to back down and cede Fashoda, accepting a negotiated settlement for the borders of their respective colonies. This occurred for one simple and telling reason that did not escape Wilhelm's notice: France, lacking naval power, could not sustain a conflict with Britain in Africa.[166]

The third development was the outbreak of the Boer War the following year. The already tense relationship between Britain and the Afrikaner Transvaal Republic, smoldering since the 1895 Jameson Raid, resulted in an exchange of ultimatums and finally in armed conflict. The war was a serious matter for Britain, not only costing lives but also humiliating the Empire as ragtag Boer militias repeatedly defeated Britain's finest troops. Britain's defense expenditure spiked, and it became tied down in a bloody, protracted guerrilla conflict that it ultimately settled, to its permanent shame, with the first use of concentration camps to isolate the Boer civilians from the guerrilla fighters. But while Britain's reputation suffered, Germany, in Wilhelm's mind, was similarly shamed. Ever since his congratulation of the Boers for repelling the Jameson Raid, Wilhelm had dreamed of a more active role in South Africa, supporting the Boers whom he regarded as ethnic kinsmen and diminishing Britain's colonial and political position. As it had been for France at Fashoda, so it was for Wilhelm: without a fleet, he could do nothing. This fact was hammered home in humiliating fashion in 1900 when Britain briefly detained three German steamships to prevent them from sending aid to the Boer guerrillas.[167] The demonstration of small-scale seapower combined with the by-now familiar problem of being unable to confront Britain for fear of what the Royal Navy might do would have served as a further impetus for Wilhelm's personal navalism.

Having followed all these developments, and excited at their success with the first Naval Law, Tirpitz and Wilhelm decided in 1900 to push through what would become known as the Second Naval Law, which upped the mandatory fleet size to 38 battleships. This time there was no difficulty: the humiliating seizure of the steamships inflamed nationalistic sentiment, and the vote in the Reichstag was near-unanimous. Although Britain was distracted by the Boer War, and although its colonial secretary, Joseph Chamberlain, was still reaching out to Germany in search of diplomatic accommodation, what in time would become known as the Dreadnought Race was officially on.[168]

German Strategic Intent: Ultimate Aims

It is therefore possible to sum up Germany's aims as of the start of the Dreadnought Race, allowing for appropriate caution in the face of considerable ambiguity. Precisely because the principal instigator of the hard-power component of the Dreadnought Race, Tirpitz, was ambiguous in his statements, at least partly for political effect, and because of the mercurial nature of Germany's top political and military leader, Wilhelm, it is difficult to pin down Germany's aims conclusively. This much appears to be known and agreed on: under Tirpitz, and sometimes over the objections of other members of Tirpitz's service, the Imperial German Navy initiated a buildup of capital ships designed, at the very least, to thwart a British blockade or invasion the event of war. At the outside, it can reasonably be assumed that Tirpitz—and Wilhelm, once convinced—were prepared to pursue a preponderance of naval power in the North Sea, cutting into and to some extent nullifying Britain's existing naval dominance there, assuming the opportunity presented itself. This would serve the varying purposes of serving as a show of national strength per se, by providing a unifying impetus to German domestic politics, forcing Britain by deterrence to allow Germany a free hand on the world stage, allowing Germany some political leverage in extracting colonies and other concessions, possibly dominating Britain to the point where it would have to settle for subordinate status, and generally increasing German prestige in the process.

How all this was to be achieved was left vague as was the question of how to satisfactorily conclude the arms race, given

Chapter Four

Tirpitz's legitimate concerns about the "danger zone" and the emergent problems of overcoming it, and likewise given the fact that no plans to protest Britain diplomatically appear to have existed and neither did plans to actually go to war. This point is driven home by the fact that, in the end, not only did arms control negotiations fail to satisfactorily conclude the Dreadnought Race, but they were stymied in part by the race's architect Tirpitz.

Finally, although in one sense the ultimate drivers of German policy are beyond the scope of this study, a word must be said concerning the ultimate intentions of Germany's leadership. Even apart from the goals just discussed, Wilhelm also sought to upend the regional naval balance of power as an end in itself, for the aforesaid psychological and personal reasons, which ultimately can be summarized as a desire for increased German international prestige, a prestige which would reflect on Wilhelm personally. Such prestige was obviously desirable to German nationalists, not least among them the thinkers who formed an enthusiastic following and propaganda machine for Tirpitz (as well as the consumers of their work), and was therefore not out of place with its times. It was also, as noted, intensely desired by Wilhelm, and this forms a key to understanding the competitive goals of his state. Wilhelm and his likeminded subordinates also sought to unify their country in the pursuit of a common objective, a fundamentally nonmaterial goal with imprecise boundaries.

To recapitulate, Germany, through its political and military leadership, may have made numerous conceptual errors in deciding to compete with Britain's naval hegemony, but its competitive goals can be summed up. First and foremost, they involved closing the naval power gap in North Sea waters, and possibly reversing the North Sea naval balance if such could be achieved. This would allow for the projection of German power outside those waters, obtaining a more favorable political rebalancing by subordinating Britain, bringing unity under a nationalist banner to their own feuding political factions at

The Origins and Aims of the Dreadnought Race

home, and obtaining increased international prestige through acquiring a powerful navy and, to a much lesser degree, more colonies.

Germany's aims were not strictly finite. They involved the establishment of a known (if expansive) quantity of hard power and thereby the acquisition of an unknown type of soft power and a nebulous political harmony at home. Whatever the case, it is clear Tirpitz and Wilhelm thought in terms similar to those described by Gary Hamel and C. K. Prahalad: they would focus their efforts and force a change, in effect requiring the creation of a "finish line" to the competition, even if they did not understand what this entailed. As the next chapter's narrative will show, however, the expansive and fuzzy nature of their goals and the flaws in the execution of the Dreadnought Race to come—and the original sin of not knowing how to apply their advantage if they ever achieved it—would ultimately ensure that they got too little and paid too much. Germany's ability to achieve these goals, and its resource allocation in their pursuit, is discussed in the analysis that follows. Next, this narrative will turn to the actual course and outcome of the Dreadnought Race.

Chapter Four

Endnotes

1. Rudyard Kipling, "Recessional," *Spectator*, 24 July 1897.
2. N. A. M. Rodger, *The Safeguard of the Sea: A Naval History of Britain, 660–1649* (New York: W. W. Norton, 1999), 2.
3. The term *Dreadnought Race* is in common usage on many websites, blogs, and news outlets, but it is not commonly used as a phrasing for the subject studied here. Scholarly literature tends to prefer "Anglo-German naval arms race" or a similar phrasing. As much as this phrasing has to recommend it, Dreadnought Race is preferred here, less for any inherent drama or evocativeness than for its brevity. Although the author has not located a previous scholarly usage of the term as a proper noun phrase, it is doubtful this is the first such instance.
4. The details of the historical controversies surrounding key points in the Dreadnought Race will be discussed in more detail as the narrative progresses.
5. Paul M. Kennedy, *Strategy and Diplomacy, 1870–1945: Eight Studies* (London: Fontana, 1983), 116–19; and Paul M. Kennedy, *The Rise of the Anglo-German Antagonism, 1860–1914* (London: Ashfield Press, 1987), 204–6, 223–28. Kennedy focuses on the role of these individuals in particular, as well as the role of individual decision making (albeit influenced by political trends) in explaining German policy. See also Ivo Nikolai Lambi, *The Navy and German Power Politics, 1862–1914* (Boston, MA: Allen and Unwin, 1984), 31–37, 57–86, 181–86. Lambi similarly places these individuals at center stage. See also Peter Padfield, *The Great Naval Race: The Anglo-German Naval Rivalry, 1900–1914* (Edinburgh, Scotland: Birlinn, 2005), 28–51; and Robert K. Massie, *Dreadnought: Britain, Germany, and the Coming of the Great War* (London: Vintage Books, 2004), 137–43, 164–85. These two narrative histories of the Dreadnought Race similarly foreground these individuals and emphasize their roles as key decision makers. See also Matthew S. Seligmann, Frank Nägler, and Michael Epkenhans, *The Naval Route to the Abyss: The Anglo-German Naval Race, 1895–1914* (London: Routledge, 2016), 1–14. These authors take a similar approach, with Tirpitz as the prime mover. As Seligmann has noted separately, however, this is as much a narrative convention as anything else, and the focus on Tirpitz and Wilhelm can obscure a more complicated picture involving more players and less unity of purpose and effort than the story of Tirpitz and Wilhelm will allow. See Matthew S. Seligmann, *The Royal Navy and the German Threat, 1901–1914: Admiralty Plans to Protect British Trade in a War against Germany* (Oxford, UK:

Oxford University Press, 2012), 23-24. This may be quite true, but it should not obscure the basic understanding that these individuals, acting largely on their own initiative and for their own purposes, did indeed set in motion the arms race in capital ships and the diplomatic maneuvers that accompanied it that are here collectively dubbed the "Dreadnought Race." The interaction of this attempt to upset the North Sea naval balance and redraw the diplomatic map with political and bureaucratic realities and structural factors is discussed in the narrative.

6. See Volker R. Berghahn, *Germany and the Approach of War in 1914*, 2d ed. (New York: St. Martin's Press, 1993), 1-14, 24-55. Berghahn lays out a case for a German foreign and military policy driven by structural forces within German society and politics. See also Hans-Ulrich Wehler, *The German Empire: 1871-1918* (Oxford, UK: Berg, 1997), 232-46. Wehler echoes Berghahn's point, arguing for an understanding of Germany's move toward war as being driven by the chaos of its social modernization. This study readily concedes these historians' points regarding the structure of the German state while nevertheless focusing on the decisions taken by its leaders and the motivations for them. In discussing a state's strategy, it is often enough to know what its leaders did and what they sought by doing so, even if the position those leaders found themselves in was not of their own making or even that of any particular person or group.

7. The story is usually presented (and admittedly is presented here) as a conflict between characters who are foils for one another: Tirpitz is pitted against Fisher and Wilhelm is pitted against a consensus-based British political system. There is no shortage of examples, all of which appear as sources in the narrative that follows. See, for example, Jonathan Steinberg, *Yesterday's Deterrent: Tirpitz and the Birth of the German Battle Fleet* (London: MacDonald, 1965), 200-7; Holger H. Herwig, *"Luxury" Fleet: The Imperial German Navy, 1888-1918* (New York: Humanity Books, an imprint of Prometheus Books, 1987), 17-20, 32-35; Kennedy, *The Rise of the Anglo-German Antagonism*; Kennedy, *Strategy and Diplomacy*, 111-26, 130-60; Padfield, *The Great Naval Race*, 29-32, 35-59, 113-16; Massie, *Dreadnought*, 150-85, 401-32; Lambi, *The Navy and German Power Politics*, 68-86, 155-70; and Patrick J. Kelly, *Tirpitz and the Imperial German Navy* (Bloomington: Indiana University Press, 2011). Steinberg treats the era as "the Tirpitz Era," while Herwig begins his narrative of the subject with a character sketch of Wilhelm and Tirpitz, who are to be the key *dramatis personae*. In Kennedy's large-scale narrative of the era, *The Rise*

of the Anglo-German Antagonism, numerous figures are treated, but where naval affairs are concerned Tirpitz looms large, while in Kennedy's other essays on British national security policy in the era, *Strategy and Diplomacy*, the Tirpitz-Fisher rivalry and Wilhelm's own intentions are foregrounded. Padfield and Massie take a similar approach. Lambi places Tirpitz at center stage but also in the context of the complex German naval bureaucracy and its interaction with the Kaiser. Kelly's book, *Tirpitz and the Imperial German Navy*, as the title makes clear, is devoted to Tirpitz. This tendency to pit great figures against each other ignores the previously discussed structural interpretations, which Kennedy has devoted an essay to analyzing. As he notes, although structural factors did help steer Germany's turn toward *Weltpolitik*, the major decisions were nevertheless taken by individual people, of whom Wilhelm and Tirpitz were the most noteworthy. See Paul M. Kennedy, "The Kaiser and German *Weltpolitik*: Reflexions on Wilhelm II's Place in the Making of German Foreign Policy," in *Kaiser Wilhelm II: New Interpretations*, ed. John C. G. Röhl and Nicolaus Sombart (Cambridge, UK: Cambridge University Press, 1982), 143–46. See also Nicholas A. Lambert, *Sir John Fisher's Naval Revolution* (Columbia: University of South Carolina Press, 1999); and Seligmann, *The Royal Navy and the German Threat*, 7–25, 65–88. There has been a push recently to break out of this frame, with Lambert foregrounding Fisher but challenging the received wisdom that Germany was Fisher's primary concern, and with Seligmann pointing out that Tirpitz had competitors within his own bureaucracy while the Dreadnought Race was but one part of a complicated naval rivalry. For a discussion of the merits and demerits of the revisionist case, see Matthew S. Seligmann, "Naval History by Conspiracy Theory: The British Admiralty before the First World War and the Methodology of Revisionism," *Journal of Strategic Studies* 38, no. 7 (July 2015): 967–68, https://doi.org/10.1080/01402390.2015.1005443. As is discussed in the narrative that follows, it is possible to accept some of the revisionist claims and to acknowledge the broader picture painted by Seligmann while still treating the matter as being, fundamentally, a competition between two states with at least some agency and whose decisions are traceable to specific leaders, who are rightly given central treatment in the historiography of this subject. No more and no less is claimed here.

8. Lambi, *The Navy and German Power Politics*; Seligmann, Nägler, and Epkenhans, *The Naval Route to the Abyss*; Massie, *Dreadnought*; and Padfield, *The Great Naval Race*.

9. Lambert, *Sir John Fisher's Naval Revolution*. For analysis and debate on these matters, see Jon Tetsuro Sumida, *In Defence of Naval Supremacy: Finance, Technology, and British Naval Policy, 1889-1914* (London: Unwin Hyman, 1989); Seligmann, "Naval History by Conspiracy Theory"; Christopher M. Bell, "Contested Waters: The Royal Navy in the Fisher Era," *War in History* 23, no. 1 (January 2016): 115-26, https://doi.org/10.1177/0968344515595330; Christopher M. Bell, "Sir John Fisher's Naval Revolution Reconsidered: Winston Churchill at the Admiralty, 1911-1914," *War in History* 18, no. 3 (July 2011): 333-56, https://doi.org/10.11177/0968344511401489; Nicholas A. Lambert, "On Standards: A Reply to Christopher Bell," *War in History* 19, no. 2 (April 2012): 217-40, https://doi.org/10.1177/0968344511432977; and Christopher M. Bell, "On Standards and Scholarship: A Reply To Nicholas Lambert," *War in History* 20, no. 3 (July 2013): 381-409, https://doi.org/10.1177/0968344513483069. See also Hew Strachan, *The First World War*, vol. 1, *To Arms* (Oxford, UK: Oxford University Press, 2001), 383; and P. J. Cain and A. G. Hopkins, *British Imperialism: 1688-2000*, 2d ed. (London: Pearson Education, 2002), 389. Strachan echoes Lambert in noting that the Royal Navy's more immediate concerns at the start of the first decade of the twentieth century involved navies other than Germany's, specifically those of France and Russia. Cain and Hopkins also take this position.

10. Seligmann, Nägler, and Epkenhans, *The Naval Route to the Abyss*, 151-81. The position is noteworthy given Seligmann's contribution to the aforementioned debate regarding the priorities of the naval planners on both sides, arguing broadly that the Dreadnought Race, whether strictly intended or a happenstance result of independent naval planning, occurred in the context of a broader naval rivalry. See Seligmann, *The Royal Navy and the German Threat*, 1-6.

11. Ronald Robinson, John Gallagher, and Alice Denny, *Africa and the Victorians: The Official Mind of Imperialism* (London: I. B. Tauris, 2015), 462-72.

12. D. K. Fieldhouse, *The Colonial Empires: A Comparative Survey from the Eighteenth Century* (New York: Delacorte Press, 1966), 178-80, 370-71.

13. Fieldhouse, *The Colonial Empires*, 370-71, 380-81, 393.

14. Cain and Hopkins, *British Imperialism*, 390-93. Cain and Hopkins do note that another major driver of the Anglo-German rivalry was disputes about trade and protectionism, a point that this study can concede but that goes beyond its scope, which concerns

Chapter Four

Germany's attempt to supplant British naval power in the North Sea and the political consequences and implications thereof.

15. Lambi, *The Navy and German Power Politics*, 427; Kennedy, *The Rise of the Anglo-German Antagonism*, 420–23; Seligmann, Nägler, and Epkenhans, *The Naval Route to the Abyss*, loc. 403–4; Padfield, *The Great Naval Race*, 343–46; and Massie, *Dreadnought*, 888–908.

16. See Padfield, *The Great Naval Race*, xiii–xvi. It is noteworthy that it is far from necessary to hew to Padfield's dire synthesis of the historiographical claims regarding German war guilt to accept the general validity of his narrative regarding the connection between the Dreadnought Race and the deterioration of Germany's international position, making such claims all the more histrionic.

17. See Fritz Fischer, *Germany's Aims in the First World War* (New York: W. W. Norton, 1967), 11–49; and Fritz Fischer, *World Power or Decline: The Controversy over Germany's Aims in the First World War*, trans. Lancelot L. Farrar, Robert Kimber, and Rita Kimber (New York: W. W. Norton, 1974), vii–ix. For a discussion of the problems, or at least the limitations, of this approach, see Strachan, *To Arms*, 3–4, 52–57, 86–87.

18. Regarding Britain's contribution, see Kennedy, *The Rise of the Anglo-German Antagonism*, 458–59. Kennedy notes that the scales had tipped in favor of British intervention in World War I simply because Britain was now exclusively focused on Germany. Although intuitively sound, the question becomes murkier when the specifics are examined, insofar as the process by which Britain moved toward war during and before the 1914 July Crisis was far from straightforward. For a countervailing analysis, see Strachan, *To Arms*, 93–98. Strachan notes that the British political system practically guaranteed a lack of consensus on whether to go to war until the last minute. This study does not dispute this, merely noting that, to the extent that the Dreadnought Race impacted Germany's political position adversely and to the extent that it played a role in shaping the environment that led to World War I, this can be accounted for in the framework proposed here as well.

19. David K. Brown, "Wood, Sail, and Cannonballs to Steel, Steam, and Shells, 1815–1895," in *The Oxford Illustrated History of the Royal Navy*, ed. J. R. Hill (Oxford, UK: Oxford University Press, 1995), 207–20.

20. See John Scott Keltie, ed., *The Statesman's Yearbook* (London: Macmillan, 1875), 236. See also Lambert, *Sir John Fisher's Naval Revolution*, 17.

21. See Arthur J. Marder, *The Anatomy of British Sea Power: A History of British Naval Policy in the Pre-Dreadnought Era, 1880–1905* (Hamden, CT: Archon Books, 1940), 5; Lawrence Sondhaus, *Preparing for Weltpolitik: German Sea Power before the Tirpitz Era* (Annapolis, MD: Naval Institute Press, 1997), 110–15; Sumida, *In Defence of Naval Supremacy*, 9, 43–44; Massie, *Dreadnought*, 167, 414; and Robert K. Massie, *Castles of Steel: Britain, Germany, and the Winning of the Great War at Sea* (New York: Random House, 2003), 123.
22. Lambert, *Sir John Fisher's Naval Revolution*, 27–29; Strachan, *To Arms*, 418; Sumida, *In Defence of Naval Supremacy*, 51–52; and Brown, "Wood, Sail, and Cannonballs to Steel, Steam, and Shells," 218.
23. Alfred T. Mahan, *The Influence of Sea Power Upon History: 1660–1783*, ed. Ellen Lyle Mahan (Boston, MA: Little, Brown, 1918; Amazon, 2011), Kindle ed., loc. 1196–1238, 2131–57, 9359–75, 9439; and Seligmann, *The Royal Navy and the German Threat*, Kindle ed., loc. 249–65.
24. Julian S. Corbett, *Some Principles of Maritime Strategy* (London: Longmans, Green, 1911; Project Gutenberg, 2005), Kindle ed., loc. 940–58, 998–1001, 1021–31, 1036–40.
25. Corbett, *Some Principles of Maritime Strategy*, loc. 940–44, 3139–73.
26. Corbett, *Some Principles of Maritime Strategy*, loc. 1946–51, 2487–91, 2513–38.
27. Corbett, *Some Principles of Maritime Strategy*, loc. 11, 2518.
28. Corbett, *Some Principles of Maritime Strategy*, loc. 1154–56, 3342.
29. Arne Røksund, *The Jeune École: The Strategy of the Weak* (Boston, MA: Brill, 2007), 1–23.
30. Røksund, *The Jeune École*, 9–12.
31. Røksund, *The Jeune École*, 210, 228–29; Lambi, *The Navy and German Power Politics*, 7–9, 164–66; Seligmann, Nägler, and Epkenhans, *The Naval Route to the Abyss*, 2–3; and Strachan, *To Arms*, 375–77. Strachan notes the long shadow that the *Jeune École* cast over French naval procurement doctrine.
32. Røksund, *The Jeune École*, 228–29.
33. Røksund, *The Jeune École*, x–xii; Lambi, *The Navy and German Power Politics*, 2, 76; Seligmann, Nägler, and Epkenhans, *The Naval Route to the Abyss*, loc. 349–56; and Seligmann, *The Royal Navy and the German Threat*, 2–6, 12–24. Seligmann's discussion of the controversy regarding the relevance of the Dreadnought Race to Anglo-German relations and to German war plans is discussed below.

Chapter Four

34. Seligmann, Nägler, and Epkenhans, *The Naval Route to the Abyss*, loc. 346–50; Lambert, *Sir John Fisher's Naval Revolution*, 38–40; Lambert, "On Standards and Scholarship," 219–20; Lambi, *The Navy and German Power Politics*, 6–8, 346; and Kelly, *Tirpitz and the Imperial German Navy*, 47–66.
35. See Lambi, *The Navy and German Power Politics*, 7–9, 65–68.
36. For the original argument for long cycles, see George Modelski, *Long Cycles in World Politics* (Hampshire, UK: Macmillan Press, 1987).
37. Karen Rasler and William R. Thompson, *The Great Powers and Global Struggle, 1490–1990* (Lexington: University Press of Kentucky, 1994), 158.
38. Rasler and Thompson, *The Great Powers and Global Struggle*, 18, 27. In theory, this can include other forms of capability that accomplish similar tasks. The U.S. Air Force's modern long-range heavy airlift and strategic bombing capability might qualify as well (an insight that this author owes to Georgetown University professor Charles Pirtle).
39. Rasler and Thompson, *The Great Powers and Global Struggle*, 1–14, 73–97, 146–54, 157–62. Note that the state that ultimately acquires global leadership status need not be the only challenger, or even consciously involved in challenging—only that it be the dominant naval power at the end of this phase.
40. An average of 2.44 percent annually between 1875 and 1898, versus just 2.1 percent for Britain for the same time period. Estimates are based on Angus Maddison's per capita gross national product (GNP) estimates and approximate population figures drawn from census data presented in *The Statesman's Yearbook*. See the appendix for methodology. For a discussion of Britain's declining relative position, see Kennedy, *Strategy and Diplomacy*, 91–92; Kennedy, *The Rise of the Anglo-German Antagonism*, 464–66; and Massie, *Dreadnought*, 134–35. For a partial concurrence, see Strachan, *To Arms*, 12–13. See also Andrew D. Lambert, "The Royal Navy, 1856–1914: Deterrence and the Strategy of World Power," in *Navies and Global Defense: Theories and Strategy*, ed. Keith Neilson and Elizabeth Jane Errington (Westport, CT: Praeger, 1995), 71–74. Lambert critiques this general picture, noting that British primacy remained assured even despite such relative setbacks.
41. Lambert, *Sir John Fisher's Naval Revolution*, 15–17; and Massie, *Dreadnought*, xiii–xvi, 373–74.
42. Røksund, *The Jeune École*, 8–12.
43. Marder, *The Anatomy of British Sea Power*, 84–87. Marder cites a London *Times* editorial of the era as condemning commerce

raiding as entailing nonstop fatal attacks on civilians, but whatever norms existed that forbade this, they were apparently not considered a sufficient deterrent by anyone involved.

44. See Seligmann, *The Royal Navy and the German Threat*, 171–72. As Seligmann has noted, the debate about whether threats to high seas commerce or of blockade by battleships were of greater concern took on increasing urgency as the Dreadnought Race ran its course, with a parallel effort being made to counter the German commerce-raiding threat.
45. Henry Kissinger, *Diplomacy* (New York: Simon and Schuster, 1994), 161–62.
46. Marder, *The Anatomy of British Sea Power*, 65–83.
47. Marder, *The Anatomy of British Sea Power*, 3–9; and Brown, "Wood, Sail, and Cannonballs to Steel, Steam, and Shells," 200–26.
48. The commitment was more general than specific. See David H. Olivier, *German Naval Strategy, 1856-1888: Forerunners of Tirpitz* (London: Frank Cass, 2004), 130–46. As Olivier notes, while Germany broadly sought to copy the *Jeune École*'s playbook, its admirals in the 1880s came up with a hybrid model that deemphasized second-order economic effects in favor of simple commerce raiding, mainly with cruisers as the weapons system of choice. For Germany, much of the *Jeune École*'s distinctive features that distinguished it from mere commerce raiding were more hypothetical than real.
49. Keltie, *The Statesman's Yearbook*, 106.
50. Marder, *The Anatomy of British Sea Power*, 7–9. See also Lambert, *Sir John Fisher's Naval Revolution*, 18.
51. Naval Defence Bill (No. 80), HL Deb, 27 May 1889, vol. 336, 1059–89.
52. Sumida, *In Defence of Naval Supremacy*, 10–16; Kennedy, *Strategy and Diplomacy*, 167–68; and Lambert, *Sir John Fisher's Naval Revolution*, 17–21. Lambert notes that the Two-Power Standard mattered more for political and electoral purposes than as a precise measurement of naval capability. The fact, however, that the Two-Power Standard was thought achievable, and what later happened to it, can serve as a useful heuristic for Britain's relative naval power. Kennedy views Britain's long-term naval arms race with France and Russia as the more dangerous one, but given that it entailed a combined effort on the part of two powers to achieve a split parity with Britain, rather than by one as was the case in the Dreadnought Race, it in the end did not pose the same unified threat that the German North Sea naval buildup posed, and was handled differently. This study does not treat the Franco-Russian

attempt to attain parity with Britain in detail, not least because, as Kennedy notes, those states' attempts ended inconclusively, with the formation of the respective ententes, but the strategy of either state's attempts to do so could be analyzed in similar fashion to the case under study here using this study's framework. The whole point of a framework such as this is to enable the analysis of a competition from the point of view of one of its participants, in terms of what it is doing and what it may do to prevail. It is important to note that, while "beta" in the aforementioned framework is typically forced to focus its resources if it is to make any initial progress, "alpha" (in all these cases Great Britain) may, in its preeminent position, face a set of challenges and have a series of priorities. As is discussed below, the challenge once competition begins in earnest is to match one's grand strategy to one's competitive strategy, focusing one's resources and avoiding taking on more than one can handle. Alpha can get away with this for a while, at least until beta becomes a real threat; beta must start out in this fashion.

53. See Fieldhouse, *The Colonial Empires*, 179–80. Fieldhouse has argued persuasively that the major impetus for African colonization was local imperatives faced by European colonial powers that made colonial expansion slightly more attractive than not in each of many unrelated cases. This general picture is confirmed in Robinson, Gallagher, and Denny, *Africa and the Victorians*, 464–72. The authors note that British imperialism was driven by ad hoc improvisation—and specifically a desire to protect access to eastern possessions—rather than a comprehensive strategy. Fieldhouse has also noted that, at least in retrospect, African colonization was hardly profitable and was as much an end as a means. He notes that this was certainly true for Germany, whose colonial acquisitions' greatest significance was as a status symbol and lagging indicator of world power status. See Fieldhouse, *The Colonial Empires*, 364, 370–71, 381, 393; and D. K. Fieldhouse, *Economics and Empire, 1830-1914* (London: Weidenfeld and Nicolson, 1973), 472–73, 476. See also Cain and Hopkins, *British Imperialism*, 389–93. The authors essentially echo Fieldhouse by connecting Germany's naval ambitions at this time to its imperial challenge to Britain but note that the real threat it posed was to Britain at home, not to the Empire. See also Kennedy, *Strategy and Diplomacy*, 20–25; and Kennedy, *The Rise of the Anglo-German Antagonism*, 410–17. Kennedy notes that Britain was quite willing to accommodate Germany on colonization—provided Germany asked nicely and was willing to negotiate. The impetus to chal-

lenge Britain's North Sea hegemony was a desire, by acquiring a satisfactory level of theater naval capability, to circumvent this proviso.

54. See Thomas Pakenham, *The Scramble for Africa: The White Man's Conquest of the Dark Continent from 1876 to 1912* (New York: Avon Books, 1991), xxvii. Pakenham dates the phrase to 1884 but notes that its originator is unknown.

55. For the origins of Britain and German's colonial stakes, including the Stanley expedition; Bismarck's initial attitude toward African colonization; and the Berlin Conference, see Pakenham, *The Scramble for Africa*, xxiii, 24–38, 207, 239–42. See also Fieldhouse, *The Colonial Empires*, 180–81, 184–90; and Robinson, Gallagher, and Denny, *Africa and the Victorians*, 462–72. Because the motivations of German (if not British and other) would-be colonialists are of relevance to understanding Germany's competitive strategy, it is worth noting briefly that no complete consensus exists as to a single set of motives that drove European African expansion. Fieldhouse notes that African expansionism was driven by a host of factors, of which ideology (imperialism) was but one of many and often not a major one, a point echoed by Robinson, Gallagher, and Denny in arguing that British expansionism in particular was driven not by ideology but by expediency. In this argument, what drove Britain's sudden increased interest in Africa was not a surge in public support or a new set of beliefs about colonies, but rather national security imperatives—specifically the need to maintain control of strategic areas that were slipping away from British influence by normal means. Fieldhouse, by contrast, assigns less importance to security concerns than to economic developments and ideology, such as the numerous factors that drove expansion in tropical West Africa and, by contrast, on the explicitly ideological roots of Belgian ambitions in the Congo. It is not necessary here to posit a single set of reasons for European powers' African adventures. What will be relevant is that, in the minds of certain senior German policy makers who moved Germany into confrontation with Britain, African colonialism, and Britain's interference with it, moved hand in hand with naval concerns.

56. Daniel Owen Spence, *A History of the Royal Navy: Empire and Imperialism* (New York: I. B. Tauris, 2015), 77–78.

57. Fieldhouse, *The Colonial Empires*, 67.

58. Spence, *A History of the Royal Navy*, 193.

59. For Germany's early recognition of the importance of seapower in African and other colonization and its early attempts to use its

60. small fleet for the purpose, see Sondhaus, *Preparing for Weltpolitik*, 154–56.
60. This exact point was highlighted in a policy memorandum for the Foreign Office by Sir Eyre Crowe at the height of the Dreadnought Race. See Eyre Crowe, "Memorandum on the Present State of British Relations with France and Germany," in *The Hidden Perspective: The Military Conversations 1906-1914*, ed. David Owen (London: Haus Publishing, 2014), 230–31, 235, 245–50. Crowe also noted German interest in East Asia at the time, although comparatively speaking the German colonial interest there was minuscule by comparison to its active and ongoing search for available African territory. See, for example, Fieldhouse, *The Colonial Empires*, 365. See also Rolf Hobson, *Imperialism at Sea: Naval Strategic Thought, the Ideology of Sea Power, and the Tirpitz Plan, 1875-1914* (Boston, MA: Brill Academic Publishers, 2002), 303–4. Hobson writes, "The 'scramble for Africa' and the prospect of a similar carve-up of East Asia in the 1890s unleashed a general rivalry in which navies were considered to play a decisive role. . . . The new navalism proper translated its colonial hopes and fears into battle fleets. Mahan's publications were enormously influential in focusing attention on the battleship and in creating the ideological link between this concentrated expression of sea power and economic and colonial growth."
61. Rudyard Kipling, "The White Man's Burden," *Times* (London), 4 February 1899.
62. Christopher M. Clark, *The Sleepwalkers: How Europe Went to War in 1914* (New York: HarperCollins, 2014), 148–49. This was the crux of the matter. Germany could have colonies and even the prestige that they represented, but to do so as it willed and not as it must, and to acquire the prestige that this status conferred, it needed capital ships. The year the demarche occurred—effectively the same time Tirpitz took over the Imperial Naval Office, as will be discussed—was therefore not without some significance. For all that, the connection between seapower and worldwide power projection acquired focus in the German official mind only at the start of the Dreadnought Race, if then. Even as of 1897, Wilhelm was still thinking of seapower in terms of deployable fleets of smaller cruisers. See Lambi, *The Navy and German Power Politics*, 34–35.
63. See Fieldhouse, *The Colonial Empires*, 364–65. As Fieldhouse notes, the German Empire was not only short-lived, but something like nine-tenths of it were concentrated on the African continent, the small remainder being located in the Pacific.

The Origins and Aims of the Dreadnought Race

64. Massie, *Dreadnought*, 61–62; and Paul M. Kennedy, *The Rise and Fall of the Great Powers: Economic Change and Military Conflict from 1500 to 2000* (New York: Vintage Books, 1989), 182–88. Paradoxically, Prussia's victory in the Franco-Prussian War is seen as a triumph of an efficient mass conscription-based system over the comparatively inefficient system employed by France at the time. As for Britain's hatred of forced service, naval historian N. A. M. Rodger attributes the usage of press gangs to "Parliament's refusal to confront the manning problem," noting that British libertarianism prevented almost any sort of forced service except in the most ad hoc fashion in emergencies. See N. A. M. Rodger, *The Command of the Ocean: A Naval History of Britain, 1649–1815* (New York: W. W. Norton, 2004), 312–13.
65. See the appendix. Data are derived from volumes of *The Statesman's Yearbook*, 1876–99.
66. See the appendix. Data are derived from volumes of *The Statesman's Yearbook*, 1876–99.
67. Massie, *Dreadnought*, 162–63; and Sondhaus, *Preparing for Weltpolitik*, 149–52.
68. Kissinger, *Diplomacy*, 132, 159, 165.
69. Gordon A. Craig, *Germany: 1866–1945* (Oxford, UK: Clarendon Press, 1978), 114–16; Kissinger, *Diplomacy*, 132–33, 204; and Kennedy, *The Rise and Fall of the Great Powers*, 188–91.
70. Kissinger, *Diplomacy*, 137–67, 204; Massie, *Dreadnought*, 76–82; and Strachan, *To Arms*, 8.
71. Massie, *Dreadnought*, 49, 56, 76–82; and Kissinger, *Diplomacy*, 121.
72. Craig, *Germany*, 131–32; Kissinger, *Diplomac*, 160, 165–66; and Massie, *Dreadnought*, 76–82.
73. Kissinger, *Diplomacy*, 160; Massie, *Dreadnought*, 82–83; and Craig, *Germany*, 121–22, 130–31.
74. Pakenham, *The Scramble for Africa*, 202–3; and Kissinger, *Diplomacy*, 146. Kissinger sees Bismarck's forays into colonialism as an instance of nationalist political pressure overriding his better judgment, which is, at the very least, a plausible hypothesis.
75. Pakenham, *The Scramble for Africa*, 188.
76. Pakenham, *The Scramble for Africa*, 200–4, 292–95; and Massie, *Dreadnought*, 84–90.
77. Pakenham *The Scramble for Africa*, 602. For the sources for the currency conversion, see the appendix.
78. Kissinger, *Diplomacy*, 180.
79. Fieldhouse, *The Colonial Empires*, 364–67, 370–71.
80. Fieldhouse, *The Colonial Empires*, 367–71.

Chapter Four

81. Fieldhouse, *The Colonial Empires*, 364, 370-71; and Fieldhouse, *Economics and Empire*, 472-73.
82. Fieldhouse, *The Colonial Empires*, 380-81, 392-93; and Fieldhouse, *Economics and Empire*, 475-76. Fieldhouse concludes his magisterial study of the subject, *Economics and Empire*, by noting that, although economics had an important role in incentivizing colonization, political factors—what this study would refer to as the security and intangibles metrics—were at least as important.
83. See Robinson, Gallagher, and Denny, *Africa and the Victorians*, 462-72.
84. See, for example, Kissinger, *Diplomacy*, 171.
85. Padfield, *The Great Naval Race*, xiv, 53. See also Kennedy, "The Kaiser and German *Weltpolitik*," 143-68. Kennedy discusses the relative merits of assigning responsibility for key decisions made by Germany to the kaiser's personal agency as opposed to structural factors, noting that in many cases the kaiser's peculiar personality must be seen as the decisive factor even when structural factors are accounted for. As noted above, because this study is concerned with strategy as conceived from the top, it focuses on the decisions of key leaders even as it can acknowledge that those leaders operated in a domestic political environment framed and shaped from below.
86. Padfield, *The Great Naval Race*, 30-33; and Massie, *Dreadnought*, 26-34.
87. Kennedy, "The Kaiser and German *Weltpolitik*," 155-56; Strachan, *To Arms*, 5-6; Padfield, *The Great Naval Race*, 53-54; and Massie, *Dreadnought*, 120-21, 143.
88. Jonathan Steinberg, "The Kaiser and the British: The State Visit to Windsor, November 1907," in *Kaiser Wilhelm II*, 121-27. Steinberg notes that the pathetically egotistical kaiser was as likely to be a victim of well-calibrated English snobbery as he was to put off his interlocutors by his manners.
89. Massie, *Dreadnought*, 6-10, 24, 26.
90. Massie, *Dreadnought*, 41-46.
91. Steinberg, "The Kaiser and the British," 121-30.
92. Lambi, *The Navy and German Power Politics*, 31-32; and Massie, *Dreadnought*, 150-58, 262-66, 302-3, 652-56.
93. See Kennedy, "The Kaiser and German *Weltpolitik*," 143-57.
94. Wehler, *The German Empire*, 62, 64, 65. The actual phrasing is "polycratic, but uncoordinated authoritarianism" and "authoritarian polycracy."
95. For the classic structuralist approaches, see Berghahn, *Germany and the Approach of War in 1914*, 1-14; and Wehler, *The German*

The Origins and Aims of the Dreadnought Race

Empire, 52-100. For a discussion, see Kennedy, "The Kaiser and German *Weltpolitik*," 143-51. See also Strachan, *To Arms*, 5-8.
96. Strachan, *To Arms*, 5-8, 11. See also Craig, *Germany*, 248.
97. Craig, *Germany*, 46-47.
98. See Kennedy, "The Kaiser and German *Weltpolitik*," 164-66; and Strachan, *To Arms*, 5-11.
99. Massie, *Dreadnought*, 97-99. Bismarck's resignation was in fact voluntary, an attempt to force the kaiser to make a policy change that backfired when the kaiser accepted it—but it was the ultimate result of an unarrested deterioration of their working relationship.
100. The term was coined by Bernhard von Bülow, before becoming official policy. Although it is closely tied to Imperial German Navy grand admiral Alfred von Tirpitz and the Dreadnought Race, and acquired specificity through this series of events, it originally simply referred to a desire to make German influence felt outside its borders. It did, however, almost of necessity dovetail with German navalism, although not originally in any specific form. See Craig, *Germany*, 275, 303; and Lambi, *The Navy and German Power Politics*, 91, 113. See also Kissinger, *Diplomacy*, 171; Massie, *Dreadnought*, 137; Crowe, "Memorandum on the Present State of British Relations with France and Germany," 231, Padfield, *The Great Naval Race*, 35, 45; and Sondhaus, *Preparing for Weltpolitik*, ix.
101. Strachan, *To Arms*, 8-9; and Craig, *Germany*, 275.
102. Craig, *Germany*, 240-47; and Strachan, *To Arms*, 8-10.
103. Lambi, *The Navy and German Power Politics*, 31-35.
104. Craig, *Germany*, 274-75.
105. Craig, *Germany*, 240-47.
106. Kennedy, *The Rise of the Anglo-German Antagonism*, 410-11; Kennedy, *Strategy and Diplomacy*, 20-25; and Crowe, "Memorandum on the Present State of British Relations with France and Germany," 230-50.
107. Fieldhouse, *The Colonial Empires*, 370-71; and Fieldhouse, *Economics and Empire*, 472-73.
108. Berghahn, *Germany and the Approach of War in 1914*, 42-45; Wehler, *The German Empire*, 94-97; Strachan, *To Arms*, 8-10; and Craig, *Germany*, 274-77, 302-3. For a detailed discussion of the connection, see Hobson *Imperialism at Sea*, 313-24. Hobson qualifies the relationship between *Sammlungspolitik*, *Weltpolitik*, and navalism, noting that ideology played at least as important a role as hard political interests in explaining Wilhelmine navalism.
109. Lambi, *The Navy and German Power Politics*, 33-35. As Lambi acidly puts it, "In the case of the navy it would have helped if

Chapter Four

Wilhelm had known exactly what he wanted. He did very much want a large fleet, but he did not really know why he wanted it and whether it should primarily consist of large ships of the line or cruisers." Tirpitz, partly for his own purposes and partly in furtherance of Wilhelm's, would later fill in the details of Wilhelm's thinking.

110. Sondhaus, *Preparing for Weltpolitik*, 184; and Massie, *Dreadnought*, 110–12.
111. Kissinger, *Diplomacy*, 179–85; Craig, *Germany*, 238; Massie, *Dreadnought*, 113–15; and Owen, *The Hidden Perspective*, 10–11.
112. Craig, *Germany*, 239; Kissinger, *Diplomacy*, 181, 194; and Owen, *The Hidden Perspective*, 6.
113. Craig, *Germany*, 238–40; Kissinger, *Diplomacy*, 185; and Massie, *Dreadnought*, 306–7.
114. Kissinger, *Diplomacy*, 185–86.
115. Kissinger, *Diplomacy*, 183, 189.
116. Craig, *Germany*, 243–47. See also Kissinger, *Diplomacy*, 184–85. To contrast, see Massie, *Dreadnought*, 218–25.
117. Kennedy, *Strategy and Diplomacy*, 112–13; Kelly, *Tirpitz and the Imperial German Navy*, 33–128; Lambi, *The Navy and German Power Politics*, 62–68; Massie, *Dreadnought*, 164–68; and Padfield, *The Great Naval Race*, 35–36, 39.
118. Kelly, *Tirpitz and the Imperial German Navy*, 79–80, 84, 110; Lambi, *The Navy and German Power Politics*, 62–65, 68; Kennedy, *Strategy and Diplomacy*, 113, 116–18; and Massie, *Dreadnought*, 168–69.
119. Kelly, *Tirpitz and the Imperial German Navy*, 71; Kennedy, *Strategy and Diplomacy*, 124–25; and Massie, *Dreadnought*, 166–69.
120. Sondhaus, *Preparing for Weltpolitik*, 176–81; quotation found in Massie, *Dreadnought*, 63. For the vagueness of Wilhelm's thinking and Tirpitz's increasing influence, see Lambi, *The Navy and German Power Politics*, 33–37.
121. Sondhaus, *Preparing for Weltpolitik*, 3–4.
122. See Sondhaus, *Preparing for Weltpolitik*, 15–17, 153. See also Kelly, *Tirpitz and the Imperial German Navy*, 19–20.
123. Keltie, *The Statesman's Yearbook* (1875), 235–36; and Brown, "Wood, Sail, and Cannonballs to Steel, Steam, and Shells," 215–17.
124. Sondhaus, *Preparing for Weltpolitik*, 179–84.
125. Craig, *Germany*, 205–6; and Padfield, *The Great Naval Race*, 15–23.
126. Barbara W. Tuchman, *The Proud Tower: A Portrait of the World before the War, 1890–1914* (London: Hamish Hamilton, 1966), 243.
127. John Keegan, *A History of Warfare* (New York: Vintage Books, 1994), 358.

The Origins and Aims of the Dreadnought Race

128. For a discussion of Treitschke's influence on Germany and its top decision makers, see Wehler, *The German Empire*, 72; Craig, *Germany*, 48–49; Padfield, *The Great Naval Race*, 16–25, 40–41, 52; and Tuchman, *The Proud Tower*, 242–43.
129. Sondhaus, *Preparing for Weltpolitik*, 224; and Kennedy, *Strategy and Diplomacy*, 124–25.
130. Kennedy, *Strategy and Diplomacy*, 157; Kelly, *Tirpitz and the Imperial German Navy*, 57, 72; Craig, *Germany*, 204–5; and Padfield, *The Great Naval Race*, 40–41.
131. Kennedy, *Strategy and Diplomacy*, 124–25.
132. See Lambi, *The Navy and German Power Politics*, 34–35, 62–68; Kennedy, *Strategy and Diplomacy*, 124–25; and Kelly, *Tirpitz and the Imperial German Navy*, 107, 110–15. See also Hobson, *Imperialism at Sea*, 202–9. Hobson gives a detailed discussion of Tirpitz's ideas and notes that he may have derived some of them independently of Mahan.
133. Sondhaus, *Preparing for Weltpolitik*, 179–80.
134. Sondhaus, *Preparing for Weltpolitik*, 176, 228.
135. Keltie, *The Statesman's Yearbook* (1875), 105–6; and John Scott Keltie, ed., *The Statesman's Yearbook* (London: Macmillan, 1890), 538.
136. Sondhaus, *Preparing for Weltpolitik*, 126.
137. Massie, *Dreadnought*, 170–71.
138. See Seligmann, Nägler, and Epkenhans, *The Naval Route to the Abyss*, 2–3; Lambi, *The Navy and German Power Politics*, 32–37, 167–68; Kelly, *Tirpitz and the Imperial German Navy*, 98–99; Padfield, *The Great Naval Race*, 38; and Massie, *Dreadnought*, 163–64. For a characterization of the British system, see Padfield, *The Great Naval Race*, 47–48. For the pitfalls of routine British naval management, see Massie, *Dreadnought*, 391–400, 462–63. This convoluted organizational structure was merely one facet of a pervasive German governance pattern famously dubbed "authoritarian polycracy" by Hans-Ulrich Wehler. See Wehler, *The German Empire*, 64. Although this narrative treats Tirpitz, the prime mover behind Germany's naval policy, as a major driver of German competitive strategy, it must be noted that this designation appears clean and tidy only in retrospect. See also Seligmann, *The Royal Navy and the German Threat*, 8–9. In particular, as Seligmann has noted, while Tirpitz tended to prevail in bureaucratic battles by means of his relationship with his monarch, this in no way means he did not have to fight them. In particular, while Tirpitz would have largely unitary control of ship procurement (and would engineer the political appropriations process to avoid

meddling), the actual use of ships—doctrine—was the responsibility of the Naval High Command, which at a number of key points attempted to interfere with Tirpitz's decisions on the basis that doctrine had to work hand in hand with procurement if a navy were to function. For a more thorough description of this bureaucratic infighting, see Seligmann, *The Royal Navy and the German Threat*, 16–19. This is treated in more detail below, but for now it is enough to note that in understanding German competitive strategy, it is necessary at times to simplify a complex decision-making narrative, and that this study merely treats the most relevant players as the sources of planning and decision making without intending to minimize the importance or the scope of a complex historical and historiographical process.

139. Lambi, *The Navy and German Power Politics*, 61–62; Massie, *Dreadnought*, 168–71; and Padfield, *The Great Naval Race*, 37–40. Tirpitz, in his previous post as chief of staff to the High Command, had been advocating for a battleship fleet for most of the preceding decade, particularly in 1895. See Hobson, *Imperialism at Sea*, 193–212.

140. Alfred von Tirpitz, *My Memoirs*, 2 vols. (New York: Dodd, Mead, 1919; Amazon, 2019), 49. In translation, the phrasing is replaced by the less-catchy "the ten years lost from 1888 to 1897."

141. Alfred von Tirpitz, memorandum dated 15 June 1897, found in Padfield, *The Great Naval Race*, 44–45.

142. Seligmann, Nägler, and Epkenhans, *The Naval Route to the Abyss*, 6; Padfield, *The Great Naval Race*, 44–45; and Massie, *Dreadnought*, 172.

143. Padfield, *The Great Naval Race*, 44–45; Massie, *Dreadnought*, 172; and Steinberg, "The Kaiser and the British," 128. For the torpedo revolution, see also Sumida, *In Defence of Naval Supremacy*, 41; and Marder, *The Anatomy of British Sea Power*, 5–6.

144. Massie, *Dreadnought*, 172; and Padfield, *The Great Naval Race*, 45.

145. See, for example, Lambi, *The Navy and German Power Politics*, 33–37; Hobson, *Imperialism at Sea*, 300–3; and Seligmann, *The Royal Navy and the German Threat*, 7.

146. See, for example, Berghahn, *Germany and the Approach of War in 1914*, 52.

147. The minimal purpose of the fleet as a deterrent to a close blockade from a position of possibly inferior strength is widely attested and agreed on. For more on this, see Steinberg, "The Kaiser and the British," 20–21; Kennedy, *Strategy and Diplomacy*, 132–33; and Hobson, *Imperialism at Sea*, 260–73, 301–7, 316–31. Hobson is perhaps the most conservative on this point, in that he argues

The Origins and Aims of the Dreadnought Race

the notion of overtaking Britain as the dominant power in the North Sea, as opposed to merely establishing an acceptable force ratio, was little more than a pipe dream for Tirpitz, and not one much found in evidence in an objective appraisal his writings. He ironically notes, with puzzlement, that the very force ratio that Tirpitz saw as a minimal deterrent was also acceptable in later arms control talks as a means of ensuring British superiority.

148. Kennedy, *Strategy and Diplomacy*, 146–54. The citation to Wegener is Wolfgang Wegener, *Die Seestrategie des Weltkrieges* [The Naval Strategy of the World War] (Berlin: E. S. Mittler & Sohn, 1929).

149. See Hobson, *Imperialism at Sea*, 263–70; and Kennedy, *The Rise of the Anglo-German Antagonism*, 422, 443–44. Although Hobson disputes Kennedy's more maximalist interpretations of Tirpitz's intentions, such interpretations have more than their share of plausibility given the personalities and implicit goals involved. See Kennedy, *Strategy and Diplomacy*, 154–56.

150. See Kennedy, *Strategy and Diplomacy*, 154–60. Kennedy notes that Germany's only sure way to subordinate Britain, which was at least under consideration as a policy goal, was to build a fleet sufficient to either defeat or contain the Royal Navy, as ambitious as such an undertaking might be.

151. "A lever and a deterrent" is the apt phrasing of Jonathan Steinberg; see Steinberg, "The Kaiser and the British," 20–21. Kennedy sees the deterrent effect of the fleet as "the first, essentially negative, stage in Germany's development as a sea power" and suggests that its ultimate purpose was to prevent Britain from interfering with German colonialism at will. See Kennedy, *Strategy and Diplomacy*, 133–35. Kennedy attributes Tirpitz's desire for the fleet to his belief in an inevitable showdown with Britain and to a general desire to acquire Britain's deference—"a middle position between war and alliance" by deterring it—and stresses its connection with Bernhard von Bülow's attempts to avoid a war, and perhaps court Russian political support, until the fleet was ready for action. See Kennedy, *The Rise of the Anglo-German Antagonism*, 223–27. Clark downplays the political angle altogether, noting that the fleet was merely the culmination of German naval ambitions and that, although it was envisioned with Britain as its adversary, it was simply using the world's most powerful navy as its yardstick. See Clark, *The Sleepwalkers*, 149. Herwig also stresses the deterrent-leverage plan but notes Tirpitz's argument that the fleet would, in ominous terms for Britain, make Germany of "value" to Britain as a partner. See Herwig, *"Luxury" Fleet*, 36–37.

Chapter Four

Herwig sums up Tirpitz's program as an "aggressive policy of forcing colonial expansion with a powerful battle fleet stationed off the British coast. . . . And if in the future Great Britain should refuse to yield to this pressure, Tirpitz was willing to stake Germany's fate on a single, decisive naval battle." See Holger H. Herwig, *The German Naval Officer Corps: A Social and Political History, 1890–1918* (Oxford, UK: Clarendon Press, an imprint of Oxford University Press, 1973), 14–15. In Herwig's earlier view, the fleet's ultimate purpose was to enable colonial imperialism, and its means could include a direct attack if necessary. Berghahn argues that it "would be wrong to deduce . . . that the Imperial Navy was designed to be no more than a deterrent . . . in fact, it was to be used for a policy of diplomatic bullying. . . . Presuming that it would retain its deterrent effect on the Royal Navy and would not have to be employed in war, Tirpitz wanted to bully the other powers into recognizing Germany's need for a colonial empire." See Berghahn, *Germany and the Approach of War in 1914*, 52–53. Epkenhans writes, "Against this background [Tirpitz's Anglophobia and belief in German national destiny] it is difficult to regard the risk theory only as 'a defensive deterrent concept'," though he subscribes to the argument that the fleet could have served a diplomatically offensive purpose from a position of military inferiority. See Michael Epkenhans, *Tirpitz: Architect of the German High Seas Fleet* (Washington, DC: Potomac Books, 2008), 33–34. By contrast, Kelly, refuting Berghahn, argues that for Tirpitz the fleet was almost an end in itself and that little documentation exists to suggest that Tirpitz had thought through the external political import of the fleet, and that Tirpitz was so narrowly focused that he ignored both external and domestic political consequences of his single-minded pursuit of a battleship fleet. He also echoes Kennedy's point that Tirpitz overlooked the Royal Navy's ability to use a wide rather than close blockade, neutralizing his fleet. He also notes the historical consensus that the fleet's purpose was primarily defensive. See Kelly, *Tirpitz and the Imperial German Navy*, 153, 200–1. Similarly, Lambi takes the most parsimonious and indeed pessimistic tack: that the fleet as originally conceived by Tirpitz was suitable for little more than a suicide run in wartime, but can be understood to have been intended to "[raise] Germany's value as an ally." See Lambi, *The Navy and German Power Politics*, 143. Craig sees the matter as essentially a spectrum: the more Germany closed the naval gap, the more political options it would have and the more respect it would be afforded. See Craig, *Germany*, 309–10. Hobson

The Origins and Aims of the Dreadnought Race

is agnostic, noting that Tirpitz "could excuse himself from the task of defining what the projected fleet could achieve in war because its political importance would ensure that there would be no war"—the fleet would be a deterrent first, with unspecified political effects as a follow-on. Hobson further argues that the fleet grew out of "the ideology of sea power"—perhaps the best way of describing what was fundamentally an open-ended set of goals for an open-ended project driven by the dreams of ambitious leaders and their navalist supporters. See Hobson, *Imperialism at Sea*, 237–38, 325. Strachan sums up the project elegantly: to "use naval power as a deterrent and as a means to a new diplomatic order." See Strachan, *To Arms*, 449. Viewing the matter holistically, and noting that the Dreadnought Race's key instigators were competitive and arrogant men with a strong sense of their own and their country's destiny, it seems fair to say that the ratio of German battleship fleet to its competitor was intended to be whatever could be achieved, and the purpose of building that fleet was simply whatever Germany could get from it. Seligmann floats the hypothesis, originally suggested by a Royal Navy officer, that the entire Dreadnought Race was a diversion or distraction to remove British attention from the real German plan to threaten British commerce. See Seligmann, *The Royal Navy and the German Threat*, 172. Intriguing though this hypothesis may be, the political context—a revisionist Germany led by an ambitious and touchy kaiser known to be jealous of his British relatives' international social position; a nebulous push for *Weltpolitik* that seemed to require a North Sea breakout first; the enthusiasm of German navalists; and, above all, the character of Tirpitz, a sometimes mendacious man who preferred ambitious goals over middling ones—should give pause.

152. Quote found in Christian Gauss, *The German Emperor as Shown in His Public Utterances* (New York: Charles Scribner's Sons, 1915; Project Gutenberg, 2013), 127.
153. Quote found in Gauss, *The German Emperor as Shown in His Public Utterances*, 162.
154. Quote found in Gauss, *The German Emperor as Shown in His Public Utterances*, 182; and Bernhard von Bülow, "Bernhard von Bülow on Germany's 'Place in the Sun' " (speech to Reichstag, 1897), *German History in Documents and Images*, accessed 2 December 2021.
155. Lambi, *The Navy and German Power Politics*, 37, 77; and Kelly, *Tirpitz and the Imperial German Navy*, 179.

156. See Kelly, *Tirpitz and the Imperial German Navy*, 47–68; Kennedy, *Strategy and Diplomacy*, 150; Hobson, *Imperialism at Sea*, 187; and Seligmann, *The Royal Navy and the German Threat*, 8–19. The German Imperial Navy was of multiple minds on this question. There had long been a broad consensus in favor of the continued development of torpedo boats for coastal defense, which would at least deter a close blockade of German shores. There was also a faction, of which Tirpitz was merely the most ambitious, that favored a Mahanian blue water battle fleet; the question was simply what form that fleet would take, what its objective would be, and when and how it might be built. Additionally, a separate faction favored a focus on cruisers and *Jeune École*-style commerce raiding. As Seligmann has noted, the dispute between these two factions was never entirely resolved, with the Dreadnought Race consuming resources that could have gone to a cruiser procurement program but not completely eradicating the commerce-raiding doctrine or its associated procurement programs, which ultimately were implemented in the form of submarine warfare and an ineffective partial liner-conversion strategy.

157. Tirpitz, *My Memoirs*, 49.

158. For Tirpitz's original and evolving conception of the danger zone, see Seligmann, Nägler, and Epkenhans, *The Naval Route to the Abyss*, 283; Kelly, *Tirpitz and the Imperial German Navy*, 196; Herwig, *"Luxury" Fleet: The Imperial German Navy*, 36–37; Massie, *Dreadnought*, 181–82; and Padfield, *The Great Naval Race*, 42–43.

159. Kelly, *Tirpitz and the Imperial German Navy*, 196–201; Kennedy, *Strategy and Diplomacy*, 150–51; and Hobson, *Imperialism at Sea*, 269.

160. Craig, *Germany*, 308; Strachan, *To Arms*, 10–12; Kelly, *Tirpitz and the Imperial German Navy*, 132–55; Steinberg, "The Kaiser and the British," 149–53; Padfield, *The Great Naval Race*, 54–56; and Massie, *Dreadnought*, 173–74.

161. Padfield, *The Great Naval Race*, 56–57; and Massie, *Dreadnought*, 173–74.

162. Herwig, *"Luxury" Fleet: The Imperial German Navy*, 37–8. See also Craig, *Germany*, 276–77; Wehler, *The German Empire*, 96–97, 167–68, 176–77; and Berghahn, *Germany and the Approach of War in 1914*, 53–54, 87–89. Craig notes that Tirpitz was able to mobilize the German nobility's fear of industrialism displacing agrarianism globally (as exemplified by the nascent Boer War) in support of the naval program, even as he in the early years made mention only of needing further tariffs. Wehler points out

that Tirpitz was capable of alternately courting and countering the German agrarian right as part of the perennial maintenance of the unwieldy anti-Social Democratic political bloc while pursuing his goals, and one does not have to subscribe entirely to Wehler's thesis that *Weltpolitik* was driven primarily by domestic politics to appreciate his understanding of Tirpitz's role (and political abilities). Wehler emphasizes Tirpitz's courting of the agrarian right by initially including protective tariffs for agriculture as the preferred means of funding the naval buildup, a point echoed by Berghahn, who deemphasizes Tirpitz's duplicity on the funding question. Whatever the case, inadequate agricultural tariffs would be the first source of funding for the Dreadnought Race. What for the nobles were the more lugubrious aspects of naval funding would wait for later.

163. Steinberg, "The Kaiser and the British," 149–200, 206; Strachan, *To Arms*, 11; Seligmann, Nägler, and Epkenhans, *The Naval Route to the Abyss*, 7–8; Massie, *Dreadnought*, 174–81; and Padfield, *The Great Naval Race*, 57–65. Steinberg in particular lauds Tirpitz political skill, which he suggests actually exceeded his skill as a strategist, given what followed.

164. Seligmann, Nägler, and Epkenhans, *The Naval Route to the Abyss*, 158–59; Strachan, *To Arms*, 11; Seligmann, *The Royal Navy and the German Threat*, 9; Kelly, *Tirpitz and the Imperial German Navy*, 6; Hobson, *Imperialism at Sea*, 261; Lambi, *The Navy and German Power Politics*, 280; and Padfield, *The Great Naval Race*, 68–70.

165. Padfield, *The Great Naval Race*, 70–71; and Massie, *Dreadnought*, 135, 138–49.

166. Pakenham, *The Scramble for Africa*, 467–69, 508–11, 516–17, 535–56; Padfield, *The Great Naval Race*, 70–72; and Massie, *Dreadnought*, 248–56, 341–42.

167. Padfield, *The Great Naval Race*, 83–87; and Massie, *Dreadnought*, 180, 271–75, 307, 553. This was far from the first time the issue had come up. As was noted earlier, the mere idea of German interference in South Africa was enough to provoke the barely veiled threat of a British naval blockade, the implications of which for German navalists were quite obvious.

168. For Chamberlain's efforts, see Padfield, *The Great Naval Race*, 87–88; and Massie, *Dreadnought*, 306–7.

Chapter Five
The Dreadnought Race

The earth is full of anger,
The seas are dark with wrath,
The Nations in their harness
Go up against our path:
Ere yet we loose the legions—
Ere yet we draw the blade,
Jehovah of the Thunders,
Lord God of Battles, aid!

~ Rudyard Kipling[1]

Britain initially responded sluggishly, if at all, to Germany's challenge. Neither the political nor the bureaucratic will to move quickly existed. The Royal Navy was mired in institutional inertia and complacency born of a lack of practice. As might be expected from a military service that had been at peace for too long, form had taken precedence over function. Captains were promoted and held to account on the basis of their ability to keep their ships in spotless condition and perform tactically useless maneuvers on parade. Gunnery tests were conducted infrequently and were not treated as important; captains who focused on gunnery to the detriment of appearances were penalized by the personnel system and not infrequently had their careers ruined, and inconvenient results of gunnery tests—which were often extremely poor—were kept out of public view. Tactics had stagnated as well. Until the 1890s, especially given the lack of actual combat experience to go on, Royal Navy officers had been taught to assume that future battles would resemble Trafalgar: ships would fall out of the line of battle to engage the enemy at close quarters, with the action decided by boarding operations; the true implications of long-range gunnery and armored hulls were slow to catch on.[2] Administratively, the Admiralty was incoherent and difficult to manage, with an almost overconfident organizational culture and with, in the historian Peter Padfield's evocative description, "tennis . . . played on the Admiralty lawn."[3]

The naval historian Arthur J. Marder famously noted that Britain took little notice of Germany as a naval threat—as opposed to a political adversary—until the intent behind the new German battleship fleet became plain by its sheer size and rapid growth. It is more likely, in fact, that the Admiralty did not fully come around to the idea of Germany as Britain's primary adversary until considerably later, and that Britain's political leadership did not internalize the German challenge until at least 1904, if not 1908.[4] When the Dreadnought Race began, the Admiralty were still committed to maintaining the Two-Power Standard. To them, acquisition policy was simply a question of maintaining a quantitative edge in fleet size over potential foreign competitors in general, not Germany specifically.[5] With the Fashoda Incident fresh in their minds, the Admiralty were more apt to expect to have to fight France than Germany.[6] When several key publications ran editorials on the rising German naval threat, the Admiralty patronizingly rejected their arguments as amateurish.[7] There was also inertia at the top. Robert Gascoyne-Cecil, Third Marquess of Salisbury, the Tory prime minister at the time the Dreadnought Race began, was an old man with but a few short years to live. He was well past his prime; he would tender his resignation to Buckingham Palace with minimal ceremony and almost no warning in 1902 and die peacefully of old age the following year.

Indeed, several misconceptions and oversights contributed to Britain's lack of a coherent response to the German challenge. Where national security was concerned, both public opinion and the government were preoccupied with the Boer War. The so-called "Khaki Election" of 1900 brought Salisbury's conservative government back into power, but this was on the basis of its conduct of the Boer War, not its ability to deal with longer-term problems.[8] About the same time, the colonial secretary, Joseph Chamberlain, who was the de facto manager of the war effort and, by this time, the government as well, was advocating for an alliance with Germany, which German chancellor Bernhard von Bülow was simultaneously advocating. The negotiations went nowhere for the very reason that Germany was willing to engage in them. As Robert K. Massie notes, Britain could not upset the balance of power by acquiescing to German domination of the continent and becoming a junior partner in a German alliance; it had instead to oppose German expansion by any means possible. Chamberlain was overruled, and despite Bülow's advocacy Britain politely turned the proposal down.[9] The fact that the question was still up for discussion, though, should illustrate just how slow Britain was to come around to the idea that Germany specifically, and not some hypothetical combination of foreign powers in general, was Britain's principal national security problem.

But the fact that the Royal Navy's leadership did not immediately treat Germany as a threat did not mean that it stood entirely idle. As noted, there was the Two-Power Standard to maintain. With the onset of the Boer War, British defense expenditure spiked, though the army received the lion's share. Still, the Royal Navy's estimates did increase, and by the end of the Boer War they had gone up almost 50 percent, from £23.8 million in 1898 to a peak of £36.9 million in 1904.[10] The Royal Navy was in the process of mothballing older battleships and building new ones with heavier guns and armor than the ones they replaced.[11] The naval modernization program begun in 1889 was therefore still in the implementation phase, even at great cost.

But the post-Boer War world was not the same for Britain as before. Queen Victoria, the stern monarch whose reign had effectively defined Britain for two-thirds of a century, had died in 1901; the *bon vivant*, playboy former Prince of Wales Edward VII was now on the throne.[12] Britain had suffered significant blows to its prestige. The Boers had proven a tough adversary, and Britain had brought them down only by employing the world's first concentration camps as a means of separating the civilian population from the Boer guerrillas, unofficially sanctioning the killing of Boer prisoners of war in the process, atrocities that not only diminished Britain's international standing but fueled German nationalist propaganda.[13] Britain had incurred an enormous war debt; despite the fact that it had been waged against a third-rate adversary over whom victory was no great achievement, the war had been no small affair.[14] Salisbury was dead. Chamberlain, the colonial secretary who had been the last hold-out in the question of a German alliance, had been discredited by the brutal tactics employed to win the war and finally brought down by his unsuccessful advocacy of protectionist policies that other members of the government did not support. The new government of prime minister Arthur J. Balfour was under attack by the liberals, who proved much more politically cohesive than Balfour's unwieldy conservative-unionist coalition.[15] The new German fleet was simply the latest of a series of problems with which Britain now had to wrestle, and it was far from the top of its priority stack. The conservative government remained of multiple minds on policy even as the British public, led by major publications, veered toward skepticism of German intentions and eventually toward hostility.[16]

Three developments occurred in tandem immediately after the Boer War that served to change Britain's strategy. The first was a diplomatic change of tack, by which Britain ended its policy of "splendid isolation" and, by a complex series of agreements, tied itself to the Franco-Russian anti-German coalition. The second was an emphasis on cost-cutting: a belated attempt by the post-Salisbury conservatives to get the budget under control, followed by the election of a Liberal government with an eye toward finding room in the budget for social spending, which led to the genesis of Britain's welfare state. And the third was the appointment of Admiral Sir John Fisher as First Sea Lord in 1904, who came in with orders to find savings in the budget, but who in short order crafted Britain's new naval procurement policy, built the battleship HMS *Dreadnought*, and brought the Royal Navy around to direct competition with Germany.

The British Awakening: Diplomatic Hawks, Cost-Cutting in the Royal Navy, and the Building of HMS *Dreadnought*

With the abortive negotiations for an Anglo-German alliance having run their course, the British government adopted an almost opposite policy. Its genesis lay in a gradual shift in British threat perception. The shift was not uniform. Not only in the government, but in public opinion as well, German actions and rhetoric—the naval program, the kaiser's inflammatory remarks regarding the Boer War, the increasingly hostile rhetoric of the German nationalist press, and the unsubtle way in which Germany had conducted the failed alliance negotiations—had created the impression of a German threat in at least certain quarters. The degree to which this was seen as important varied greatly. The First Sea Lord, Admiral Sir Walter T. Kerr, repeatedly warned the British government regarding the German capital ship program, but he viewed it as one of several problems. Several prominent editors were prepared to go further, openly considering in print a "Copenhagening" of Imperial German Navy grand admiral Alfred von Tirpitz's nascent High Seas Fleet: a (hopefully bloodless) preventive strike to seize and then sink all of the new battleships launched or in production—an action that, despite the obvious risks, was within the Royal Navy's capabilities. The expression recalled an earlier action during the Napoleonic Wars. Conversely, several more obscure ministers were deeply ambivalent about the degree of threat that Germany posed, while many civil servants were not sold on the idea of a German threat at all. But at the highest levels, there was a growing consensus in favor at least of doing something—a consensus that included the First Lord of the Admiralty, William W. Palmer, 2d earl of Selborne, as well as, to a lesser degree, the heads of the War and Foreign Offices, H. O. Arnold-Forster and Henry C. K. Petty-Fitzmaurice, 5th marquess of Lansdowne, respectively, as well as Balfour. Overshadowing this, in spirit if not through official power, lay Edward VII's own personal animus toward not only Germany but his impulsive relative on the German throne.[17]

What emerged was the Entente Cordiale, officially a mere shakeup of diplomatic relations with France, but one with far-reaching implications. In 1903, the British foreign secretary, Lord Lansdowne, began a series of wide-ranging and open-ended negotiations with France, with the stated, modest objective of sorting out some longstanding colonial issues. A series of agreements, collectively known as the Entente Cordiale—a mere "understanding," as British public opinion and even the government would not admit of anything so drastic as an actual foreign partnership—were formalized in 1904. Although the aforementioned policy makers were not entirely of one mind, as Paul M. Kennedy notes, a consensus emerged in favor of a diplomatic move to reduce tensions with France so that Britain could deal with Germany, and each time negotiations stalled (as they frequently did), there was sufficient impetus from within the government to persevere. This may be taken, insofar as it is possible to generalize, as an approximate indication of the Balfour government's mood, even if, as noted, opinion was far from uniform. For all this, a major distinction still existed among proponents of the entente between those who simply saw it as a way of managing a general (and hopefully temporary) decline in Britain's political position and those who saw it actually as a means of clearing the decks to oppose Germany specifically. This ambiguity of purpose would be paralleled by the military developments of the same period (discussed below), which even now are seen either as a direct attempt to compete with Germany or simply as a means of resolving resource pressures. What looks ambiguous in retrospect was no clearer at the time, and the one aspect of Britain's posture by 1904 that is clear is that it still lacked a clarity of purpose.[18] It would not acquire such clarity until the end of the decade.

On paper, the Entente Cordiale looked like little more than (badly needed) diplomatic housekeeping. Even this was not easy, and the entente explicitly comprised what were seen as several fairly dramatic agreements. In effect, France would guarantee noninterference in Egypt, which was under a de facto British protectorate, in return for a British pledge of noninterference in Morocco, which, being opposite the Strait of Gibraltar, was a longstanding British concern. Since these two areas were key geopolitical chokepoints, this represented an agreement to put aside two major areas of regional competition in their own right. Other matters were similarly cleared up, such as fishing rights off the coast of Newfoundland and designating spheres of influence in Southeast Asia. The concessions made by both sides were numerous, and they required considerable political finesse both at the negotiating table and with their respective publics, not to mention substantial political risk-taking given the way each side's electorate might react. A royal visit by Edward VII to Paris—which initially was warily received—also helped mend fences. However, on its face, the entente represented the settling of imperial disputes far from home, nothing more.[19]

The spirit of the entente mattered far more than the letter. Britain was patching up its differences with the state that had historically been one of its most feared adversaries, even as it declined an alliance with the state alongside which it had fought Napoléon. In practice, the entente signified a far more wide-ranging agreement. As Henry Kissinger describes, while a series of governments disavowed any sort of military or national security commitment that the entente might be claimed to signify, the general impression it created led in turn to what Kissinger refers to as "moral obligations," a set of soft promises that impacted the political situation and could be seen as a soft alliance, even if its proponents disavowed this. In effect, sorting out colonial disputes meant the end of Anglo-French enmity. Put together with the end of Anglo-German negotiations and the already-existing Franco-Russian alliance, the entente meant that Britain had chosen its side in European political disputes.[20]

There was no going back. Under subsequent governments, the entente resulted in a dialogue on defense matters that led to mutual preparations for war against Germany as a common adversary—linking French and British policy on Germany even in the absence of a formal mutual security guarantee and ending the Anglo-French naval rivalry while ensuring naval cooperation.[21]

The exact motivations for the Entente Cordiale, as well as its precise nature and extent, are not only the subject of dispute now but were subject to evolution at the time. Dovetailing with Britain's ambivalence toward confronting Germany in this time, the entente promised little as of its formation. It would subsequently assume more of a role in British policy even as, right up to the end, British policy makers would deny that it committed Britain to anything at all. As Samuel R. Williamson Jr. has noted, the German naval challenge gave the entente more significance as the decade wore on, again without any official admission that such was the case.[22]

While the practical effects of the Entente Cordiale took a while to register, they were wide-ranging. Militarily, the entente effectively removed the French naval threat to Britain, thereby allowing Britain the freedom—which it was admittedly slow to exercise with full intent—to focus on competing with the new German naval program. In practice, as has been noted, the Royal Navy was more apt to view its French counterpart with arrogant contempt, but the removal of the need to worry about it did allow for the switch, toward the end of the decade, toward naval competition with Germany alone and specifically. Moreover, as Arne Røksund has noted, the entente effectively terminated the French *Jeune École*, which would survive as a school of thought but not as a procurement doctrine. With Anglo-French relations thawing, and ultimately with the commencement of military coordination between the two powers, the need on France's part for a comprehensive set of plans for strangling the British economy in the event of war faded away.[23]

Chapter Five

In Germany, at the highest levels, the Entente Cordiale was seen as having significantly altered the Anglo-German relationship. It also eliminated a favorite Bismarckian ploy—though admittedly one that, post-Bismarck, Germany had been unable to pull off—namely exploiting diplomatic divisions regarding colonies and other matters between Britain and France to prevent a total diplomatic containment of Germany. Nor, as Kennedy has noted, did any solutions suggest themselves. Bülow and Kaiser Wilhelm II discussed the possibility of a rapprochement with Russia, which, because of the Russo-Japanese War and Britain's partnership with Japan, was a potential adversary to Britain even as it was a formal ally of France. This would have been a necessary step to allow Germany the necessary strategic focus to compete effectively with Britain; sadly, for Germany and Bülow, it failed.

The idea of a Russian rapprochement went nowhere. As will be discussed when dealing with the Royal Navy's own preparations, the idea had circulated among British policy makers, and ultimately in the British press, that an obvious solution to the problem of Germany's naval program was a preventive strike against it, an event that Tirpitz, in his "danger zone" hypothesis, had anticipated. The German naval leadership, rarely of one mind on anything, were therefore united in opposition to making any hostile moves against Britain, even diplomatically; paradoxically, this included Tirpitz, who opposed a rapprochement with Russia for these reasons. There was little evidence at the time that Russia would be ultimately amenable to a thaw in relations with Germany as opposed to Britain—notwithstanding the Dogger Bank crisis, in which the Russian Navy inadvertently fired on British fishing vessels, which the British press ultimately spun into a conspiracy theory involving Germany—and, in fact, Anglo-Russian rapprochement ultimately occurred instead. But perhaps the biggest contributing factor to the Entente Cordiale—Germany's heavy-handed diplomacy combined with its too-obvious intention of supplanting Britain as the dominant naval power in European waters and farther afield—was also the biggest contributing factor to the lack of options for dealing with it.[24] The only option was to attempt to undermine it, at which Germany, under Wilhelm's leadership, would fail spectacularly.

The entente saw its first test with the Algeçiras crisis of 1905–6. In March 1905, Wilhelm, at the urging of Bülow, made a state visit to Tangiers in French-controlled Morocco, at which he pledged support for the sovereignty of the sultan of Morocco, then under French dominion. This overt threat to French control of an important colony would serve as an attempt to divide the entente, but it would fail. The implicit threat to French control of Morocco led in turn to the implicit counterthreat of war in France, despite French policy makers' acute understanding of France's weakness relative to Germany. Britain backed France diplomatically, effectively reaffirming the Entente Cordiale. At the Algeçiras Conference the following year, which was called to settle the crisis, Germany received the face-saving concession of being allowed to support the nominal transfer of the Moroccan police force to the sultan's control, after which the crisis dissipated. Germany was unable to obtain the diplomatic support of any great power apart from Austria-Hungary at the conference, which left it in a diminished international political position.[25]

The second and third developments came in quick succession. Britain had incurred a substantial war debt fighting the Boers, and financially the Royal Navy was flailing. Maintaining the Two-Power Standard had led to a more than 50-percent increase in naval estimates; at least from the point of view of a government that had to justify its existence to an electorate, this was an unsustainable trend.[26] As will be discussed in the following chapter, Britain's fiscal conservatism would come to be its undoing in the Dreadnought Race for this very reason. The man brought in to do it in 1904 was the Balfour government's First Sea Lord—the empire's chief naval officer—Admiral Sir John Fisher.[27]

Fisher was a perfect foil for his counterpart, Tirpitz. A man of humble origins with an aggressive and eccentric personality and a gift for both religious allusion and profanity, who had worked his way up the Royal Navy's bureaucracy by hard work and bravery under fire, Fisher would come to influence the Royal Navy in everything from force posture to personnel policy to procurement and doctrine, even as he occasionally had to bow to consensus and political reality.[28] Much of the historiographical dispute surrounding Fisher concerns the degree to which he was able to put his more *avant-garde* ideas into practice, as well as the numerous purposes to which those ideas were to be put and their relationship to the political environment that swirled around them. This study is prepared to accept the "revisionist" interpretation of Fisher's tenure of office, which holds that he was concerned with revolutionizing the Royal Navy for reasons of his own, and not merely Germany—as noted several times above, it is possible and even necessary to analyze a state's strategy even if that strategy is incoherent or its decision makers distracted. This study can, however, suggest that Germany was indeed among Fisher's top priorities and that this basic intuition was not as off-base as it is sometimes portrayed.[29]

Fisher's Reforms

Fisher had intuited the German threat earlier than many. His extant writings suggest a preoccupation with countering Germany as early as 1902, and he came into office with the intention of doing something about the German naval threat—if necessary, by taking preventive military action to eliminate it, a position he reiterated at the height of the Algeçiras crisis.[30] Although Fisher is best known as the father of HMS *Dreadnought* and its successors, his first acts were administrative. Ostensibly as a cost-cutting measure, Fisher began a wide-ranging shake-up of the Royal Navy's deployment.

The Royal Navy's ships were distributed according to the strategic conceptions of an earlier, more congenial era. In better times, when the Royal Navy's control of the seas had been undisputed and its main task was imperial gunboat diplomacy and simply showing the flag, the navy had been divided into numerous squadrons spread across the surface of the globe. Fisher's initial review of the situation pointed out the obvious: all of these squadrons (many of which contained obsolete ships) could be defeated in detail; they were weaker than the local fleets they opposed. At the same time, ships would be needed to face the German High Seas Fleet in due course—and disturbingly soon. Fisher therefore reorganized the navy's squadrons. Overseas squadrons were consolidated into larger units, obsolete ships were mothballed or decommissioned even at the cost of having a smaller fleet, and ships that could fight were brought progressively closer to home. The China fleet was effectively disbanded. The Mediterranean Fleet, which had responsibility for maintaining access to the Dardanelles and containing potential Russian moves to seize them, was downsized. Several of these ships were then used to augment the Channel Fleet, the fleet closest to home that was responsible for the security of the British Isles; Fisher later consolidated all of the squadrons around the North Sea into a single Home Fleet. The latter effectively doubled in size. Meanwhile, ships of lesser fighting value were given skeleton "nucleus" crews so as to keep them in partial readiness, rather than alternate between being fully manned and being in drydock. In a sign of what was to come, Fisher experienced significant pushback on these decisions from active duty naval officers and retired flag officers. He ignored the latter and, where necessary, fired the former.[31]

There is continued controversy as to whether Fisher pursued these reforms—and, perhaps equally important, was allowed to pursue these reforms—primarily as a cost-cutting measure and a chance to give British naval policy a badly needed shake-up, or whether he did so with the actual intention of confronting Germany. On balance, it would appear that Fisher was of the more hawkish faction within the British government where Germany was concerned, and that, as noted, his pronouncements regarding policy went well beyond the existing policy consensus.[32] Despite this, neither Fisher nor his political masters appear to have been exclusively concerned with Germany until well into the second half of the decade.[33] For all this, however, the redeployment of a large chunk of the Royal Navy to home waters after the entente and the deterioration of Anglo-German relations is best understood for what it was: the beginnings of a policy shift toward direct competition with Germany for control of these waters and the political influence that came with it. The fact that Britain had other concerns and was not yet single-mindedly fixated on Germany, and the fact that Fisher was hardly without other things to do, does not fundamentally alter this basic interpretation.

In his more radical moments, Fisher appears to have envisioned a drastic change in naval warfare predicated on long-range weaponry, smaller target profiles, and speed. Observing the rapid pace of naval technological development, Fisher had come to a simple but radical thesis: future naval engagements would be decided by the ability to outdistance the enemy. In particular, as Fisher saw it, battles would be won or lost on the basis of a fleet's ability to consistently hit an enemy from a longer range than that at which its own guns could respond, which in turn would require superior speed to maintain that distance. Armor would become irrelevant: not only was it becoming useless in the face of bigger naval guns firing more powerful shells, but its weight would slow down a ship and prevent it from outrunning a lighter-armed opponent. In Fisher's conception, the ideal ship would be an "all big-gun" ship, carrying 12-inch or bigger guns that could hit at the longest range possible in any direction, not least directly forward during a chase. It would have the best available powerplant to maximize its speed, and it would have just about enough armor—almost as an afterthought or concession—to survive in combat, and no more. In Fisher's conception, the concept of the cruiser, a ship that relied on speed at the expense of armor and firepower, would merge with the concept of a battleship, a ship of the line that sacrificed speed for firepower and the ability to take as much as it gave.[34] Fisher would ultimately become famous for the design and procurement of two major new types of battleship: HMS *Dreadnought* and its successors, which would revolutionize the concept of the battleship, effectively the apogee of the battleship as far as Fisher's preferred design concept would allow; and the genuine battlecruiser HMS *Invincible* (1907) and its successor classes, which were the more faithful expression of Fisher's vision of a light, fast, up-gunned ship. Such a design ran counter to the more conservative design concepts and tactical doctrines then-prevalent in the Royal Navy.

Doctrinal questions regarding the prescribed use of Fisher's lighter battlecruisers remain unresolved even now, in that documentation on Fisher's intended use for the battlecruisers remains both spotty and subject to multiple interpretations. The most radical of these has been advanced by Matthew S. Seligmann, who has made a persuasive case that a number of senior Admiralty officials from Fisher on down saw the new light-armored battlecruisers as an answer to the problem of countering German cruiser attacks—specifically, armed German merchant vessel attacks—on British shipping in the event of war. The case, in brief, lies not merely in the fact that a number of British admirals explicitly advocated this, but in the genuine mystery of what the battlecruisers were actually supposed to do otherwise, as they were not suitable for line-of-battle service (a fact that would be made clear during the 1916 Battle of Jutland). Fisher, who for all his verbosity sometimes did not lay out his thoughts in detail, appears to have neglected to explain this to the entire Admiralty.[35]

It is nevertheless possible that Fisher's vision really was as radical as it has been made out to be: that the battlecruisers were intended to run rings around heavier capital ships in open battle. Where Seligmann's argument is concerned, he is no doubt quite correct, but his argument misses the more fundamental question of whether countering commerce raiding was *Invincible*'s only or primary purpose. The up-gunned, underarmored *Invincible* appears to have grown out of Fisher's vision, imperfectly realized the first time around with HMS *Dreadnought*, for a ship that would use speed almost as a form of armor while pummeling slower enemy vessels that could not respond in kind. As such, *Invincible* reflected Fisher's personality as much as any specific mission, and it was more likely to have been designed as something worthwhile, and then put to a specific use, rather than the other way around. The iconoclastic and rebellious Fisher was an extremely aggressive individual—not only at war but in person—who thrived on motion (literally in his personal manners as well as organizationally and at sea). The highly innovative, fast-moving, hard-hitting *Dreadnought* and *Invincible* were, quite simply, exactly the kind of vessels Fisher would have designed—and, indeed, perhaps the only kind he could have imagined. This, perhaps, best explains a point that Seligmann glosses over: the large caliber of *Invincible*'s armament was fundamentally unnecessary if its mission was simply to hunt commerce raiders, and an expensive luxury to boot (simply going from 9-inch to 12-inch guns raised the cost of a battlecruiser from £1 million to £1.75 million). Even more evidence of the lack of clarity about doctrine—suggesting that countering commerce raiders was only part of the battlecruisers' mission—is the simple fact that the battlecruisers were put to use as line-of-battle ships when World War I broke out.[36]

Allowing, however, for the assumption that Fisher's more ideal vision was corrupted in execution by committee, the concept of the state-of-the-art surface warship as a hyperpowered, heavily gunned ship did indeed come to fruition under Fisher's supervision, initially as HMS *Dreadnought* and subsequently as a series of ships and ship classes.[37]

Fisher was, moreover, highly sensitive to larger technological trends, to a greater degree than his contemporaries, even if some of their expressed skepticism was performative. During his tenure of office as First Sea Lord, he proposed assigning roles to submarines that had previously been reserved for capital ships, and he pushed through the rapid technological development and large-scale procurement of submarines and torpedo craft.

Fisher is also credited with at least speculating about replacing battleships for blockade purposes (their primary role in the event of war) with a "flotilla defence" concept in which fleets of submarines and torpedo boats would substitute for battleships, freeing the latter for other operations.[38] As one would expect from a man generally regarded as a genius in such matters, not all of Fisher's ideas came to fruition or were even fully baked. His idea for flotilla defense would be adopted, after his tenure of office had ended, as a stopgap measure for defending the Mediterranean, freeing up battleships to defend home waters—but no one was going to gamble the safety of British home waters or the effectiveness of a wartime blockade on an unproven technology.[39] In the following chapter, some analysis will be offered concerning the balance of forces in this area, as it was of at least speculative relevance to the naval balance that Germany was trying to overturn, but the focus of the competition would remain the building of capital ships.

The Royal Navy's problems with form as a substitute for substance did not begin or end with technological wizardry, however. Throughout his short career as First Sea Lord (1904–10), the polarizing Fisher had to contend with insubordination from Lord Charles William de la Poer Beresford, the admiral in command of the Channel Fleet, which because of its geographic position was the most important and politically visible combatant command in the Royal Navy. A ruthlessly and remorselessly political officer who could not be removed from command because of his political connections (and because he would be even more dangerous as a politician than as a subordinate), Beresford repeatedly demanded resources that the Royal Navy did not possess and threatened to ruin Fisher's political reputation when the latter refused to change policy. More importantly, however, Beresford developed a bureaucratic and ultimately personal feud with the one officer who Fisher managed to put in a position to revolutionize the Royal Navy's gunnery doctrine, Admiral Sir Percy M. Scott. Although Scott ultimately succeeded in updating the Royal Navy's doctrine and training practices to account for new long-range gunnery, Beresford's meddling slowed the process at every turn.[40] For all Fisher's brilliance, he never succeeded in systematically overhauling the Royal Navy's personnel system or bending it to his will, a deficiency attributed by some modern analysts to his failure to develop and cultivate a modern strategic planning staff. It is fair to say that even at the time he retired as First Sea Lord, the Royal Navy's people were not as up to date as its technology.[41]

The year 1905 saw the election of a liberal government, to be led by Henry Campbell-Bannerman as prime minister, with the future prime minister Herbert H. Asquith as chancellor of the exchequer.[42] The liberals shared the previous government's interest in cost-cutting, but for different reasons. The rationale for the Campbell-Bannerman government was to rapidly pay down the national debt (particularly the part of it that had risen during the Boer War) to free up funds for the liberals' proposed social programs—notably the Old Age Pensions Act (1908) and unemployment insurance—which, though modest by modern standards, were revolutionary in their time and represented an enormous increase in government spending.[43] The liberals planned to pay down the debt and ultimately raise taxes to support social programs.[44] Although this general policy gave Fisher the support he would need for a naval program that was ostensibly a cost-cutting measure, it would also serve to hamstring Britain's efforts to stay ahead of Germany in the Dreadnought Race.[45] The moment of truth would come three years later, when, in 1908, the German High Seas Fleet almost caught up.

This, however, was far off at the time HMS *Dreadnought* was launched. *Dreadnought* had a short development span. Although Fisher preferred a more radical design, when forced by his civilian superior, First Lord of the Admiralty William Palmer, to adhere to a ship-of-the-line concept, he designed *Dreadnought* in 1905; its keel was laid in October of that year; and it was finished and christened on 9 February 1906. Following

Fisher's concept of future tactics and ship design, *Dreadnought* was built for long-range fighting. The ship mounted five turrets with two guns each: one amidships on the bow; two abeam, with one on each side; and two astern, normally facing rearward. Each gun was of 12 inches' diameter. In keeping with the new tactical concept, *Dreadnought* finally did away with a vestigial feature of previous warships: the beaked ram on the bow. What made *Dreadnought* revolutionary, though—indeed, what made it possible at all, and what allowed Fisher to square the circle of armor versus speed—was its powerplant, which allowed for a maximum speed of 21 knots, compared with a previous theoretical maximum of 18–19 knots and a practical maximum of 14 knots. Recognizing that he could not have everything and, like Tirpitz, worried about the political ramifications of his work, Fisher kept *Dreadnought*'s size within the general limits of previous ships, and in so doing kept costs down. The new ship cost only £181,000 more than a previous ship of the line—an acceptable increase given the increased capabilities and weighed against a £31,000,000 naval budget. With the Boer War over and the liberals in power, Fisher had to be careful not to be seen to spend money, a built-in cost-controlling factor that probably helped to make *Dreadnought* a revolutionary design and its descendant ships sustainably affordable.[46]

The term *dreadnought* hereafter refers to ships designed roughly to these specifications, as it came to be an internationally recognized term of art for this new kind of capital ship.[47] In practice, as the design was copied abroad and, as new designs were put forward, there was considerable variation. In particular, the tension between armor and speed was never entirely resolved. The ensuing years would see both lighter battlecruisers of a design closer to that originally envisioned by Fisher, as well as heavier battleships with upgraded powerplants that allowed them to compete acceptably with the lighter ships on speed. What united these disparate design concepts was an acceptance of Fisher's tactical concept: long-distance engagements fought with large guns by ships that were designed for the purpose, that could survive in this type of tactical environment, and that were much faster than anything previously launched.

Tirpitz's Countermove

At a stroke, HMS *Dreadnought* ruined Tirpitz's plans. Although it is often remarked that *Dreadnought* leveled the playing field by rendering all capital ships in all navies obsolete at the same time, the fact was that the Royal Navy had a momentary technical edge (before the technology diffused) and a lasting financial edge. Because of its relative competence in shipbuilding, Britain could turn out ships of whatever quality at a lower unit cost than Germany—a situation that Tirpitz never succeeded in remedying. Fisher had forced the issue and in so doing forced Tirpitz to compete on a new set of terms.[48] Tirpitz, who was caught so off-guard that he nearly resigned his office, was now faced with the challenge of suspending his building program, reverse-engineering Fisher's design, and then improvising a new building program for dreadnoughts.[49]

With this in mind, Tirpitz in 1906 sought an amendment (or *novelle*, in the German legislative parlance of the time) to the existing Naval Laws that would allow him to catch up once more. The novelle, which became known as the Third Naval Law, reinterpreted the permission given in the preceding two laws for 38 battleships and 20 cruisers as permission for 58 battleships or battlecruisers of similar type to *Dreadnought* that, as Fisher had intended, had muddied the distinction between the two classes of ship. A successor novelle in 1908 also permitted a quicker pace of production for new ships and a shortened planned obsolescence for new builds. Ships would now be built with the intention of being commissioned for 20 years instead of a previous 25, effectively authorizing a slight but significant increase in shipbuilding.[50]

However, Tirpitz's dockyards were still not up to the pace of the Royal Navy's. The German shipyards could put out a ship in 42 months, and a crash program could bring that time down to 35 months—but *Dreadnought* had been built in 12. And while

Dreadnought had represented a modest increase in the cost per ship, Tirpitz's new planned ships were expected to be twice as expensive as his old ones. Nor, in fact, would the new ships be in quite the same technological league as *Dreadnought*: Tirpitz struggled to build ships that could pull 20 knots, while *Dreadnought* could manage 21.[51] Worse yet, Tirpitz was not in a position to make meaningful gains in the one indisputable area of his own intentions: altering the balance of forces with Britain. The most he could hope for was to build two ships to Britain's three, making narrowing the capital ship gap in the North Sea a tenuous proposition.[52] Germany was not playing to its strong suit; it was, in fact, struggling even to keep pace with an adversary that, despite its internal inertia and mismanagement, was quite used to handling these sorts of challenges.

Tirpitz, moreover, was embroiled in multiple bureaucratic battles merely to implement his own vision. As an inevitable outgrowth of Kaiser Wilhelm's purely personal leadership style and fractured organization, it was difficult to sustain consensus to almost any degree without constant bureaucratic jockeying. For this reason, Tirpitz's radical vision—eroding Britain's naval primacy in North Sea waters as a means of increasing political leverage—was never uniformly adopted by the multiple heads of the Imperial German Navy. The upshot was a policy problem that manifested in at least two major forms: the persistence of the *Jeune École* as a doctrine not only for tactics but impacting procurement as well (despite Tirpitz's theoretical control of these matters), and repeated challenges to Tirpitz's basic plan on matters of budgeting.

Where the first was concerned, the Admiralty Staff—who would actually be ordering ships into combat in wartime and who answered not to Tirpitz but (if to anyone) to the kaiser personally in his role as commander in chief—continued preparations for *Jeune École*-style commerce raiding operations throughout this period. As Matthew Seligmann has painstakingly documented, the adoption of Tirpitz's plans for the High Seas Fleet did not end the thinking in the operational arm of the Imperial German Navy's leadership that commerce raiding would be of primary importance once war broke out. Moreover, with the High Seas Fleet consuming an ever-increasing portion of the Imperial German Navy's resources, the High Command turned to the German merchant marine as a force multiplier, making elaborate plans to commandeer and outfit civilian ocean liners (which were built for speed) as commerce raiding cruisers, complete with instructions to carry disassembled guns in peacetime in readiness for wartime operations. Throughout this period, the Admiralty Staff, which had replaced the High Command, feuded continuously with Tirpitz's office about the devotion of scarce resources to these projects, even as they also pressured the Imperial Naval Office to procure submarines for more standard commerce raiding missions.[53]

The second problem was simple bureaucratic feuding. Although the kaiser's dissolution of the High Command had left its replacement, the Admiralty Staff, in a relatively neutered bureaucratic position, it was still in a position to obstruct Tirpitz and was, at the least, completely uncoordinated with him. As of 1905, the Admiralty Staff had no plans for war with Britain (with or without French involvement), despite the Algeçiras crisis forcing the issue. Throughout the latter half of the decade, the Admiralty Staff kept Tirpitz at a distance in its war planning. The Admiralty Staff in particular waged a bureaucratically wasteful rearguard action against Tirpitz's building program, which because of resource constraints had to sacrifice certain technical advantages for the sake of fiscal solvency; the Admiralty Staff repeatedly requested resources the state simply did not have.[54] Indeed, in Hew Strachan's apt phrasing, with "only brief exceptions, relations between the two departments [Tirpitz's office and the Admiralty Staff], in peace and war, verged on the fratricidal." Tirpitz won most of his bureaucratic battles with the Admiralty Staff, but this came at the cost of the latter's effectiveness: during the course of the decade, Tirpitz succeeded in seeding the Admiralty Staff with his own loyalists whose abilities as naval officers were second to their loyalty to Tirpitz and their ability to imitate the latter's capacity for back-biting. But over time, this was not sufficient to prevent Tirpitz's ultimate failure to carry his building program to a successful conclusion, and lobbying from the Admiralty Staff played a role in Tirpitz's gradual sidelining immediately prior to the outbreak of World War I and subsequent forced retirement in 1916.[55]

For all that, Tirpitz had his legal authorization for new ships. The new 1908 Naval Law ultimately allowed for a building tempo of four ships per year.[56] Fisher, meanwhile, was under the same pressure to cut costs as before: based on the estimates as they stood in 1906, the Royal Navy would be adding only three new *Dreadnought*-class warships.[57] This left the Royal Navy for a time still in the lead—insofar as *Dreadnought* represented a unique development in naval technology that only Britain then possessed, and that it would take several years for Tirpitz to catch up—but the race was headed for its most competitive phase.

Tirpitz also lobbied successfully for the widening of the Kiel Canal, which connected the North and Baltic Seas via the Jutland peninsula, to admit the new dreadnoughts, which would otherwise be useless. The existing canal imposed a practical upper limit to the size and speed of ships, which would doom any attempt to compete with the Royal Navy unless the canal were modified.[58] Fisher forecast that the canal would be fit for the passage of the big ships by October 1914 and predicted that war would follow immediately after its completion, likely on a bank holiday. He was, in the event, right in all the particulars except for the month: the canal was completed early in June 1914, with war following in August.[59]

Although *Dreadnought* remains Fisher's best-known naval innovation, Fisher was perhaps even more dedicated to another ship class authorized at the same time: HMS *Invincible* and a series of class ships. *Invincible* would carry eight 12-inch guns and, with less armor than *Dreadnought*, could steam at 26 knots or even 28 knots for short distances. The *Invincible*-class ships were true battlecruisers of the type Fisher preferred to build, relying on speed and long-range gunnery to the practical exclusion of armor. There is an apocryphal story that, knowing his adversary, Fisher carefully referred in public statements to *Invincible* having 9.5-inch guns; when Tirpitz laid the keels for a new class of German heavy cruiser that would imitate the *Invincible*, he relied on this information and built them with similar weaponry. The new German cruisers were therefore obsolete from the start, since the ships they were designed to counter outgunned them. It appears, however, that Tirpitz had the correct information from the start and simply was unable initially to duplicate *Invincible*'s armament. Whatever the case, Fisher's new ship design effectively put Tirpitz's building program back on its heels once again.[60]

Consensus Forms: British Democracy Does the Right Thing after Trying Everything Else[61]

While Britain had gained an edge in military technology, diplomatically, the British government was already losing its customary cohesion, due again to an inability to handle human resources. While Campbell-Bannerman sought to cut costs, his energetic but introverted foreign secretary, Sir Edward Grey, with whom he never developed a congenial working relationship, was beginning a long correspondence with his French counterparts regarding plans for a hypothetical outbreak of hostilities between France and Germany. France repeatedly sought assurances that Britain, in the spirit if not the exact letter of the Entente, would come to its aid in the event of a German invasion; moreover, it wanted troops, not merely naval assistance. Although Grey did not formally commit his government, he allowed himself and his office—and ultimately the British Army—to be drawn in to a series of discussions (now simply known as the "military conversations") of a hypothetical British commitment of an expeditionary force in the event of a war, along with a parallel series of naval discussions that assigned the French fleet responsibility for the Mediterranean and in so doing allowed Britain to concentrate forces in the North Sea.[62]

Grey may not have formally agreed, but he did not disagree, either, and the fact that the discussions were ongoing, with Grey's blessing and encouragement, amounted to what Henry Kissinger refers to as a "moral obligation." The British Foreign Office, by habit, was now in the process of encouraging France to believe that Britain would send an expeditionary force, an encouragement that was diplomatically as good as a promise. The military conversations' practical effect was a slow but noticeable sclerosis of British policy options. In particular, they made and encouraged more specific assumptions about Britain's commitment to the integrity of Belgium, now universally understood to be the likely front line in the event of a Franco-German war. Under the Salisbury government, contrary to the widespread impression of an ironclad British guarantee of Belgian territorial integrity under Britain's treaty obligations, there had been considerable variation of opinion. Although British policy makers were in broad agreement that Britain had some sort of obligation to Belgium and that there were reasons for this, opinion ran the gamut from a mere naval show of force (favored by military intelligence) to the actual dispatch of an expeditionary force to the continent (favored by Lord Lansdowne's Foreign Office)—although even here the details were murky and only a small force was envisioned—to a general lack of interest in specifics (professed by Salisbury, who suggested that policy would guide treaty interpretation rather than the other way around). All of this would change under Grey, insofar as the military conversations led, beginning as early as 1905, to serious discussions about the inevitable dispatch of several British Army divisions to Belgium after the outbreak of hostilities.[63]

Exactly how much this influenced British policy—and, critically, its ultimate entry into World War I—is debatable. By 1911, military-to-military dialogue endorsed by the Foreign Office between the British Army and its French counterpart had led to detailed plans for the deployment and logistical support of a British expeditionary force in France if war broke out. With this kind of hands-on engagement, it became more difficult for Britain to opt out of reinforcing France on land in the event of war. However, the process was chaotic and its impact is disputable. The British Army would not even coordinate operational planning with the Royal Navy. When, at a famous meeting of the Committee of Imperial Defence in 1911 in response to the Agadir crisis, the British Army and the Royal Navy presented conflicting war plans to the government, the army's recommendation was adopted over the navy's. However, as various historians have noted, it is more plausible to state that Anglo-French naval talks, which created soft commitments for Britain to defend France while the latter's fleet was deployed to the Mediterranean, had more to do with Britain's "moral obligation" than the land forces' coordination, which does not seem to have entered into the debate in Britain during the July Crisis of 1914 regarding whether to intervene. As Hew Strachan and others have noted, although the existence of the military conversations influenced some British decision makers, the question of whether and how to intervene was left open up to the very end of the July Crisis. It is perhaps permissible to conclude that the need to coordinate with a de facto partner had something to do with Britain's entry into World War I, but it is difficult to speak more definitely than this.[64] This study claims no more than this. What can be stated definitively, however, is that from Germany's perspective, the staff talks led to closer cooperation between Britain and France and a corresponding dearth of diplomatic options for dealing with Britain.

    The staff talks did, however, contribute to the British government's chaos. Astoundingly, once Asquith succeeded Campbell-Bannerman as prime minister in 1908, Grey did not immediately perceive a need to inform him of the exact nature of these discussions, only belatedly briefing Asquith on his activities in this area in 1911.[65] The British government was beginning to factionalize over the issue. Perhaps counterintuitively, Grey and the Foreign Office stood on the more hawkish end of the spectrum. Effectively at Grey's side during these deliberations was an influential policy aide, Sir Eyre Crowe, who performed for the Dreadnought Race a similar function to that performed by George F. Kennan for the U.S. State Department at the start of the Cold War.[66] In a long memorandum (the similarities to the "Long Telegram" of Cold War fame may again be noted) to Grey dated 1 January 1907, Crowe summed up the Anglo-German relationship:

> The general character of England's foreign policy is determined by the immutable conditions of her geographical situation on the ocean flank of Europe as an island State with vast oversea colonies and dependencies, whose existence and survival as an independent community are inseparably bound up with the possession of preponderant sea power. . . . The colonies and foreign possessions of England more especially were seen to give to that country a recognized and enviable status in a world where the name of Germany, if mentioned at all, excited no particular interest. . . . Here was a distinct inequality, with a heavy bias in favour of the maritime and colonizing Powers. Such a state of things was not welcome to German patriotic pride. . . . [Over] and beyond the European Great Powers there seemed to stand the "World Powers." It was at once clear that Germany must become a "World Power." . . . A few fresh possessions were added by purchase or by international agreement. . . . On the whole, however, the "Colonies" have proved assets of somewhat doubtful value. Meanwhile the dream of a Colonial Empire had taken deep hold on the German imagination. . . . If . . . Germany believes that greater relative preponderance of material power, wider extent of territory, inviolable frontiers, and supremacy at sea are the neces-

sary and preliminary possessions without which any aspirations to such leadership must end in failure, then England must expect that Germany will surely seek . . . ultimately to break up and supplant the British Empire.[67]

Crowe's memorandum, which Grey approved, was hardly concise—it ran to more than 40 pages of florid prose—but it accurately reflected the essentially nonmaterial nature of Germany's competitive goals.[68]

On the other side of the debate lay the prevailing opinion of the British government. Although Campbell-Bannerman was astute enough to recognize the need to take precautions, the radical faction within the liberals headed by David Lloyd George, who became chancellor of the exchequer when Campbell-Bannerman died and Asquith became prime minister in 1908, wanted little to do with the national defense.[69] Lloyd George possessed wide-ranging ambitions for poverty relief and social reform in an already reformist government that passed the Old Age Pensions Act in 1908 to be followed at the beginning of the next decade by the National Insurance Act, and for this reason regarded defense spending as a political distraction and a fiscal waste.[70] He was joined in these objections by an ardent and influential admirer, namely the young Winston Churchill, at the time a mere member of British Parliament, whose own opinion was that there simply were not sufficient points of contention between Britain and Germany to be worth a fight—in effect, that the threat was minimal because the requisite harmful intentions could not possibly be there.[71] Churchill, at the time a member of the liberal coalition, would become president of the Board of Trade in 1908 before becoming First Lord of the Admiralty (the civilian overseer of the First Sea Lord) in 1911. By that time, he had completely reversed his views.[72]

While the dispute was underway, the dynamic Grey was already seeking to augment Britain's political position. Negotiations begun in 1907 with Russia concluded in 1908 with the signing of the Anglo-Russian Convention, which, combined with the Entente Cordiale with France and the preexisting Franco-Russian alliance, formed the Triple Entente by which Britain would ultimately enter World War I on the side of France and Russia. As with the Entente Cordiale, the Anglo-Russian Entente was formally a piece of colonial housekeeping—it settled British and Russian spheres in Afghanistan, Iran, and elsewhere as a means of building goodwill. Formally, it was not meant to be interpreted as a binding military alliance (which Britain would not enter); likewise, the Triple Entente, taken together, was not intended as an agreement to act in concert on all issues. As with the Entente Cordiale, however, the Anglo-Russian agreements cleared up issues that might previously have hindered cooperation in order for Britain and Russia to cooperate.[73]

The British internal dispute came to a head during the period from late 1908 to early 1909, when the Royal Navy's intelligence sources and the British naval attaché in Germany acquired a much more disturbing picture of German fleet-building, which suggested the possibility of near-parity in fleet size by 1914, with 16 British dreadnoughts arrayed against 13 German ones. There was also intelligence, which was initially close-hold information even within the German government, that Tirpitz was accelerating the contracting process for the ships slated for construction in 1909, a technical violation of the Naval Laws. Britain was not moving quickly enough. The British naval estimates for the years from 1905 (when the keel of *Dreadnought* had been laid) through 1908 had initially budgeted for four new ships per year, for a total of 16 new hulls, but this had been cut to 12 as of 1906, with only two ships slated for construction in 1908. Shocked at the latest naval intelligence, the Admiralty, led by Fisher, demanded that the estimates be revised to allow for the construction of six ships. The Lloyd George faction of the government, which included Churchill and commanded considerable support from the back benches, agitated for only four ships.[74]

Chapter Five

The disputes reached the press, in which a predictable storm of name-calling broke out between liberal and conservative publications. With the debate having gone public, Asquith succeeded in brokering a compromise. The radical faction would get its limit of four ships, but a contingency authorization of four additional hulls would be given subject to confirmation of the latest intelligence regarding German plans and intentions, which it was understood would be forthcoming. Churchill summed up the compromise wittily: "The economists wanted four, the Admiralty wanted six, and we compromised on eight."[75]

The so-called "Naval Scare" had a number of longer-term effects. It forced the liberal government, with Lloyd George now serving as chancellor of the exchequer, to levy new taxes to fund both its social programs and the new ships, which in turn provoked a legislative showdown with the reactionary House of Lords. It also led to a flurry of diplomatic exchanges between the Foreign Office and Germany, with German chancellor Bernhard von Bülow for the first time floating the idea of a negotiated settlement, which would also contain the Imperial German Navy's growing fiscal burden. This in turn led to Tirpitz concealing his intentions regarding the aforementioned accelerated contracting process even from Bülow, so as to enable the latter to answer honestly that he had no information regarding the issue to his British interlocutors. Most importantly, however, it led to a sea of change in Britain's understanding of the German naval challenge, building a consensus around the concept of Germany as Britain's chief competitor.[76]

In effect, as several historians have noted, the Naval Scare, which awakened the British government and public to the true balance of naval forces near British home waters, amounted to the end of the Two-Power Standard and the beginning of a "one-power standard" directed at Germany alone. Although in concrete terms it merely led to an accelerated naval acquisition process, psychologically the Naval Scare put the government, the Admiralty, and the Foreign Office on the same political page for the first time and led to a focus of effort directed at

Germany to the exclusion of other concerns.[77] By 1910, the British government had unofficially adopted a new standard for naval dominance: the Royal Navy would be maintained at 60 percent greater strength than the German High Seas Fleet, and as Germany was expected to have 24 dreadnoughts in short order, Britain would have 38.[78] The next year, 1910, saw authorization given for five additional new dreadnoughts with an option for four more, one more ship even than the previous year's "four and four" compromise.[79]

The practical effect of the new naval policy was the creation of prototypical "superdreadnoughts," the new *Orion*-class battleships whose keels were laid beginning in 1909. The *Orion*-class ships had 10 13.5-inch guns—an improvement over *Dreadnought*'s 10 12-inch guns—and for the first time deployed all their main armament along the centerline of the ship (*Dreadnought* had located 4 of its 10 guns broadside), which enabled a wider field of fire for each gun. Ironically, this innovation was a copy of an American design: the U.S. Navy's first dreadnought, USS *South Carolina* (BB 26), was the first ship to have this feature. This innovation yet again put the German building program behind the technological curve.[80]

All this notwithstanding, it took another two years and another crisis to finally remove all doubt from British policy makers' minds—notably Churchill and Lloyd George's—and to change their policy stances about the need to counter German naval dominance. It was not until 1911 that a crisis in France's colonial territory brought about a complete consensus on the German threat.

German Actions: Bureaucratic Infighting, Lost Consensus, and Premature Aggressiveness

In Germany, meanwhile, the resolve to build the High Seas Fleet was being sustained almost exclusively by Tirpitz and Kaiser Wilhelm. Within the kaiser's immediate circle, there was alarm—notably expressed by Chancellor Bülow—over the possibility of Britain launching a preventive war to stop Germany's shipbuilding program. Legislatively, meanwhile, support was collapsing, as the financial strain imposed by the shipbuilding program forced legislators in favor of it to contemplate a shift towards funding it with direct taxation on wealth, a program sure to be bitterly opposed. A domestic consensus in favor of the naval project had always been elusive. Big business—at least those parts of it that could profit from shipbuilding or use it as a lobbying tool for subsidies—was the mainstay of domestic support for the fleet. Arrayed against it were the German Social Democrats, who saw it as the route to war and a tool of domestic repression, as well as, oddly enough, the traditional Prussian Junker nobility and their allies in the other states. For this latter group, the High Seas Fleet was an unwelcome innovation and an aberration from the familiar and traditional policy of strength on land. Not only did it threaten the traditional Prussian way of life, the fleet also threatened the Junkers' finances and land holdings. As of 1909, desperate to find money to finance the fleet at the tempo of shipbuilding that Tirpitz wanted, Wilhelm had to propose an inheritance tax, the unpopularity of which with his traditional political support base can well be imagined. The tax was defeated in the Reichstag. Wilhelm used the defeat as an excuse to dismiss Bülow, with whom he was gradually growing weary. Bülow accelerated his own end by his failure to adequately manage the increasingly erratic kaiser's public image, though he was afforded the rare privilege of nominating his successor.[81]

The chancellor who replaced Bülow, Theobald von Bethmann Hollweg, was a dutiful civil servant with minimal ambitions and (unusually for a Prussian high official) a modest ego. He had little interest in war except, perhaps, for its value for *Sammlungspolitik*, and he favored negotiation over the arms race.[82] In 1909, therefore, he began a series of attempts to reach an arms limitation agreement with Britain, which continued for two years. Bethmann Hollweg called on his British interlocutors for an unspecified mutual drawdown in naval construction combined with a neutrality agreement that would reduce the significance of the Entente Cordiale.[83]

During the next two years, Bethmann Hollweg was stymied at every turn. Wilhelm backed him in this endeavor at best unenthusiastically. Tirpitz, who had Wilhelm's ear and was on far better terms with him than was Bethmann Hollweg, argued against any agreement that would make crossing the (ever-widening) "danger zone" more difficult, refused to suspend his building program, and demanded tangible British concessions before considering any change of course. Strangely, the German Social Democrats, and particularly their leader, August Bebel, made overtures to Britain to argue for a hawkish British naval policy. They feared that if Britain showed weakness or otherwise slackened the pace of its naval buildup, the kaiser might succeed in leading Germany to war, which would crush the Social Democrats' political position by ruining their reputation and provide further political justification to their opponents. Ironically, they were in exactly the opposite political position from the left wing in British politics, which, as noted, had every reason to oppose British navalism in all its forms. Bebel corresponded with the British Foreign Office on this issue, repeatedly urging Britain to continue its arms buildup with sufficient aggression as to deter Wilhelm from pursuing the race further. And where formal channels were concerned, Bethmann Hollweg's initiative was dead on arrival. Grey, as always, preferred a hard line, arguing that Britain would be getting nothing substantive in return by agreeing to a nebulous slowdown in construction and a neutrality agreement. This time, the government was with him. Bethmann Hollweg, undermined at home and refused abroad, reached no agreement.[84]

In 1911, the Agadir Incident served to eliminate the British government's last political reservations about Germany's intentions. After rebellion broke out in French Morocco, the thuggish new German foreign minister, Alfred von Kiderlen-Wächter, persuaded Wilhelm, over Bethmann Hollweg's head, to send a fleet to Morocco to threaten intervention. Wilhelm, without informing Tirpitz, sent the small gunboat SMS *Panther* (1901). The move was a pathetic bluff, in the absence of any serious German war plans for such a situation, but it was enough to raise the hackles of the Asquith government. Britain diplomatically backed France; France, in turn, bribed Kiderlen-Wächter to back down by offering a tract of politically and financially near-worthless land in the French Congo to Germany in exchange for ceasing its military intervention in Morocco. Kiderlen-Wächter, satisfied that he had gotten the better end of the deal, had *Panther* sent home.[85]

British Reaction: New Leadership, Strategic Consensus, and Rapid Arms Buildup

The so-called *Panthersprung* (panther-leap, in reference to the *Panther*'s long-distance deployment) further cemented British policy toward Germany, particularly in its effect on the last two major political holdouts in the British government, Lloyd George and Churchill. During the crisis, at some political risk to himself, Lloyd George made a carefully worded (and Foreign Office-approved) statement during a dinner speech before a financial audience that was being copied in shorthand by journalists that subtly reminded anyone who might hear that Britain did not back down in the face of threats. Grey privately praised Lloyd George's political courage. The speech finally cemented Lloyd George's public support for a harder line against Germany. Churchill, meanwhile, after meditating briefly on the meaning of it all, decided that he had been wrong in his earlier skepticism and that (in his customary style) courage and resolution were now called for. Meanwhile, after the aforementioned fracas in the meeting of the Committee of Imperial Defence in which the Navy had been humiliated by the Army, Asquith, perhaps sensing an opportunity, appointed Churchill, who at the moment was Home Secretary, to a new position: First Lord of the Admiralty.[86]

By this time, Fisher had retired; the old sea dog submitted his resignation in 1910. A combination of factors had influenced the decision. Fisher wanted his protégé, Admiral Sir Arthur K. Wilson, to succeed him as First Sea Lord and did not want to obstruct the younger man's professional advancement; he was also finally tired of feuding with Beresford about the latter's attempts to undermine his reforms to advance his own career. Not least of all, Fisher, at 69, was simply growing too old for the job. He accepted a barony from King Edward VII in gratitude for his services and stepped down.[87] Although Wilson was adequate to the task of carrying on the Royal Navy's reforms, he was not a dynamic leader, and he was responsible for having humiliated his service in the aforementioned war council. Wilson was sidelined by Churchill about bureaucratic disputes; his undistinguished successor, Admiral Sir Francis Bridgeman, fared little better; and Bridgeman's successor, Prince Louis of Battenberg, another Fisher disciple, was similarly marginalized. From this point on it would be the navy's civilian leadership—in the form of Churchill, unofficially advised by Fisher—that drove policy.[88] Edward VII was likewise not long for the world, and he died in May 1910, with his son George V assuming the throne.[89] At the royal, civilian, and military levels of decision, the guard had changed.

Churchill at the Royal Navy's Helm

With Fisher advising him from retirement, Churchill successfully pushed for the construction of new classes of superdreadnought battleships. The first two classes, the *George V*-class and *Iron Duke*-class, deployed 10 13.5-inch guns (all at centerline) and had a maximum speed of 21 knots (as with the previous classes). However, the latest class, the *Queen Elizabeth*-class, which were finally commissioned in 1915, bore eight enormous 15-inch guns, all at centerline, and had a maximum speed of 24 knots, 3 knots faster than *Dreadnought*. They also introduced a new innovation with enormous geopolitical implications: oil-fired boilers, a new design feature made necessary by the added weight required to stabilize the ship while firing its gigantic armament. Churchill approached the last class with his

Chapter Five

customary energy and an almost chastened sense of urgency, pushing through the early construction phases without sufficient testing or construction of a prototype, on the basis that there was no time to waste. Fisher was sufficiently enthusiastic (as ever) about the new design as to earn the title of "oil maniac" in the media. In addition to solving some of the weight problem, oil-fired boilers were more energy-efficient relative to space constraints. An equivalent amount of energy could be carried in a much smaller space, with correspondingly less weight, with oil-fired engines than with coal-fired ones. Once appropriate measures had been taken to secure a fuel source for them (the Persian Gulf), oil-powered vessels were also easier to refuel and could be refueled at sea, which would theoretically give the ships a greater combat radius.[90] The new ships were therefore qualitatively superior both to anything Tirpitz had built and to their Royal Navy predecessors. They are, in fact, regarded as rendering the previous classes obsolescent in the quantitative analysis that follows in the next chapter.[91] Yet again, Tirpitz was behind the curve.

Tirpitz's Last Chance

During this time, Tirpitz was hard at work devising a new novelle that would authorize still more capital ship construction. His building tempo had been reduced to a mere two ships per year as a result of the inability of the German government to find financing (notwithstanding the broad authorization the law gave him to build new ships up to the maximum 58 allowed). As of 1912, Tirpitz was hoping that Britain, which relied on well-paid, long-service volunteer crews (unlike Germany's conscript crew) for the Royal Navy (ironically, in stark contrast to the impressment system resorted to during the Royal Navy's more famous era a century before), would be unable to staff its ships even if it continued building beyond a certain point, which Tirpitz thought Britain would reach around 1918 or 1919. Bethmann Hollweg, in turn, was still hoping, albeit in the face of contrary evidence, that Britain might be amenable to some type of arms limitation agreement and political coexistence. In fact, the salient political issues of the previous decade were largely irrelevant at this point. Edward Grey noted publicly that if maintaining peace were merely a matter of accommodating German colonial ambitions, it might be possible to make room for German colonization in Africa, but that the issue was now one of survival for Britain, and that all other issues were irrelevant at this point. Tirpitz achieved passage of the novelle (the fifth, and final, Naval Law) in 1912, which authorized three more capital ships, bringing the final total authorized capital ship count to 61, which in turn, by Grey's own declaration, rendered arms control talks pointless. Ironically, to pass the novelle and create the necessary parliamentary and official consensus, Tirpitz had to acquiesce in the removal of a battlecruiser from the fleet bill and a reduction in building tempo to three ships in alternating years and two otherwise, which pushed his goal further out of reach.[92]

Churchill decided to build faster than Tirpitz. Tirpitz was now building at an alternating rate of three ships one year and two ships the following year; Churchill proposed to build at a 4:3 and then 5:4 alternating annual rate (thereafter to be sustained at a continuous 4 per year), a pace of production that, although fiscally onerous, the British government agreed to with comparative ease, now that opposition to naval rearmament amounted to a fringe position. Lloyd George and his faction offered ineffective opposition. From 1912 onward, Churchill officially sanctioned a 60-percent superiority standard above Germany alone, prioritized competing with German strength in the North Sea over other theaters, and publicly announced his intention to build two ships for every one that Tirpitz produced. In the political arena, beginning in 1912, Churchill, along with the British war secretary Richard Burdon Haldane, began a diplomatic initiative to propose a temporary shipbuilding freeze that (conveniently) would only apply to dockyards in Britain, allowing for construction elsewhere in the dominions. The proposal failed—although success, as opposed to sowing discord among German decision makers, may not have been the objective. Substantively, the dominions proposal also failed, as Canada, the principal member expected to contribute, declined to do so. This same year, Bethmann Hollweg renewed his attempts to obtain a mutual freeze in the naval arms race combined with a British guarantee of neutrality in any continental conflict, a proposal that was as good as dead on arrival in view of Britain's increasing commitment to the Entente Cordiale and the military conversations. As noted, Grey peremptorily rejected the German proposal.[93]

Although the question of what might have happened had Grey been willing to consider the proposal provides an interesting counterfactual, in the end, the neutrality agreement sought by Germany—essentially, an agreement by Britain to act favorably toward German interests—was not materially different from the Entente Britain already had with France and Russia, insofar as it guaranteed nothing on paper but imposed soft obligations. Given all this, it was perhaps inevitable that the search for such an agreement died as readily as the German proposal for an alliance, with Britain as a junior partner, did a decade previously. In sum, Germany was, rather ham-fistedly, seeking to legitimize by agreement Britain's acquiescence in Germany's new world power status, and in the absence either of any provision to sweeten the pill for Britain or a decisive naval advantage (the danger zone still not having been fully crossed), such an agreement was neither in Britain's interest nor likely to be seen as such. This was as close as Germany ever came to a settlement with Britain—creating a "finish line"—that would alter the political status quo in its favor and in line with its objectives. It failed utterly.

In broad terms, the political tides were moving in an altogether different direction. Beginning in 1912, German policy makers, influenced by leading opinion makers in the German press (notably the editor Walther Rathenau), deemphasized *Weltpolitik* as the major focus of German foreign policy. In its place, beginning with Rathenau's suggestion to Wilhelm and

Bethmann Hollweg, Germany would pursue a policy of tighter European trade ties, to culminate in a customs agreement, to be dubbed *Mitteleuropa* (middle Europe). This, as Hew Strachan has noted, would accord more reasonably with Germany's trade, which was becoming unbalanced, and with its allies Austria-Hungary and Italy, which stood little to gain from Germany's naval policy. As Strachan notes, this was delaying the inevitable: a mere focus on trade and a corresponding lack of emphasis on naval power and colonies would not resolve the more fundamental structural problems Germany posed for the great power system or suppress German nationalist ambitions forever. It was, however, an admission made official by Bethmann Hollweg in a speech to the Reichstag that Germany had to put first things first and that it could not devote unlimited resources to its navy when there was work to do closer to home. In so doing, it had effectively decided to fold its hand, even if Tirpitz and Wilhelm would continue to try in vain to continue the naval arms race for a while longer. As will be discussed, this would require more funding than they could obtain.[94]

Even as he made these overtures, Churchill continued the base consolidation begun under Fisher, transferring the bulk of the Mediterranean Fleet to the Channel Fleet so as to keep scarce naval resources close to home. The move was seen both in Britain and elsewhere as a final indignity and an indicator of how desperate the situation had become. In doing so, he employed a mix of creativity, pragmatism, and conservatism. Recognizing the promise of submarines—a point on which Fisher had come to hold some rather avant-garde views, even if he did not always follow them up—Churchill sought to speed up the development of long-range fast submarines that could, at least theoretically, take on capital ships on the high seas. He also made plans to base submarines in the Mediterranean as part of a "flotilla defence" strategy that would employ the smaller craft in the event of war against threats to strategic chokepoints, while in turn freeing up battleships for service closer to home. In this, however, Churchill was more conservative. Although both he and Fisher may at times have understood that the battleship's status as a capital ship might well be transient, Churchill confronted the basic fact that, for the moment at least, battleships remained the dominant high seas vessel and acted accordingly. He proceeded to do exactly what Tirpitz had gambled would be too difficult for the Royal Navy to accomplish all at once, moving the majority of the Royal Navy's capital ships to the Home Fleet to defend North Sea waters, in effect staring down Tirpitz' High Seas Fleet. The move completed the work begun by Fisher of moving the Royal Navy's capital ships close to home, an obvious blow to British prestige that created the appearance of a siege mentality but also ensured that the German High Seas Fleet would be in no position to take offensive action.[95]

With neither more funding nor more time, however, Churchill was forced to make the best of the situation. He therefore doggedly resolved to continue ship construction. By this point, the maintenance costs for the ships and crews in both the German Imperial and British Royal Navies were nearing the breaking point, as the sizes of the two fleets were approaching ludicrous levels with no end in sight. Tirpitz believed that he could force an end to the race once Britain became unable to crew its new ships, which cost more to crew than their German counterparts because of the latter's use of conscription. In the event, no satisfactory method of either forcing a political resolution or winning a great sea battle presented itself, and Tirpitz was forced to face the fact that the German Army always took the first cut of defense funding. The Imperial German Navy was nearing the upper limit of its available funds.[96]

By 1914, Wilhelm wanted yet another Naval Law or novelle to increase fleet size still further. Tirpitz demurred: because of Germany's financial straits and opposition in the Reichstag, further authorizations for shipbuilding simply could not be obtained. Tirpitz believed that a new law might be possible in the indefinite future, as Germany's economy was growing faster than Britain's and would eventually allow for more construction. As of 1914, a de facto shipbuilding pause was fast approaching. The fact that even the latest ships in the High Seas Fleet at this point employed design concepts that the Royal Navy's ship designers had discarded several years previously (such as broadside-mounted turrets and coal-fired engines) merely reinforced the sad fact that Tirpitz and his fleet could not catch up.[97] During the two-year period from 1912 to 1914, Germany and Britain avoided war over the Balkans and appeared to have reached something resembling détente, with Grey even acting as a peace broker to prevent escalation of the Balkan wars.[98] In this sense, with the status quo having been expensively maintained, Britain had achieved something like a victory in the Dreadnought Race.

In June 1914, the newly widened Kiel Canal, whose completion Fisher had predicted would presage the outbreak of war, opened. That same month, Archduke Franz Ferdinand was assassinated, and 39 days later Britain, France, and Russia were at war with Germany.[99]

The War that Followed

The two sides' naval operations throughout the long war that followed were famously inconclusive. As might well have been predicted, the very first step the Royal Navy took was to blockade German ports, effectively tying down the incongruously named High Seas Fleet. It avoided a close blockade, however, which would have allowed for a decisive encounter with the High Seas Fleet, in favor of a distant blockade enforced by a series of patrol lines and quick-reaction forces.[100]

Ironically, both Fisher and Churchill saw their reputations tarnished by an only tangentially related matter, the abortive and disastrous plan to seize control of the Dardanelles prior to the Gallipoli offensive, after the Ottoman Empire joined Germany's coalition. Fisher had been brought out of retirement to serve again as First Sea Lord at the onset of war. He resigned in protest against Churchill's handling of the Gallipoli operations, ending what had been a very friendly working relationship between the two men. Churchill was ultimately forced to step down as First Lord of the Admiralty as a result of the same debacle.[101]

Tirpitz, meanwhile, had gradually lost influence before the war due to his ongoing feuds with an increasing number of fellow officers and high officials, including Bethmann-Hollweg, who was horrified by Tirpitz's support for what would become Germany's unrestricted submarine warfare program. Having less of a role to play now that success depended on competence in combat and not merely in force design, and lacking his former bureaucratic influence, Tirpitz ultimately had to resign his position in March 1916. Though he would become a player in various German political factions during and after the war, his official contribution to German wartime decision making was effectively over.[102]

The High Seas Fleet, or elements of it, made a series of attempts to break the stalemate, none of which succeeded. In 1915, in the Battle of Dogger Bank, the aggressive Royal Navy admiral David Beatty succeeded in intercepting a squadron of the High Seas Fleet attempting to run the blockade and chasing it back to harbor. One German ship, the cruiser SMS *Blücher* (1908), went down, while Beatty's flagship, HMS *Lion* (1910), a fast battlecruiser, sustained serious damage. Beatty's success demonstrated the value of the high-speed ships that Fisher had lobbied so successfully to construct, but it was marred by his squadron's inability to finish the job.[103]

The next year, at the Battle of Jutland, the High Seas Fleet made a break for the North Sea and was confronted by the Royal Navy's Battle Cruiser Fleet, under Beatty, and the Grand Fleet, composed of heavy battleships and led by one of Fisher's favorite officers, Admiral John Rushworth Jellicoe. Ironically, HMS *Dreadnought* was not present, instead undergoing a refit in drydock. A series of deadly maneuvers followed, but despite the Royal Navy's squadrons' success in "crossing the T" of the German formations on two occasions—cutting across their line of movement in perpendicular fashion at close range, so as to make it near-impossible to miss—only one German capital ship was sunk. Beatty's battlecruisers, however, sustained heavy punishment. Not intended for use at close range in the tight confines of a major battle, having traded armor for speed per Fisher's earlier designs, and being steamed directly at the enemy as if they were heavy battleships, the battlecruisers suffered heavily: three were hit and exploded, one of them Fisher's favorite, HMS *Invincible*. Depending on one's point of view, the fast battlecruisers' performance can be seen as a vindication of Fisher (because they performed as expected and were being used incorrectly) or a condemnation of his whole plan (because, in the end, it was deemed necessary to put the comparatively light, fast ships into the line with the heavy battleships, whose design was a compromise with Fisher's ideal design). Churchill was furious with Jellicoe for failing to follow up his success, noting ruefully that he "was the only man who could lose the war in an afternoon." Although Jellicoe was hardly an aggressive commander, his larger strategic instincts were arguably right. Britain had little to gain and more to risk by taking offensive action against the High Seas Fleet that, as long as it could not leave the blockaded waters of the North Sea, was no threat to anyone.[104] Even Churchill admitted that, as the High Seas Fleet had failed to accomplish anything, British naval strength had been reaffirmed.[105]

A few other, smaller actions followed, all of them having the same result: the High Seas Fleet stayed where it was. In 1917, as the German economy began to fail and Marxism-Leninism was taking hold in Russia, the High Seas Fleet's enlisted personnel mutinied while the ships were in dock. At war's end, as part of the peace terms, the remainder of the High Seas Fleet was finally "Copenhagened" as Fisher had demanded so long earlier, being escorted under guard to Scapa Flow in Scotland and there scuttled and sunk.[106]

As for the military technology that had driven the entire competition: precisely because of the dearth of actual combat between battleship and battlecruiser fleets after Jutland, it is difficult to determine who was right on the speed versus armor question. It is enough perhaps to note that, as is the larger lesson of the entire affair, the relative merits of any strategic choices depend on the degree to which they fit into the broader plan. Used correctly, either Fisher's compromise battleships, of which *Dreadnought* was the original one, or his ideal fast battlecruisers, such as the *Invincible*-class, might have served well, but right up to Jutland training and operational doctrine for the new ships had never been satisfactorily agreed on, nor was there much experience before (or after) to draw on.[107] Although it is tempting to argue that the entire battleship-battlecruiser dispute, which involved debate about everything from ship design to tactical doctrine, made a mess of the performance of both categories of ships in the one great action in which they were involved, as recent research by Nicholas A. Lambert places the blame for the loss of the battlecruisers at Jutland on ammunition storage rather than either design or doctrine.[108] The most that can be said with certainty is that the Royal Navy muddled through.

As for Wilhelm and Tirpitz, despite the turn the war ultimately took and the High Seas Fleet's ineffectiveness and sad fate, neither expressed serious remorse for the futility of their decisions.[109]

Endnotes

1. Rudyard Kipling, "Hymn Before Action," in *The Seven Seas* (London: Methuen, 1896), 103.
2. Robert K. Massie, *Dreadnought: Britain, Germany, and the Coming of the Great War* (London: Vintage Books, 2004), 391–400; Peter Padfield, *The Great Naval Race: The Anglo-German Naval Rivalry, 1900–1914* (Edinburgh, Scotland: Birlinn, 2005), 47–51, 94–95, 143–46; Jon Tetsuro Sumida, *In Defence of Naval Supremacy: Finance, Technology, and British Naval Policy, 1889–1914* (London: Unwin Hyman, 1989), 9; Arthur J. Marder, *The Anatomy of British Sea Power: A History of British Naval Policy in the Pre-Dreadnought Era, 1880–1905* (Hamden, CT: Archon Books, 1940), 519; and Arthur J. Marder, *From the Dreadnought to Scapa Flow*, vol. 1, *The Road to War, 1904–1914* (Oxford, UK: Oxford University Press, 1961), 10. This view of the Royal Navy's culture has been called into question more recently, with revisionists noting that it was not as anti-intellectual as it has been made out to be. See Christopher M. Bell, "Contested Waters: The Royal Navy in the Fisher Era," *War in History* 23, no. 1 (January 2016): 5, https://doi.org/10.1177/0968344515595330. In fact, it is hard to square this impression with the Royal Navy's considerable attention to technological development during this same period—in particular, its quiet attention to submarine warfare, while maintaining a façade of disinterest to detract from their popularity in open sources. For more on this, see Nicholas A. Lambert, *Sir John Fisher's Naval Revolution* (Columbia: University of South Carolina Press, 1999), 39–41. But anecdotes of a lackadaisical attitude toward combat readiness, at least among officers serving aboard capital ships, do abound, as the above-cited sources suggest, as does the simple reality that HMS *Dreadnought* (1906) was, indeed, a major technological innovation compared to what went before it. It is probably accurate to say that, while the Royal Navy most certainly did contain gifted people who as individuals possessed demonstrated intellectual and innovative potential, the institution does not seem to have shown much interest either in confronting Germany or in a thoroughgoing renovation for its own sake. This study is also willing, albeit cautiously, to adopt the now-classic view, first promulgated by Arthur J. Marder, that Adm Sir John Arbuthnot Fisher, who features prominently in these pages, was indeed an innovator who left a larger mark on the institution than most of his contemporaries and gave it a direction it had previously lacked, even if he did not succeed in hammering through all of his ideas. See Arthur J. Marder, *From*

the Dreadnought to Scapa Flow, vol. 1, *The Road to War, 1904–1914* (Annapolis, MD: Naval Institute Press, 2013), 13. Christopher M. Bell notes that Marder's criticism of the Royal Navy's administration in the pre-Fisher era has since come to be regarded as somewhat hyperbolic. See Bell, "Contested Waters," 5. The Royal Navy's officer corps may not have been composed entirely of dullards, but the mobilization of its intellectual gifts under Fisher made more use of its latent talent than had so far been the case. It is possible to hold this view even if one accepts, as this study does, that Fisher's impact was neither wholly consistent nor wholly complete. Further details will emerge as the narrative progresses.

3. Padfield, *The Great Naval Race*, 47–48. For the Royal Navy's administrative habits, see Massie, *Dreadnought*, 395–96. Sumida offers an in-depth look at the proficiencies and shortcomings of the Royal Navy's administration, which he explicitly contrasts from its tactical leadership capabilities, in which he offers the balanced conclusion, which may anticipate the later parts of this narrative, that the "failures of Fisher's radical policies on capital ship design, the related poor performance of the British battle cruisers at Jutland, and defective Admiralty leadership during the war, were in large part the result not simply of prewar policy disagreements within the Admiralty, nor of random glitches in the functioning of the bureaucracy, nor again of personal idiosyncrasy with regard to the delegation of work, but of financial limitations, structural defects in the administrative system, and what might be called the Admiralty's culture of counterproductive work practices that were the outgrowth of chronic prewar administrative undermanning." Sumida, *In Defence of Naval Supremacy*, 25–26. In short, the Royal Navy began the Dreadnought Race as a capable but muscle-bound bureaucracy in which a set of external factors created an adequate but incoherent administration, and it never entirely overcame these limitations even under its most famous and gifted First Sea Lord of this era.

4. Arthur J. Marder notes that Britain's suspicion of Germany prior to 1905 stemmed from the latter's political activities rather than its naval program. See Marder, *The Anatomy of British Sea Power*, 543–45. Later, he partly reverses this judgment, noting briefly that the German fleet was objectively a more pressing matter post-1900 than equivalent building programs in the United States and Japan, although he appears to be generalizing at this point. See Marder, *The Road to War*, 11. Massie notes that

it took until 1901 for the Admiralty to pay attention to German ship construction as part of a plan to oust the Royal Navy from its dominant position; until this time, the overall distribution of naval capability was the Admiralty's main concern. See Massie, *Dreadnought*, 184. This study is prepared to go further, agreeing in part with the "revisionist" line of argument that even the construction of HMS *Dreadnought*, and many of the reforms that accompanied it and thereby framed the Dreadnought Race, may have been motivated by a multiplicity of concerns besides the German challenge, even if by that point some British decision makers had accepted that Germany was beginning to become a problem. For these arguments, see Sumida, *In Defence of Naval Supremacy*; Matthew S. Seligmann, *The Royal Navy and the German Threat, 1901–1914: Admiralty Plans to Protect British Trade in a War against Germany* (Oxford, UK: Oxford University Press, 2012), 3–6; Matthew S. Seligmann, Frank Nägler, and Michael Epkenhans, *The Naval Route to the Abyss: The Anglo-German Naval Race, 1895–1914* (London: Routledge, 2016), 103–8; and Bell, "Contested Waters," 2–12. Hew Strachan is prepared to accept that the German naval buildup "provided the thrust to British naval policy" post-1901, although arguing that *Weltpolitik* alone cannot be considered a cause of World War I, and that many of the diplomatic developments of the first half of the decade were driven by other factors. See Hew Strachan, *The First World War*, vol. 1, *To Arms* (Oxford, UK: Oxford University Press, 2001), 10, 13–14, 17. As Seligmann, Nägler, and Epkenhans note, "the perception that there was a German naval threat did not have to emerge in one go with everyone recognising this fact simultaneously." See Seligmann, Nägler, and Epkenhans, *The Naval Route to the Abyss*, 106. This study is prepared to adopt the sensible synthesis offered by Seligmann, Nägler, and Epkenhans that at least some British policy makers had caught a whiff of a threat from Germany as early as 1901 or 1902, even if the notion of Germany as Britain's chief challenger was slower to catch on and did not become a consensus until the end of the decade. On this point, see Paul M Kennedy, *The Rise of the Anglo-German Antagonism, 1860–1914* (London: Ashfield Press, 1987), 251–65. This study emphatically does not need to accept some of the more speculative conclusions of the revisionists regarding subsequent British naval dispositions and doctrine, however, as will be discussed. But in any event, the fact that a multitude of concerns faced British policy makers at all times does not negate the basic truth that even a bad or incoherent strategy—or one blindly made—is still in a crucial

Chapter Five

sense a strategy, and that a competition in which one side is not running at full speed is still a competition.

5. Padfield, *The Great Naval Race*, 94. Sumida notes that French and Russian naval modernization programs (of more modest scope) preoccupied the Admiralty. See Sumida, *In Defence of Naval Supremacy*, 18–20. See also Massie, *Dreadnought*, 184.
6. Padfield, *The Great Naval Race*, 94; Sumida, *In Defence of Naval Supremacy*, 229–37; and Massie, *Dreadnought*, 184.
7. Padfield *The Great Naval Race*, 94. Massie, *Dreadnought*, 190, 310.
8. Massie, *Dreadnought*, 288–89.
9. Massie, *Dreadnought*, 290, 304–7, 312; Henry Kissinger, *Diplomacy* (New York: Simon and Schuster, 1994), 186–87; and Kennedy, *The Rise of the Anglo-German Antagonism*, 230–46. Kennedy notes mordantly that the "years 1897 to 1901 are characterized, so far as Anglo-German political relations are concerned, by two contradictory trends: on the one hand, by the inauguration of German external and fleet-building policies which . . . would soon lead to a worsening in . . . relations; and on the other, by the attempts of a group of British politicians and by certain German officials—none of whom fully appreciated the implications of the turn to *Weltpolitik*—to explore the possibility of an Anglo-German alliance. In retrospect, the historian cannot fail to agree with Bülow's statement that these two tendencies were contradictory." Kennedy, *The Rise of the Anglo-German Antagonism*, 233. A more concise and poignant depiction of both Britain's confusion about whether to view Germany as a threat and Germany's lack of consistency in its competitive goals could scarcely be written.
10. Data on the British Army and Royal Navy are from Frederick Martin and John Scott Keltie, eds., *The Statesman's Yearbook* (London: Macmillan, 1875–1914). See the appendix for more information. Another essentially similar assessment is given in Sumida, *In Defence of Naval Supremacy*, 20–22. By Sumida's data, the peak expenditure in 1904 was actually £38.3 million, up from £21.8 million in 1897, an even more drastic increase.
11. Sumida, *In Defence of Naval Supremacy*, 18–22.
12. Massie, *Dreadnought*, 13–20, 296–303.
13. Massie, *Dreadnought*, 271–75, 307–8, 553–54. For a note on the extent to which the German nationalist press capitalized on Britain's atrocities to drum up anti-British sentiment, see Kennedy, *The Rise of Anglo-German Antagonism*, 247.
14. Sumida, *In Defence of Naval Supremacy*, 22–23. Research conducted for this study based on *The Statesman's Yearbook* data for the relevant years and the historical economic research of the economist

Angus Maddison (see the appendix) indicates that Britain had reduced its outstanding public debt to an historic low of 31.9 percent of gross national product in 1899 prior to the Boer War, only to see it rise more than 7 percentage points to 39.1 percent in 1903. The resulting "sticker shock" may have had as much to do with the subsequent government reaction as the nature of the debt.

15. Kennedy, *The Rise of Anglo-German Antagonism*, 261–66; and Massie, *Dreadnought*, 307–12, 325–36.
16. Strachan notes in particular that British diplomatic attention was focused more on events on the imperial periphery than on the European continent. See Strachan, *To Arms*, 13–14. For all this, a wave of public hostility to Germany over the latter's stance on the Boer War had gripped Britain, even if the government was not unified as to how to manage Germany. See Kennedy, *The Rise of Anglo-German Antagonism*, 223–71.
17. Kennedy, *The Rise of Anglo-German Antagonism*, 252–55.
18. Kennedy, *The Rise of Anglo-German Antagonism*, 252–53, 266–74; and Samuel R. Williamson Jr., *The Politics of Grand Strategy: Britain and France Prepare for War, 1904-1914* (London: Ashfield Press, 1990), 1–29.
19. See Kennedy, *The Rise of Anglo-German Antagonism*, 266–74; Williamson, *The Politics of Grand Strategy*, 1–29; Massie, *Dreadnought*, 344–48; and Kissinger, *Diplomacy*, 189. The political reasoning behind the entente on the British side was hardly uniform or even coherent. Williamson notes that a multitude of imperial concerns having little to do with Germany drove the formation of the Entente Cordiale. On this point, Kennedy notes that the government—which a few years previously could not even decide if it wanted an alliance or to diplomatically protest Germany over interference in the Boer War—had little unanimity on the subject, with some officials at the cabinet level and below seeing the entente as serving the purpose of containing Germany and others seeing it as imperial house cleaning. See also Strachan, *To Arms*, 14. Strachan is emphatic that the entente served the latter purpose and was not driven by naval developments. As noted above, the fact that Britain was in competition with Germany did not obligate it to have a coherent strategy, nor does this fact prevent an analysis of what strategy it can be said to have had. It is nevertheless possible that an increasing awareness of German ambitions was at least in the background throughout these events.

20. Kissinger, *Diplomacy*, 189, 191, 197–98, 212–13. Kissinger cites the subsequent military coordination between France and Britain, not the entente, as the beginning of substantive, if soft, guarantees by Britain that entailed credibility problems if they were not honored, but the entente certainly paved the way for the ever-closer relationship. As an aside, although Kissinger is often cast as a realist, his understanding of the human nuances of international dialogue and their effects on policy is consistent with a more constructivist reading of the situation as well.

21. See Kissinger, *Diplomacy*, 212–13; and David Owen, *The Hidden Perspective: The Military Conversations, 1906-1914* (London: Haus Publishing, 2014), 27–28. See also Strachan, *To Arms*, 26–27, 62, 93; and Williamson, *The Politics of Grand Strategy*, 264–83, 358, 367–72. Williamson sees the naval side of the talks, more than that of the land, as conferring a "moral commitment," while admitting that the talks were not the driving factor behind Britain's decision to declare war in August 1914. Strachan insists that the staff talks were a factor but did not force Britain's hand, and that Britain arrived at its decision to declare war deliberately. An analysis of Britain's strategy can accounts for the inherent ambiguity both of Britain's commitments and their historical result. Owen notes the effect that the entente had on German threat perception as well, in that German officials saw the Anglo-French relationship as more rigid and less open to diplomatic suasion. This point is echoed by Kennedy, who notes that Tirpitz and his operational counterparts in the German Naval High Command opposed taking diplomatic action to counteract the entente—specifically, seeking a rapprochement with Russia as a way out of the encirclement—precisely because it might lead to a British preventive strike on the German naval dockyards and the nascent High Seas Fleet, even as Wilhelm and the German diplomatic corps wanted the exact opposite. It is safe to say that the entente was an ominous and highly unwanted development from the German perspective, even if it is strange in retrospect that it was not anticipated. It is admittedly curious that the entente—which imposed no formal legal obligations—was seen as a threat by Germany, even as Britain's equally noncommittal acquiescence in the preceding decade was seen as inadequate by those German policy makers who wanted a formal alliance with Britain.

22. Williamson, *The Politics of Grand Strategy*, 367. This study hews to the orthodox view most prominently advocated by Kennedy and Lambi, and most recently reaffirmed by Seligmann, Nagler, and Epkenhans that the German decision to persist in building

capital ships had political implications that led to the "encirclement" of Germany, and that this, in turn, created an environment in which war between Britain and Germany was more likely and in fact precipitated by diplomatic developments (the military talks, discussed below). See Kennedy, *The Rise of Anglo-German Antagonism*; Ivo Nikolai Lambi, *The Navy and German Power Politics, 1862-1914* (Boston, MA: Allen and Unwin, 1984), 427; and Seligmann, Nägler, and Epkenhans, *The Naval Route to the Abyss*, 404. It is not necessary to adopt some of the more fringe views of German war guilt and the causes of World War I—in particular, the notorious "Fischer thesis" regarding early German decisions to go to war—to note that the souring of Anglo-German relations, the tightening of Anglo-French military relations, and the defensive stance of both Germany and Britain as a result of the Dreadnought Race made war more likely. Nor is it necessary to posit that the Dreadnought Race caused World War I *per se* (a view held by few, if any) to note that it was a contributing factor and perhaps a necessary condition for it. The Fischer thesis was originally promulgated in Fritz Fischer, *Germany's Aims in the First World War* (New York: W. W. Norton, 1967). Lambi devotes a chapter to discussing and ultimately refuting Fischer's arguments for a more conspiratorial German decision to seek war prior to 1914. See Lambi, *The Navy and German Power Politics*, 361-84. Briefly, the German decision to challenge Britain in this regard soured relations and accelerated diplomatic and military developments in such a way as to push the two states closer to open conflict—even if it is true that other factors ultimately caused the war, or that, as is obvious, a different path might have been taken at any point.

23. Arne Røksund, *The Jeune École: The Strategy of the Weak* (Boston, MA: Brill, 2007), 221-22.
24. Kennedy, *The Rise of Anglo-German Antagonism*, 267-75; Paul M. Kennedy, *Strategy and Diplomacy, 1870-1945: Eight Studies* (London: Allen and Unwin, 1983), 151; and Patrick J. Kelly, *Tirpitz and the Imperial German Navy* (Bloomington: Indiana University Press, 2011), 200-1.
25. Kennedy, *The Rise of Anglo-German Antagonism*, 275-85; Williamson, *The Politics of Grand Strategy*, 28-52; Strachan, *To Arms*, 16-17; Kissinger, *Diplomacy*, 190-91; Padfield, *The Great Naval Race*, 127-31; Massie, *Dreadnought*, 351-69; and Owen, *The Hidden Perspective*, 19-20. Williamson notes that the initial intent may not necessarily have been to divide the entente, but that the implications of the German policy were clear.

26. Sumida, *In Defence of Naval Supremacy*, 20–27.
27. Sumida, *In Defence of Naval Supremacy*, 26–27; Massie, *Dreadnought*, 459; and Padfield, *The Great Naval Race*, 113–15.
28. Lambert, *Sir John Fisher's Naval Revolution*, 73–76; Ruddock F. Mackay, *Fisher of Kilverstone* (Oxford, UK: Clarendon Press, an imprint of Oxford University Press, 1973), 1–6, 32, 92–93; Marder, *The Road to War*, 14–19; Kennedy, *Strategy and Diplomacy*, 111–20; Sumida, *In Defence of Naval Supremacy*, 26–27; Massie, *Dreadnought*, 401–32, 437; and Padfield, *The Great Naval Race*, 75–76, 95–99. Another more controversial comparison to Tirpitz is suggested by Lambert, who argues that Fisher's memoranda represented an attempt to manipulate and harness trends in public opinion rather than Fisher's genuine views. See Nicholas A. Lambert, *Planning Armageddon: British Economic Warfare and the First World War* (Cambridge, MA: Harvard University Press, 2012), 76–77. For a criticism, see Matthew S. Seligmann, "Naval History by Conspiracy Theory: The British Admiralty before the First World War and the Methodology of Revisionism," *Journal of Strategic Studies* 38, no. 7 (July 2015): 968–71, https://doi.org/10.1080/01402390.2015.1005443. This view is not universally held, and this study does not entirely subscribe to it, but it is beyond dispute that Fisher, like Tirpitz, was an expert at attaching himself to powerful people and harnessing political winds to serve his own purposes. See Kennedy, *Strategy and Diplomacy*, 117–20.
29. For these arguments, see Seligmann, "Naval History by Conspiracy Theory," 966–84; Christopher M. Bell, "Sir John Fisher's Naval Revolution Reconsidered: Winston Churchill at the Admiralty, 1911–1914," *War in History* 18, no. 3 (July 2011): 333–56, https://doi.org/10.11177/0968344511401489; Christopher M. Bell, "Contested Waters: The Royal Navy in the Fisher Era," *War in History* 23, no. 1 (January 2016): 115–26, https://doi.org/10.1177/0968344515595330; Nicholas A. Lambert, "On Standards: A Reply to Christopher Bell," *War in History* 19, no. 2 (April 2012): 217–40, https://doi.org/10.1177/0968344511432977; and Christopher M. Bell, "On Standards and Scholarship: A Reply to Nicholas Lambert," *War in History* 20, no. 3 (July 2013): 381–409, https://doi.org/10.1177/0968344513483069. The basic source material includes Marder, *The Road to War*; Sumida, *In Defence of Naval Supremacy*; and Lambert, *Sir John Fisher's Naval Revolution*.
30. Marder, *The Road to War*, 116–17; Padfield, *The Great Naval Race*, 120–22, 129–30, 134; and Massie, *Dreadnought*, 360, 462.

31. Marder, *The Road to War*, 36-43; Sumida, *In Defence of Naval Supremacy*, 27-28; Lambert, *Sir John Fisher's Naval Revolution*, 97-115; Strachan, *To Arms*, 17, 376, 378; Massie, *Dreadnought*, 461-65; and Padfield, *The Great Naval Race*, 116-17.
32. Again, note Fisher's "Copenhagen" remark in Marder, *The Road to War*, 112-13. Fisher also was a fatalist regarding war with Germany. See Marder, *The Road to War*, 26.
33. By 1906, the Admiralty collectively anticipated war with Germany, and by 1907, they were making plans to reduce the Royal Navy's Mediterranean presence to consolidate forces in home waters. But opinion was far from uniform on this matter before this time. See Strachan, *To Arms*, 376.
34. Sumida, *In Defence of Naval Supremacy*, 37-38, 41-46, 50-54; Massie, *Dreadnought*, 398-400, 468-74; Padfield, *The Great Naval Race*, 125-26; and Marder, *The Road to War*, 43-44, 57-66. A perfect example of Fisher's constant harping on trading armor for speed is found in a memo he wrote in retirement to Winston Churchill after the latter had become First Lord of the Admiralty: "I. There MUST be the 15 inch gun . . . II. There MUST be sacrifice of armour . . . IV. There must be further VERY GREAT INCREASE OF SPEED." Randolph S. Churchill, ed., *The Churchill Documents*, vol. 5, *At the Admiralty, 1911-1914* (Hillsdale, MI: Hillsdale College Press, 2019), 1545-46. Once one generally understands Fisher's mindset in this, much of the questions regarding his intended usage of the ships he designed and built are resolved.
35. See Seligmann, *The Royal Navy and the German Threat*, 65-88. There is no question that other naval planning was occurring even as Britain wrestled with the issue of how to deal with the German High Seas Fleet, and this study makes no attempt to minimize such questions.
36. For the increased cost, see Seligmann, *The Royal Navy and the German Threat*, 84. See also Nicholas A. Lambert, " 'Our Bloody Ships' or 'Our Bloody System'?: Jutland and the Loss of the Battle Cruisers, 1916," *Journal of Military History* 62, no. 1 (January 1998): 29-55, https://doi.org/10.2307/120394. The plot thickens when one considers Lambert's contention that the battlecruisers' dismal performance at Jutland could well have had nothing to do with inadequate armor or poor tactics, as is commonly suggested, and instead can be put down simply to improper ammunition storage necessitated by rapid fire. About all that can be stated with certainty is that Fisher's vision was hopelessly muddled in committee, both by the fact that other minds and hands were involved and by his famous and demonstrated cageyness in

handling bureaucratic disputes, which has muddied the historical waters still further.

37. Sumida, *In Defence of Naval Supremacy*, 37–38, 55–57, 59–61. See also Marder, *The Road to War*, 69–70.
38. Lambert, *Sir John Fisher's Naval Revolution*, 121–26, 182–83.
39. Bell, "Sir John Fisher's Naval Revolution Reconsidered," 333–37, 355–56.
40. Padfield, *The Great Naval Race*, 145–56, 167–70, 175–78, 194–98; Massie, *Dreadnought*, 498–543; and Marder, *The Road to War*, 88–104.
41. See Paul M. Kennedy, *The Rise and Fall of British Naval Mastery* (New York: Charles Scribner's Sons, 1976), 234.
42. Massie, *Dreadnought*, 558, 575.
43. Padfield, *The Great Naval Race*, 140–41.
44. Officially, the Liberal platform implied a tax reduction, but once it became clear that a guns-butter dilemma existed, the Liberals, led by David Lloyd George, came to favor tax increases. The Tories in the House of Lords regarded the entire Liberal platform as "socialist" from the outset. See Massie, *Dreadnought*, 625, 640, 642, 646–47.
45. Lambert, *Sir John Fisher's Naval Revolution*, 129–32. Lambert, who views the Dreadnought Race as being at most peripheral to Fisher's concerns, nevertheless notes the intention of the Liberal government to extract fiscal concessions from the Royal Navy, on the basis that the Two-Power Standard was all with which the Navy needed concern itself. The seesaw effect of official threat perception—from an unevenly hawkish public mood early in the decade to a frantic official mood regarding the German naval challenge later, with complacency in between, and with the Royal Navy's uniformed leadership preoccupied in multiple directions until it was not—can readily be seen.
46. Massie, *Dreadnought*, 477; and Sumida, *In Defence of Naval Supremacy*, 55, 59–60. Naval budget figures are drawn from Keltie, *The Statesman's Yearbook* (1905–6). The "revisionist" consensus now holds that Fisher's unveiling of the *Dreadnought* design was driven by matters other than the German challenge, including not only the aforementioned need for cost-cutting and his own preference for innovation but also the decision to disrupt the global naval technology curve before other navies—notably those of the United States and Japan—did so. See Lambert, *Sir John Fisher's Naval Revolution*. See also Kelly, *Tirpitz and the Imperial German Navy*, 253n105. Lambert, in particular, notes that HMS *Dreadnought* was at best a compromise, in that Fisher's preference was for an even less well-armored fast battlecruiser and that he

believed battleships were obsolescent. Nevertheless, there is no question that *Dreadnought's* design was a basic, if imperfect, realization of Fisher's ideas for a faster, all-big-gun ship. See Lambert, *Sir John Fisher's Naval Revolution*, 9, 108. See also Marder, *The Road to War*, 43–69. Sumida essentially concedes this point even while noting *Dreadnought* was not Fisher's ideal preference. See also Kelly, *Tirpitz and the Imperial German Navy*, 253. As Kelly notes, from the German perspective it scarcely mattered what Fisher intended, only what he did. This can serve to reaffirm the point noted earlier that, with a competition underway and multiple minds involved on each side, a state can be said to have a strategy even if it does not have a coherent one or even intend to. As has been noted already, Britain was increasingly aware collectively of a German challenge, but its "official mind" was in multiple places at the same time. This study can readily concede this point.

47. Padfield, *The Great Naval Race*, 136. This study, in its subsequent analysis, is prepared to accept the more radical arguments regarding the light battlecruisers' capabilities. This includes, in particular, Fisher's original contention that heavily armed and fast ships could indeed take on larger vessels and hold their own—a fact demonstrated by the mere observation that their performance at Jutland, while perhaps not ideal, was far from abysmal, as well as by Lambert's very plausible contention that there may have been nothing wrong with the battlecruisers at all except for the need to keep large amounts of propellant in gun turrets to facilitate rapid firing. See Lambert, "'Our Bloody Ships' or 'Our Bloody System'?," 53–55. On that basis, and in view of the general doctrinal murkiness and the inevitable counterfactual reasoning associated with imagining the full range of use of these ships that did not occur but easily could have, this study counts battleships and battlecruisers together when analyzing the ebb and flow of the naval arms race. To do otherwise would in fact be presumptuous: it is simply too difficult to tell at this distance how these weapons systems might have performed given even slightly different conditions or in hypothetical engagements.

48. Strachan, *To Arms*, 12, 17; and Padfield, *The Great Naval Race*, 142. Strachan notes that the additional cost of a *Dreadnought*-equivalent ship for Germany was 7 million marks that, at the exchange rate of 20 marks per pound that fluctuated slightly but not significantly, ran to £350,000 more per ship, as opposed to £181,000 additional for HMS *Dreadnought*. This was despite Tirpitz's legendary ability to browbeat dockyards into producing ships at cost. See Strachan, *To Arms*, 407. The German ships were

inferior in quality as well, as would become painfully apparent as the Dreadnought Race wore on, and while the British ships eventually evolved into "superdreadnought" designs with turrets at centerline, their German counterparts never did. See Richard Hough, *Dreadnought: A History of the Modern Battleship* (London: Michael Joseph, 1965), 120-34. Strachan notes that in certain respects—notably gunnery—the German ships were inferior to their British counterparts, although in other ways—such as armor—they ran approximately even. See Strachan, *To Arms*, 390.

49. Padfield, *The Great Naval Race*, 136; and Massie, *Dreadnought*, 484-86.
50. Kelly, *Tirpitz and the Imperial German Navy*, 240-63; Lambi, *The Navy and German Power Politics*; Strachan, *To Arms*, 20-21; and Padfield, *The Great Naval Race*, 112, 124-25, 132-34, 136-37, 163-66, 173-75. Kelly sees the real arms race as beginning after the 1906 law and the decision to counter the *Dreadnought* program, a point echoed by Strachan, even if HMS *Dreadnought* had been built for other reasons. He also amusingly notes Wilhelm's repeated attempts to interject himself into the ship design and procurement process, which achieved nothing but provide further evidence of Wilhelm's amateurish conduct of office.
51. Kelly, *Tirpitz and the Imperial German Navy*, 255; Massie, *Dreadnought*, 485-86; and Padfield, *The Great Naval Race*, 163-64.
52. Strachan, *To Arms*, 20-23.
53. Seligmann, *The Royal Navy and the German Threat*, 9-45; and Lambi, *The Navy and German Power Politics*, 167-68.
54. Lambi, *The Navy and German Power Politics*, 257, 333, 424-26; and Kelly, *Tirpitz and the Imperial German Navy*, 316-19, 357-61, 371-74, 378, 385-86, 403-9, 464.
55. Strachan, *To Arms*, 409. See also Kelly, *Tirpitz and the Imperial German Navy*, 316-19.
56. Strachan, *To Arms*, 21; Kelly, *Tirpitz and the Imperial German Navy*, 278-92; and Padfield, *The Great Naval Race*, 163, 166.
57. Keltie, *The Statesman's Yearbook* (1906), 64; and Keltie, *The Statesman's Yearbook* (1907), 66-68.
58. Although Tirpitz seems to have believed that this was deliberate, there is a broad consensus that neither Fisher nor his superiors actually intended for this fortunate coincidence to occur when HMS *Dreadnought* was launched. See also Marder, *The Road to War*, 67; Kelly, *Tirpitz and the Imperial German Navy*, 254; Padfield, *The Great Naval Race*, 124, 127, 133, 187; and Massie, *Dreadnought*, 486. Here again, it is possible to say that Britain's competitive strategy, understood as its decisions rather than its official awareness, was

advantageous, even if only by accident—although it might have achieved much more in this area had Fisher and others had better information.

59. Padfield, *The Great Naval Race*, 269-70, 313-15; and Massie, *Dreadnought*, 407.
60. Sumida, *In Defence of Naval Supremacy*, 37, 58, 60-61. The story is recounted in Massie, *Dreadnought*, 494-97; its counterpoint, with more reliable sourcing, is found in Kelly, *Tirpitz and the Imperial German Navy*, 269-73.
61. With apologies to Winston Churchill.
62. Williamson, *The Politics of Grand Strategy*, 370; Massie, *Dreadnought*, 587-88; Padfield, *The Great Naval Race*, 159; and Owen, *The Hidden Perspective*, 23-26, 27-32, 128, 82-83, 97, 134. Williamson notes that the discussions merely formalized a series of decisions Britain would probably have made anyway. For the size of the British Army at the time, see Keltie, *The Statesman's Yearbook* (1907), 61.
63. Williamson, *The Politics of Grand Strategy*, 21, 64-65, 89-101.
64. Strachan, *To Arms*, 26, 93-97. See also Hew Strachan, "The British Army, Its General Staff, and the Continental Commitment, 1904-1914," in *The Schlieffen Plan: International Perspectives on German Strategy for World War I*, ed. Hans Ehlert, Michael Epkenhans, and Gerhard P. Gross; English translation ed. MajGen David T. Zabecki, USA (Ret) (Lexington: University Press of Kentucky, 2014), 313; and Williamson, *The Politics of Grand Strategy*, 191-92, 370. Williamson stresses just how little this famous committee meeting actually decided; he also notes the relative effect of the naval, vice land, commitments.
65. Padfield, *The Great Naval Race*, 257-59; Owen, *The Hidden Perspective*, 82-101; Kissinger, *Diplomacy*, 191, 197, 212-13; and Massie, *Dreadnought*, 590-91.
66. Kissinger, *Diplomacy*, 192-93; and Owen, *The Hidden Perspective*, 78-79. Owen notes that comparisons of Kennan and Crowe, and the outsized influence of their policy memoranda, are found in Margaret MacMillan, *The War That Ended Peace* (New York: Random House, 2013); the comparison appears to be quite apt.
67. Eyre Crowe, "Memorandum on the Present State of British Relations with France and Germany," in *The Hidden Perspective: The Military Conversations 1906-1914*, ed. David Owen (London: Haus Publishing, 2014), 226, 230-31, 235.
68. The entire memorandum is reproduced in Owen, *The Hidden Perspective*, 216-61. Grey's statement of approval is found in Owen, *The Hidden Perspective*, 261-62.

Chapter Five

69. Kennedy, *The Rise of Anglo-German Antagonism*, 333–35; Massie, *Dreadnought*, 562–63, 640, 644–47; and Padfield, *The Great Naval Race*, 140–41, 175–76.
70. Kennedy, *The Rise of Anglo-German Antagonism*, 333–34; Massie, *Dreadnought*, 609–10, 644–47, 709; Padfield, *The Great Naval Race*, 141, 204; and Sumida, *In Defence of Naval Supremacy*, 188–89, 266n17.
71. Padfield, *The Great Naval Race*, 178, 183, 205–6; and Massie, *Dreadnought*, 819.
72. Kennedy, *The Rise of Anglo-German Antagonism*, 334; and Massie, *Dreadnought*, 766–69, 819–20.
73. Owen, *The Hidden Perspective*, 6–8; Kissinger, *Diplomacy*, 182; Massie, *Dreadnought*, 594–602; and Padfield, *The Great Naval Race*, 120, 137–38, 160, 163, 186.
74. Seligmann, Nägler, and Epkenhans, *The Naval Route to the Abyss*, 348–52; Padfield, *The Great Naval Race*, 198–209; Massie, *Dreadnought*, 609–16, 618–22; and Marder, *The Road to War*, 151–59.
75. Seligmann, Nägler, and Epkenhans, *The Naval Route to the Abyss*, 349; David Stevenson, *Armaments and the Coming of War: Europe, 1904–1915* (Oxford, UK: Clarendon Press, an imprint of Oxford University Press, 1996), 167–69; Marder, *The Road to War*, 159–70; Massie, *Dreadnought*, 616–18; and Padfield, *The Great Naval Race*, 213–21. The Churchill quotation is found in Marder, *The Road to War*, 151; and Massie, *Dreadnought*, 618. Stevenson notes in particular that the Admiralty's intelligence somewhat overstated Germany's true building capacity, but that the political consequences were such that Britain was now permanently motivated to action.
76. Massie, *Dreadnought*, 618–25; and Stevenson, *Armaments and the Coming of War*, 170–72.
77. Strachan, *To Arms*, 376; Stevenson, *Armaments and the Coming of War*, 169; and Massie, *Dreadnought*, 624–25. It is important to qualify this statement: Marder notes that other navies—notably those of Austria-Hungary and Italy—featured as items of concern in British parliamentary and cabinet debates about naval strength in mid-1909. See Marder, *The Road to War*, 170–71. Nevertheless, these concerns were much more transient than the thereafter ongoing concern with staying ahead of Germany. Strachan notes that the arms race between these navies and that of France was of equal or greater consequence for European stability before World War I. See Strachan, *To Arms*, 377. This assertion can be readily granted by this study without taking its

focus off the question of German competition with Britain over the North Sea.
78. Strachan, *To Arms*, 376; Stevenson, *Armaments and the Coming of War*, 170; and Padfield, *The Great Naval Race*, 237-38, 248.
79. Padfield, *The Great Naval Race*, 238-39.
80. Hough, *Dreadnought*, 36-38, 55-58; Massie, *Dreadnought*, 473, 782; and Strachan, *To Arms*, 388. Strachan notes that the German battleships' inferiority in this regard—which amounted to a nearly 50-percent lag in the weight of shells fired per broadside—was never remedied before the outbreak of war.
81. Strachan, *To Arms*, 21; Kelly, *Tirpitz and the Imperial German Navy*, 298-99, 305, 341; Holger H. Herwig, *"Luxury" Fleet: The Imperial German Navy, 1888-1918* (London: George Allen and Unwin, 1980), 72-73; Padfield, *The Great Naval Race*, 26-27, 67-68, 211-12, 225-26, 235, 242-43, 259-60; and Massie, *Dreadnought*, 688-95. As Padfield and Kelly note, the idea of raising funds through a new inheritance tax resurfaced in 1912, but it did not come to a vote. The extreme sensitivity of the issue may serve as an indicator of just how opposed Wilhelm and Tirpitz's vision was to the German national security establishment's predilections. For Bülow's demise, see Kelly, *Tirpitz and the Imperial German Navy*, 295-96; and Massie, *Dreadnought*, 684-91.
82. Kelly, *Tirpitz and the Imperial German Navy*, 306; Padfield, *The Great Naval Race*, 233-35; and Massie, *Dreadnought*, 702-7.
83. Kelly, *Tirpitz and the Imperial German Navy*, 306-9; Stevenson, *Armaments and the Coming of War*, 172-73; Strachan, *To Arms*, 23; *The Great Naval Race*, 236-37; and Massie, *Dreadnought*, 706-11.
84. Kelly, *Tirpitz and the Imperial German Navy*, 308-12; Stevenson, *Armaments and the Coming of War*, 173; Lambi, *The Navy and German Power Politics*, 301, 363-71; Massie, *Dreadnought*, 706-11; and Padfield, *The Great Naval Race*, 236-49, 259-67, 284-86.
85. Strachan, *To Arms*, 24; Kelly, *Tirpitz and the Imperial German Navy*, 322-26; Stevenson, *Armaments and the Coming of War*, 181-95; Lambi, *The Navy and German Power Politics*, 315-22; Herwig, *"Luxury" Fleet*, 73; Padfield, *The Great Naval Race*, 249-65; Kissinger, *Diplomacy*, 196-97; and Massie, *Dreadnought*, 715-13. As Strachan notes, Germany was in no position to back up its threats, but the mere presence of a single gunboat was enough to alert Britain—a kind of demonstration of the intuitive, as opposed to strictly effect-based, grip that sea power and its exercise had on the minds of decision makers.
86. Stevenson, *Armaments and the Coming of War*, 181, 184-85, 213; Christopher M. Bell, *Churchill and Sea Power* (Oxford, UK: Oxford

University Press, 2013), 15; Strachan, *To Arms*, 25; Padfield, *The Great Naval Race*, 252–59; and Massie, *Dreadnought*, 731–33, 737–38, 744–49, 767.

87. Massie, *Dreadnought*, 540–43; Padfield, *The Great Naval Race*, 229–30; and Marder, *The Road to War*, 204–47.

88. Marder, *The Road to War*, 213–14; Strachan, *To Arms*, 380; Bell, *Churchill and Sea Power*, 14–15; Massie, *Dreadnought*, 770–71, 778; and Padfield, *The Great Naval Race*, 269.

89. Massie, *Dreadnought*, 655–57. Massie notes that Wilhelm used the funeral as an opportunity to seek the favor of his English royal relatives, while preening over his status as a member of the royal family—one more reminder of the essentially intangible nature of Wilhelm's priorities.

90. Hough, *Dreadnought*, 120–31; Marian Kent, *Moguls and Mandarins: Oil, Imperialism and the Middle East in British Foreign Policy, 1900–1940* (London: Frank Cass, 1993), 34–35; Massie, *Dreadnought*, 781–85; and Padfield, *The Great Naval Race*, 270–71. The details of Britain's new oil policy are documented in Kent, *Moguls and Mandarins*, 34–59.

91. See Keltie, *The Statesman's Yearbook* (1912), 57, as well as the appendix. *The Statesman's Yearbook* began documenting superdreadnoughts as a separate ship class in 1912. The series also documents the complete absence of a German equivalent that same year. See Keltie, *The Statesman's Yearbook* (1912), 857. For the superiority of the new ships, see Padfield, *The Great Naval Race*, 271.

92. Kelly, *Tirpitz and the Imperial German Navy*, 326–47; Lambi, *The Navy and German Power Politics*, 369–73; Stevenson, *Armaments and the Coming of War*, 195–210; and Padfield, *The Great Naval Race*, 273–93.

93. Stevenson, *Armaments and the Coming of War*, 213–14; Bell, *Churchill and Sea Power*, 17–19, 28–30; Padfield, *The Great Naval Race*, 276–86, 299; and Massie, *Dreadnought*, 790–804. Stevenson puts the matter succinctly: "After vast expenditure the outlines of a modus vivendi were emerging although one that the two sides were unable to embody in a formal accord." Stevenson, *Armaments and the Coming of War*, 214. To call this an achievement of Tirpitz's original goals requires an extremely minimalist interpretation of them—as Rolf Hobson has noted, regarding this matter, a 3:2 ratio might have been just about what Tirpitz could have lived with for deterrence purposes, and this was bizarrely a ratio that also satisfied Britain to a minimal level. See Rolf Hobson, *Imperialism at Sea: Naval Strategic Thought, the Ideology of*

Sea Power, and the Tirpitz Plan, 1875-1914 (Boston: Brill Academic Publishers, 2002), 264-67. However, the fact that Tirpitz did not seek to formalize this arrangement and in fact sought to press on with his building program—and apparently was still dreaming of naval parity—suggests that, in the final analysis, his goals were not met. "There is however," Hobson writes "no compelling evidence that he had thought through these matters systematically, as opposed to hoping that things would turn out alright." Hobson, *Imperialism at Sea*, 265. This point the study here can readily concede. A better epitaph for Tirpitz's dreams could scarcely be written. Regarding Churchill's plan for the dominions, it must be said that the proposal did involve a substantial shift in the imperial social contract and, for this reason, a shift along the intangibles metric for the empire writ large, if one were assessing the security of Canada or Australia, it would be a straightforward shift in security. Churchill's proposal in effect held that the British Empire would protect its vitals (British home waters—the region being competed for here) and that a flying squadron of ships operated by the dominions would provide their security ad hoc. In a draft memorandum to Asquith written on 14 April 1912, Churchill proposes a "division of labour" in which "you [the Dominions] shall patrol the Empire" and concludes, "We will cope with the strongest [combination] in the decisive theatre." See Churchill, *At the Admiralty*, 1538-40. Suffice it to say, this left many concerns unaddressed.

94. Strachan, *To Arms*, 46-47. It must be noted that here, again, it is not necessary to posit consistency across the "official mind" to admit the existence of a strategy—even a bad one. Germany had therefore partly abandoned the Dreadnought Race even as its key leadership did not make consistent decisions in this regard.
95. Bell, *Churchill and Sea Power*, 18-39. See also Lambert, *Sir John Fisher's Naval Revolution*, 182-83; and Bell, "On Standards and Scholarship," 2, 18-19. Lambert holds that Fisher's plans for overhauling the Royal Navy were so wide-ranging as to include plans to eliminate plans for the defensive use of battleships in home waters altogether in favor of a "flotilla defence" concept employing torpedo craft. This argument turns on semantics—the extent to which Fisher, particularly in his later tenure as advisor to Winston Churchill, was serious in his efforts to accomplish this. Lambert takes note of Fisher's enthusiasm for the concept. Bell merely argues that battleships remained the preferred means of defending the North Sea. For the scholarly dialectic on this subject, see Bell, "Sir John Fisher's Naval Revolution Recon-

sidered," 333–56; Lambert, "On Standards," 217–40; and Bell, "On Standards and Scholarship," 381–409. Though is obviously preferable for a study such as this to avoid wading into individual historical disputes to the extent possible, so as to avoid drawing conclusions dependent on singular interpretations contentious points of historical dispute, it is not always possible to do so, and in this particular instance this study must take a firm stand. This study sides with Bell and Seligmann in rejecting the thesis that the Royal Navy, first under Fisher and then, after he retired, under the strong guidance of Churchill as its civilian First Lord, seriously considered implementing a "flotilla defence" concept for the North Sea that would rely on small torpedo craft on the assumption of the obsolescence of battleships. Such ideas may have entered the "official mind," and particularly the mind of Fisher, but they appear to have not gone beyond the brainstorming stage. Whatever the Royal Navy's plans, it treated battleships as capital ships with a prominent role in the defense of home waters throughout the Dreadnought Race. See Bell, "Sir John Fisher's Naval Revolution Reconsidered," 355–56. However much Fisher may have envisioned a post-battleship era—and there is no question that Fisher was forward-thinking in this regard—the battleship remained the capital ship of the era through the outbreak of war, and the Royal Navy's procurement and deployment policies reflected this. See, for example, Lambert, *Sir John Fisher's Naval Revolution*, 182–83.

96. Massie, *Dreadnought*, 818–37; and Padfield, *The Great Naval Race*, 210, 276, 287–312. Projections in 1913 held that the Royal Navy would equip and staff upward of 79 capital ships by 1920, of which 62 would be battleships.
97. Padfield, *The Great Naval Race*, 311–13; and Hough, *Dreadnought*, 131–34.
98. Kennedy, *The Rise of Anglo-German Antagonism*, 455–56. Kennedy notes that tensions remained high during this time nonetheless, and that the Balkan wars acted as a dress rehearsal or harbinger for the July Crisis, which ignited a conflagration for which Europe was merely not quite ready a year or two before.
99. Padfield, *The Great Naval Race*, 315, 317–18; and Massie, *Dreadnought*, 849–53, 855–60. The opening of the Kiel Canal was celebrated by an optimistic show of diplomatic courtesy, with the Royal and Imperial German Navies performing joint maneuvers. In view of what followed later that month, the celebration was a tragedy. As is discussed in more detail in the following chapter, this study does not draw a direct connection between the Dread-

nought Race and the outbreak of World War I but rather sees fit to point out a possible indirect connection between the two, in that the decision by Britain to intervene in the war was predicated on its preexisting antagonism with Germany in which the Dreadnought Race had played a part. The implications of this, or lack thereof, are considered more fully in the following chapter.

100. Padfield, *The Great Naval Race*, 329–32, 334, 338–39; and Kennedy, *Strategy and Diplomacy*, 137.
101. Robert K. Massie, *Castles of Steel: Britain, Germany, and the Winning of the Great War at Sea* (New York: Random House, 2003), 287–95, 431, 483–90, 498.
102. Kelly, *Tirpitz and the Imperial German Navy*, 371–74, 378, 385–86, 403–12, 419–22, 430–43, 464.
103. Massie, *Castles of Steel*, 373–425.
104. Massie, *Castles of Steel*, 579–684. The Churchill quotation can be found in Massie, *Castles of Steel*, 681. In his own account of the war, *The World Crisis*, Churchill praised Jellicoe, but then sniped, "But the Royal Navy must find in other personalities and other episodes the golden links which carried forward through the Great War the audacious and conquering traditions of the past." Winston S. Churchill, *The World Crisis*, vol. 3, *1916–18* (New York: Charles Scribner's Sons, 1927; New York: Rosetta Books), Kindle ed., loc. 2319. With his love of the dramatic, Churchill ignored the imperatives that an established power ("alpha" in this framework), both regionally and globally, must follow—maintaining one's position is not exciting, but frequently what must be done.
105. Randolph Churchill, *The Churchill Documents*, vol. 6, *At the Admiralty, July 1914–April 1915* (Hillsdale, MI: Hillsdale College Press, 2020), 1511–12.
106. Padfield, *The Great Naval Race*, 340–41; and Massie, *Castles of Steel*, 682–84, 778–88.
107. Massie, *Dreadnought*, 496–97; Massie, *Castles of Steel*, 672–74; and Sumida, *In Defence of Naval Supremacy*, 338–39. Massie notes in *Castles of Steel* that poor gunnery training and signals discipline played as much a role in the loss of the battlecruisers at Jutland as improper tactics and inadequate armor. This roughly confirms Sumida's assessment that the Royal Navy had all the correct technology except for fire control.
108. Lambert, "'Our Bloody Ships' or 'Our Bloody System'?," 54–55.
109. Padfield, *The Great Naval Race*, 341–45.

Chapter Six
The Dreadnought Race in Strategic Perspective

Far-called our navies melt away;
On dune and headland sinks the fire:
Lo, all our pomp of yesterday
Is one with Nineveh and Tyre!
Judge of the Nations, spare us yet,
Lest we forget—lest we forget!
                        ~ Rudyard Kipling[1]

When your weapons are dulled and ardour damped, your strength exhausted and treasure spent, neighboring rulers will take advantage of your distress to act. And even though you have wise counselors, none will be able to lay good plans for the future.
                        ~ Sun Tzu[2]

If not in the interests of the state, do not act. If you cannot succeed, do not use troops. If you are not in danger, do not fight. . . . Therefore, the enlightened ruler is prudent and the good general is warned against rash action. Thus the state is kept secure and the army preserved.
                        ~ Sun Tzu[3]

Having overviewed the history of the Dreadnought Race as it is relevant to this study, this book now turns to analysis of the strategies of the competitor states. This analysis will proceed in three stages, following the framework established in chapter 3. First, it will address in general terms the outcome of the race in light of the strategic goals of both sides. Second, it will consider the race from the perspectives of both sides as measured against the three basic strategic metrics discussed in chapter 2, namely those of national security, finance and welfare, and intangible goals. Finally, it will consider the competition in light of the strategic principles discussed in chapter 2, particularly competitive advantage, strategic intent, and decision-cycle analysis, showing that such principles are in fact applicable to this scenario and offer some insight into the competitors' performance against one another—insights that can be extrapolated to similar competitive scenarios.

A few important notes are required before this analysis begins in earnest. The first concerns the meaning of strategy as used in this study and as commonly understood, as it relates to the question of analysis of each state's competitive strategy. It will be recalled that *strategy* has been defined here as "a plan of action and process of decision making for the allocation of resources in anticipation of a contingent event, orchestrating simultaneous and sequential engagement, to achieve an organization objective, in the context of a contest with other organizations." Crucially, such a plan and process may not exist on paper, nor may they exist in any single individual's mind. They are said to exist only in the abstract sense in which any particular organization may be said to have agency, exactly in the same way that one may say that "Britain perceived x" or "Germany intended y." Organizations such as states, of course, do not perceive anything, except in an almost metaphorical sense, and similarly do not plan or decide anything in the sense of having the cognition or consciousness of an individual human being. They are treated as acting like individual human beings for purposes of analysis or simplicity, and when this occurs generalization is inevitable. As such, there is no reason to assume that a state's strategy will ever be entirely coherent, as it might be (or might not be) if a single person made it. It is certainly true that strategy is traditionally understood to be made by an organization's leadership, and it is equally true that that leadership may involve more than one person or even change over time. It is simply posited here that a good strategy is a coherent one, tightly linking means and ends and aligning available resources behind a well-understood strategic intent. Failure to do so, in the end, is understood, deductively, to be detrimental to the goal that the strategy is supposed to serve—though, of course, in any particular instance, it is still possible for one to simply be lucky.

For this reason, this study should not expect—and will not find—that either competitor in the Dreadnought Race played brilliantly, well, or even coherently according to the framework employed here and its underlying principles. At any given time, not to have a coherent strategy, or any strategy at all, is still, in an important sense, a strategy, as assuredly as not to decide on any matter is to choose the status quo. Although it has the potential to cloud the issue as much as to illuminate, Henry Mintzberg's important point that a strategy can be a "pattern"—a consistent resource allocation policy over time—as assuredly as a "plan" or "process" is relevant here: a state such as Britain or Germany has a strategy if it can be said to be following a plan for resource allocation or to be involved in an active process of doing so, whether any given individual is at the helm and whether that person might be asleep at the wheel.[4] A general tendency in a given direction is as good as a decision, and a decision to allocate resources in a given direction is effectively a plan.

As to what that strategy was at any given time, it can only be understood by looking at the available history. Although such matters are quite murky when looked at closely, this study has endeavored, when faced with controversy, to evaluate the major points of contention and navigate through them, either by extracting a consensus where possible or, where not, choosing from among differing views. Because of the gravity of the subject matter and the numerous questions that surround it, every effort has been made to be modest in assumptions. For all that, certain assertions can be confidently made regarding what each competitor appeared to want to do at any given time and what it actually decided to do to achieve it, and from this to synthesize a general analysis of each competitor's strategy, according to the framework presented in this study. To that end, the framework is followed here as a line of inquiry into the strategies of the two competitors in this case. In each section following, the elements of the framework being employed are noted at the beginning.

General Analysis: A German Loss; a Qualified British Victory

1. Determine the competitive objective—what is at stake in the competition. Within the limits of this study, the competitive objective refers to regional hegemony, but exactly what that entails will vary in each case.
2. Determine which competitor is "alpha"—the reigning hegemon—and which is "beta," the challenger.

The framework first asks what the competition is about—not simply its regional boundaries (though of course this defines the competition's scope) but also its implications, to include what each side wants out of it and what each side is in broad terms willing to put in. It asks concomitantly which competitor is the reigning regional hegemon (alpha) and which is the revisionist attempting to achieve that status (beta).

As noted, the exact strategic goals of the two competitors can be understood with some precision. The nature of competition is a zero- or negative-sum relationship, and therefore within the context of the Dreadnought Race it was not possible for both Germany and Britain to "win." For certain purposes, however, it was possible for both to lose.[5]

Germany's strategic goals, in the end, were to achieve something close to naval parity with Britain in the North Sea (the aforementioned "floor" of a 2:3 ratio in capital ships), if not naval superiority over it, to acquire equal political status to Britain as an Atlantic hegemon and overseas colonialist.

Chapter Six

These goals have to be understood in tandem. Nevertheless, in their details they were quite murky, and specificity was lacking. In essence, Germany's goals—summed up in the never-quite-defined term *Weltpolitik*—boiled down to something like:

- Germany wanted, at minimum, a deterrent to a British blockade of German ports in the event of war, which was to be achieved by closing the North Sea capital ship gap to an acceptable level—at least a 2:3 ratio in ships, and ideally closer.
- Germany wanted to build up its forces in North Sea waters and the seaward approaches to Germany and the European continent, effectively altering or potentially even reversing a naval balance by which Britain could threaten to blockade or invade the continent at will, to one in which Germany could do so to Britain. Whether Germany would ever have exercised such privileges if it had them was beside the point, and a century later this can only be speculated on. Although at numerous points, Germany, via Alfred von Tirpitz, Theobald von Bethmann Hollweg, and others, would claim a willingness to seek a less ambitious naval balance, there was no upper limit to German ambitions in this regard except the constraints that Germany confronted in the course of the competition.
- Germany wanted the international prestige that came with such dominance, which would signal the rise of Germany from a mere continental land power to a world power.
- Germany in particular sought to leverage that prestige to acquire new colonies, most notably in Africa, the major imperial playground of the era, notwithstanding the fact that such colonies were scarcely profitable and there was little experience of their being so in its own case. In effect, Germany sought to leverage a prestige instrument (a new navy) to acquire more prestige instruments.

- As Kaiser Wilhelm's frustration with regard to German inability to affect the outcome of the Boer War suggests, Germany wanted a place at the top of the international naval hierarchy that would ensure that it—and not Britain—could intervene at will in the colonial periphery. This last is more speculative and is based on a few chance remarks, but as Paul M. Kennedy, Gordon A. Craig, and others have noted, it is clear that *Weltpolitik* required Germany first to break out of European waters before the question of smaller-scale gunboat diplomacy in Africa could be seriously discussed.
- Germany's leadership—Wilhelm and others below him—viewed all of the above as a solution to their own domestic political problems, and insofar as they were the major decision makers, *Weltpolitik* was supposed to serve their purposes.

On this basis, the Dreadnought Race was a competition for regional hegemony, the region in question being the waters surrounding Europe and especially the North Sea. What was sought was not merely a preponderance of hard power, but also the soft power and idiosyncratic national goals that came with it. This competition had global implications, and its ultimate intentions form part of any assessment of its competitors' strategies—but it was bounded geographically in its substance, and in the end it sought control of a region of the map.

Britain's hegemony is usually referenced in global terms, as described by Karen Rasler and William R. Thompson, but it was also of a regional kind, insofar as it dominated the naval approaches to northern Europe and therefore was able to contain European great powers and their conflicts. This hegemony was not total: even under the Two-Power Standard, Britain maintained only enough naval forces to deter a fight with most conceivable combinations of great powers, not any and all such combinations. With this regional naval hegemony came concomitant international prestige among the region's powers—the very prestige that Wilhelm most coveted and by which he was personally affronted. In brief, Britain sought to maintain this state of affairs. Although, as is extensively documented, Britain does not appear to have taken Germany seriously as a specific threat to its hegemony at first, Britain clearly preferred a continuation of the regional status quo to the extent possible.[6] As Paul Kennedy has noted, Britain would make appropriate adjustments—even entailing what in a later era would be called "appeasement"—to its policies when fiscal and other constraints required it to do so, but it could not and would not surrender its regional naval dominance *in toto*.[7]

The competition can therefore be framed in very simple terms. Germany sought to approach naval parity in the North Sea as a means to achieving the level of political prestige, regional military power, and access to colonies enjoyed by Britain, amounting to a challenge to Britain's regional hegemony. Britain sought to prevent this from occurring. From the first German Naval Law onward, beginning about 1898, the die was cast. That Germany failed in its attempt is quite obvious. It never was able to challenge the Royal Navy on equal terms, much less achieve any of its more ultimate goals. Britain succeeded in preserving its hegemony on these terms.

Despite this, the outbreak of World War I, which might have been foreseen but which, in the end, neither competitor wanted, must be seen as a limit to Britain's success. Britain succeeded in preventing Germany from winning the competition, but a devastating world war intervened that, regardless of the complex questions of causation it has always posed, may have had something to do with the Dreadnought Race and its course. Not only does this temper Britain's success, but all further analysis of the competition must be viewed through a counterfactual lens. If there are strategic lessons, they can only be drawn on the assumption that better decisions would have led to better results. Appropriate caution and humility must be exercised here, but certain conclusions can be drawn.

The Dreadnought Race in Strategic Perspective

It is a truism that any linkage of the Dreadnought Race to the war's outbreak is likely to be indirect and arguable. The war began in the Balkans, not in the North Sea, and the need for Britain's entry into it was a matter of debate at the time, to say nothing of after the fact. Moreover, the traditional argument linking the Dreadnought Race to Britain's entry into the war involved assumptions about the primacy of structure over agency that are now heavily questioned by historians. Briefly, it can no longer be said that the war began entirely accidentally or over the objections of those involved: high-level decision makers in all the major powers consciously made decisions that were likely to lead to war, or even to go to war, without force or fatalism.[8] For these reasons, the argument that the Dreadnought Race "caused" Britain to go to war with Germany, or even set Germany on a collision course with Britain, cannot be taken for granted, as intuitive as it may seem.

With that said, it is safe to take certain propositions as more or less demonstrated. There is a line of argument, which this study is prepared to adopt, that the Entente Cordiale (less so the Triple Entente) came about at a time of German hostility to Britain, if not strictly because of it; that Germany's challenge to British naval primacy in European waters was a crucial and inextricable part of that hostility; and that the entente, in turn, led to the military conversations, creating Henry Kissinger's "moral obligations" that made British involvement in a continental war very likely once war broke out—even if, as Samuel R. Williamson Jr. persuasively argues, it was less Britain's military than its naval commitment to France that spoke most persuasively.[9] *Likely*, of course, does not mean *certain*, and it can be said that there were plenty of opportunities for Britain to avoid war, albeit perhaps at perilous diplomatic cost, and even that the process by which it ultimately went to war paid only partial heed to previous commitments and existing plans. The decision to go to war was taken at the last minute, and it was not even necessarily precipitated entirely by the German violation of Belgium, even if some members of the British government were in favor of war for that reason.[10]

But this comes second to the more basic point that the Dreadnought Race created an encirclement of Germany and a corresponding environment of hostility that made war more likely and would most likely have had dire consequences even if a completely different chain of events had occurred either in June 1914 or at some other time. The ultimate argument accepted here is not that the Dreadnought Race led to World War I or Britain's involvement in it—in fact, a strong case can be made that just the opposite was true, that the war broke out for extraneous reasons and that, after all, the Dreadnought Race had been mitigated by the time of war's outbreak.[11] Nor is it necessary to subscribe to theses arguing for German "war guilt" to see the connection between the Dreadnought Race and the ultimate outbreak of war between Britain and Germany—and this study does not subscribe to them.[12] All that is accepted here is that the Dreadnought Race led to two linked developments that contributed to the war: the diplomatic encirclement of Germany, in which Britain played a key role, and the linkage of British military and foreign policy to continental developments in a way neither previously seen nor necessarily likely to have been the case. As Matthew S. Seligmann, Frank Nagler, and Michael Epkenhans have argued, echoing an old debate most comprehensively laid out by Paul Kennedy and Ivo Lambi, the Dreadnought Race may not have caused World War I, but it unquestionably created the "environment" for British involvement in it.[13] This commonsense proposition is difficult to dispute, even if the actual causes of the war and the decisions involved in beginning it are far from settled and never likely to be.

There is a reasonable counterargument, to the effect that Britain's involvement in World War I was precipitated by nothing more or less than an inability to remain aloof from continental politics. Briefly, as Zara S. Steiner and Keith Neilson have argued, from a very broad perspective it may well have been impossible for Britain to remain neutral once the July Crisis occurred, insofar as there was no outcome favorable to Britain that could result from neutrality. Either Germany would dominate the continent, or the rest of the entente, having eked out a victory, would not only regard British credibility as suspect but revert to hostility toward Britain in the absence of any reason to maintain good relations.[14] This elegant argument in this way avoids becoming bogged down in documentary details and rests on a simple intuitive understanding of the situation that Britain faced in 1914—one which it might well have faced regardless of whether the Dreadnought Race had ever occurred. But while one can readily concede this point, and even admire its lucidity, one nevertheless must fall back on the aforementioned and equally obvious and lucid point made by Kennedy: that Britain during the July Crisis was debating war with Germany on the side of France and Russia and no other proposition, and that this situation had its origins at least in part in the Dreadnought Race.[15]

Such being the case, this study is prepared to consider the cautious but necessary judgment that each of the competitors in the Dreadnought Race, to the extent that they were responsible for their own welfare and that of their own people, fell down when it came to the question of managing the competition in such a way as to ensure that their desired objectives were met. During an approximately 15-year period, Britain succeeded in out-competing Germany. The next year, it fell into war. Something similar can be said of Germany, even allowing that it did not necessarily intend for war to break out or that, when it decided to go to war, it did so for reasons of its own. The fact that both Britain and Germany's decisions to go to war were made for a complicated set of reasons, many of them quite valid, or that Britain had other options, does not entirely remove the links between their decisions during the Dreadnought Race and the advent of the war that followed, with all its consequences. It is enough here to note that the narrative of the Dreadnought Race must end with the outbreak of war between the competitors that, notwithstanding their having reached a kind of resolution with the German abandonment of *Weltpolitik*, had seen their diplomatic relationship eroded in the course of a high-stakes competition for control of a key piece of ocean. The Dreadnought Race contributed to the conditions that led to war's outbreak—conditions that the various participants were unable to manage successfully—as well as to the conditions at the outbreak of war. This is all that is claimed here, but it is not a claim without implications or consequences. These implications will be considered as the three metrics are examined for each competitor.

The Three Metrics

3. Determine how that objective is manifested in terms of changes in the three metrics for each of the competitors. Determine what resources alpha and beta each have in terms of the three metrics. In so doing, also assess in particular which of the three metrics it is willing to trade for the others, and whether it has an abundance or a scarcity along this metric, to determine whether it is utilizing a strength or a relatively weak area.

By the metrics of security, financial well-being, and intangible goals, Germany can still be said to have been roundly beaten in the competition, and well before World War I removed any doubt. On the other hand, Britain's performance becomes more ambiguous, even as the specific costs and benefits are more precisely tallied. Germany remains a strategic failure, whereas Britain's resource management appears excellent until the context of the impending world war is considered. In this context, a more nuanced picture emerges.

Britain
Security
It must be said that analysis of the overall security of a state from military invasion or attack is a daunting prospect, for a number of reasons. Although approximate calculations of the

military balance between states are quite possible, such calculations can never be exact. As many from Carl von Clausewitz onward have noted, war is the realm of uncertainty, and absolute certainty about the outcome of a war would mean it would never be started.[16] For competitors who are somewhat equal in terms of their ability to harm one another—the very concept of security competition implies a level of equality or competitiveness—there can be no complete certainty. Although an arms race (for that is what the Dreadnought Race was) implies an attempt by at least one side (if not both) to gain military dominance over the other, realistically, the very fact that competition exists means that such dominance will be difficult to obtain. Moreover, all security is not created equal. A threat to one's capital city or home territory is graver by nature than a threat to a far-flung imperial province that, under some set of diplomatic conditions, might be ceded to avoid a fight. As Clausewitz came to realize late in the writing of his magnum opus, *On War*, although wars may tend toward "the maximum exertion of strength," it is often the case that they are started and even ended with the aim of achieving limited objectives.[17] Political factors also come into play: a state may control and rely on its own forces, but control of an ally's forces is far less likely.

For all that, it is both possible and necessary for states to make some judgments about how secure they are and what risks they can afford to run, and it is possible for an outside analyst to examine such judgments. In the case of the Dreadnought Race, the metrics for success, where security is concerned, are actually fairly straightforward. For although the Dreadnought Race had profound political implications and was run for reasons that initially had little to do with apprehended threats, it was, in the end, simply a naval arms race into which other means were brought. It can therefore be understood by analysis of the naval balance between Britain and Germany on the one hand, and the military and security implications of the overall political balance on the other.

Chapter Six

The Dreadnought Race may have begun as a prestige gambit on the part of Germany, but it had serious implications for Britain's vital security interests. Britain had a choice (at some level) as to whether to compete to hold onto its global hegemony, but it had rather less of a choice about the security of its home waters. Fundamentally, Germany's decision to pursue naval, rather than land-based, strength amounted to a threat to Britain's home territory. Maintaining naval preeminence, and specifically naval supremacy in northern European waters, had therefore long been a vital security interest for Britain; without it, Britain was vulnerable to blockade or even (in theory) to invasion. For Britain, naval primacy was not simply what the Dreadnought Race was about; it was all that mattered, and matching Germany's capital ship fleet in North Sea waters was of vital importance.[18]

The naval balance between Britain and Germany before, during, and after the Dreadnought Race can be fairly easily assessed, given the basic symmetry in technology involved, though the quality of the ships that each side deployed could vary and was also relevant. A complication arises from the need to accurately (for statistical purposes) categorize ships of varying technological advancement and capability. This was difficult even at the time, given the aforementioned debates about speed versus armor, advances in gunnery, and disputes over the merits of creating a separate category of battlecruisers as opposed to battleships. This study tends to adopt British admiral Sir John Arbuthnot Fisher's view that the distinction was minimal for the macro level of military analysis—"simply *armoured* ships"—despite the speed-versus-armor tradeoff between the two and longstanding legitimate arguments over the ultimate performance of these ships in the 1916 Battle of Jutland.[19] For this reason, this study counts both battleships and battlecruisers as capital ships.[20] Further difficulties arise in characterizing a "second-class" or "second-rate" ship as a capital ship, particularly as, with changes in technology, the previous year's state-of-the-art capital ship could become second-rate.

The Dreadnought Race in Strategic Perspective

In general, this study counts only ships specifically styled "first rate" in any given year as capital ships in that year. There was considerable rotation, localized to three specific periods during the competition: the initial decision by Tirpitz to build new, technologically advanced battleships under the First Naval Law, from 1898 onward; the launching of HMS *Dreadnought* and the subsequent scramble to replace obsolete ships on both sides, beginning around 1905–7; and the creation of the first oil-fired "superdreadnoughts" by the Royal Navy in the last few years preceding the outbreak of World War I, a move that relegated the old dreadnoughts to second-rate status and that was not copied by Germany. Strictly for counting purposes, this study treats dreadnoughts as battleships prior to the advent of the superdreadnought and as battlecruisers thereafter. This does not appear to have a substantive effect on the overall count of ships in each competitor's fleet, since repurposed dreadnoughts are still counted as capital ships.

To get a full picture of each side's naval strength, it is useful to count not only capital ships, but fighting ships in general, as well as overall fleet size. The first category, as noted, counts first-rate battleships and battlecruisers; the second category, any and all seaworthy and combat-ready large vessels, excluding small torpedo boats, destroyers, and submarines; and the last category, the total number of ships of any kind deployed by each navy. The strengths and weaknesses of relying on any individual one of these metrics should be obvious.

Moreover, a word must be said about torpedo boats and submarines, which were coming to be seen as the wave of the future and a possible technological paradigm shift in naval warfare. As noted, Fisher and others anticipated a day in which these craft might supersede surface ships with conventional armament; they may, in their more radical moments, have contemplated replacing battleships with submarines. For all that, their actions did not match their speculations: when hard decisions had to be made, battleships were built and relied on. A word will be said about the balance of small craft between the two competitors, but the choice to compete to build fleets of large capital ships rather than torpedo boats was a carefully considered, well-taken decision: there was a reason for it.[21]

At the beginning of the Dreadnought Race, in 1898, Britain's fleet of capital ships—39 all told—was more than five times the size of Germany's capital ship fleet (figure 2), which had only 8 (rapidly obsolescent) ships. This ratio (487.5 percent) was actually historically high. The historical average from 1885 to 1898 was 343.9 percent, although prior to 1885 Germany had possessed essentially no capital ships afloat that could compete with Britain's top ships. Although this study treats the Dreadnought Race as having begun with Tirpitz's naval laws, the earlier modernization of the German fleet can nevertheless serve as a reminder that Tirpitz's and Kaiser Wilhelm's ambitions did not come as a bolt out of the blue. During the Dreadnought Race—from 1899 through 1915, when several ships planned before the war were finally launched—the ratio of British to German capital ships averaged 409 percent. This figure is misleading, however, and largely derives from the temporary boost to Britain's position offered by the "reset" that accompanied the launching of HMS *Dreadnought*. In actuality, the capital ship ratio during the Dreadnought Race was quite a bit lower than this: it averaged 165.75 percent between 1906 and 1915, finishing at 147.6 percent in 1914 and 156.5 percent in 1915, when World War I was already on. The highest it rose consistently after 1905, let alone after *Dreadnought* was launched, was 200 percent in 1910.[22]

It is clear from these figures, which tell the story more plainly than accounts of parliamentary debates ever can, that Britain struggled to maintain its naval advantage over Germany during this time period. By historic standards, Britain really was in actual danger of losing its preeminence and with it its national security. The one thing Germany succeeded in accomplishing as a result of the Dreadnought Race was producing a modern fleet that was closer to parity with Britain's than ever before, even if it did not match the Royal Navy ship for ship.

The Dreadnought Race in Strategic Perspective

Figure 2. Ratio of British capital ships to German capital ships, 1875–1915

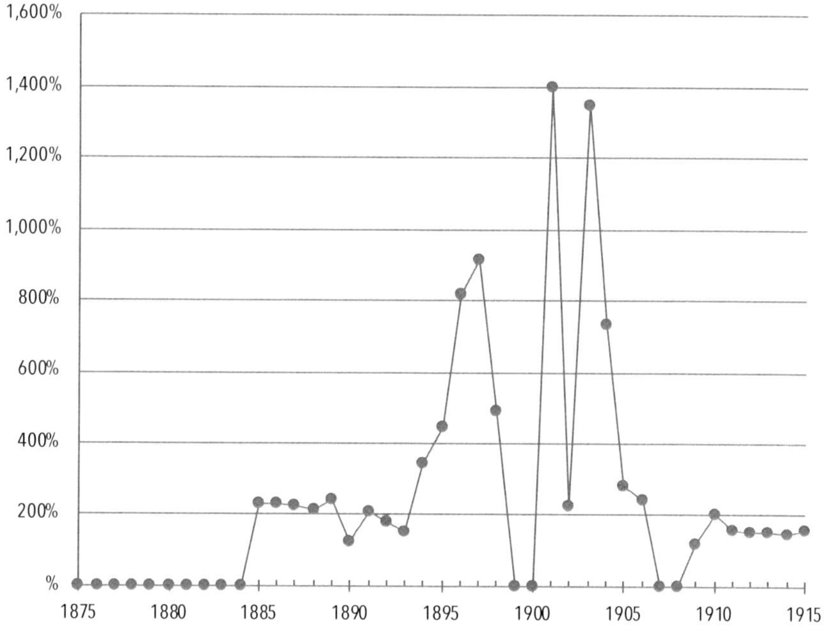

Source: compiled by the author, adapted by MCUP. Points that show 0 percent are a divide-by-zero error indicating the absence of German capital ships qualitatively understood as such. See appendix for sources and methodology.

Despite this, and with the benefit of hindsight, the Imperial German Navy never really acquired the resources with which to challenge the Royal Navy. The vaunted and feared High Seas Fleet spent World War I bottled up in port, unable to do anything about Britain's wide blockade, much less threaten Britain. In that sense, it just barely served its hypothetical role as a deterrent to invasion (which would have been a long shot anyway) but utterly failed to do anything else. The great historical test of both sides' national security strategy *vis à vis* each other is the laboratory of combat. Although the Battle of Jutland was a tactical and operational disappointment (even embarrassment) for the Royal Navy, it proved that the High Seas Fleet could not win a favorable operational decision. Britain's national security, as measured against a historic high, may have been reduced but not unacceptably so.

Chapter Six

Other measurements tell a similar story. The rapid, half-planned obsolescence of older capital ships—with the introduction of HMS *Dreadnought*, *Invincible*-class battlecruisers, and *Orion*- and *Queen Elizabeth*-class superdreadnoughts—did have a silver lining for Britain's overall naval posture: it increased the total number of combat-ready vessels available to the Royal Navy even as it meant that most of these vessels were of lesser value. Britain's average ratio of major combatant vessels (i.e., large surface ships, whether capital ships or of lower rates) to Germany's actually increased during the Dreadnought Race (figure 3).[23] The historical average from 1875 to 1898 had been 274 percent, allowing for serious caveats regarding what counted as a combat-ready ship in an age of rapid innovation and obsolescence. From 1899 to 1915, the average was 331 percent. This figure is somewhat misleading, as from 1904 onward the average was merely 177 percent—proof that Germany was catching up. Still, from this perspective, Britain was well ahead of Germany by a comfortable margin. Even accounting for the fact that the ratio declined from a peak of 628 percent in 1902 just after the start of the Dreadnought Race to a lesser 175 percent in 1913, Britain still finished well ahead of Germany.

In terms of the sheer number of vessels in each navy (figure 4), the ratio of Britain's overall fleet size to Germany's averaged 211.6 percent from 1875 to 1898 but only 203.5 percent from 1899 to 1915, peaking in 1903 at more than 401 percent with another peak in 1908 (after *Dreadnought* was built) at 245.45 percent. Once again, the intensity of the competition is reflected in the statistics. Overall, Germany was gaining on Britain, though not to the point where it was close to overtaking it.[24]

Another area in which Britain's position improved during this timeframe was that of small attack vessels, including submarines, torpedo craft, and small destroyers. As noted, the utility of these vessels for control of the high seas rather than coastal defense remained unproven at the time, though they had some utility as a force multiplier given their capabilities for flotilla defense, as previously discussed. It is therefore inter-

Figure 3. Ratio of British major combatant fleet to German major combatant fleet, 1875–1915

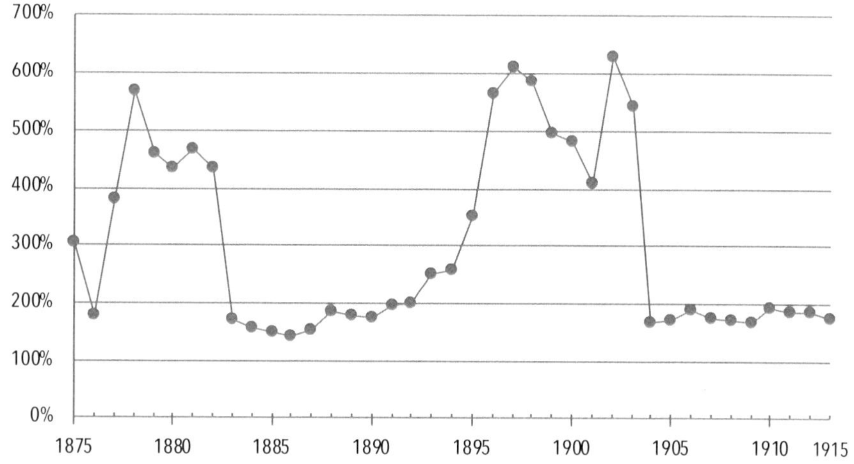

Source: compiled by the author, adapted by MCUP. See appendix for sources and methodology.

esting to count such craft, insofar as their hypothetical utility for such purposes was at least under discussion at the time.

Germany had initially possessed a small advantage in this area, in that it was the first to deploy any torpedo boats, during a time when such craft were more of a curiosity than anything else. From 1875 to 1884, Germany possessed a pair of such craft, which were used to impress international onlookers, while Britain had possessed none.[25] By the 1890s, however, both Britain and Germany were making serious efforts to acquire small assault craft. Britain acquired its first in 1888, and by 1894 the two powers were essentially at parity in this area, with Britain's torpedo-craft fleet at a 101.5 percent ratio to Germany's (figure 5). The average ratio of British to German small assault craft fleets from 1893 to 1898 was 104.9 percent. Subsequently, from 1899 to 1915, the average ratio was 199.9 percent. Effectively, therefore, during the Dreadnought Race, and even while racing to build dreadnoughts, Britain doubled its strength in small craft relative to Germany's. As noted above, the potential uses and effectiveness of such craft and the associated doctrine

Figure 4. Ratio of British fleet size to German fleet size, 1875–1915

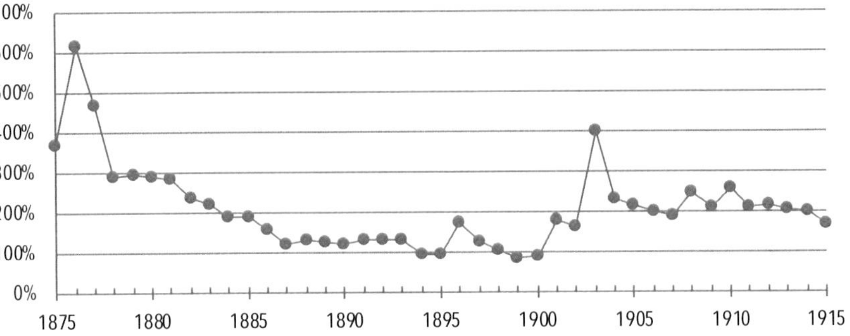

Source: compiled by the author, adapted by MCUP. See appendix for sources and methodology.

were all evolving and in question at the time. Whatever the case, Germany lost out even in the balance of unconventional craft, regardless of their ultimate utility. Germany's inherent disadvantage in naval procurement meant that even if it competed with Britain in one area, it lost in others. In any case, this was another area in which Britain made net gains *vis à vis* its German competitor during the Dreadnought Race.

Matthew Seligmann's more intriguing arguments regarding Germany's bureaucratic preference, reflected in the deliberations of the Naval High Command even as they were shunted aside by Tirpitz and Wilhelm, for waging a *handelskrieg* (trade war) against British shipping using refitted ocean liners have been discussed in the last chapter. Because the Dreadnought Race was an attempt by Germany to outdo Britain in conventional battleships and dominance of the North Sea rather than in commerce-raiding capabilities, these arguments are largely outside the scope of this study. However, Seligmann's arguments raise a very important point: to the extent that Germany and Britain were on a collision course for war (about which more will be said shortly), the Dreadnought Race imposed bureaucratic and fiscal constraints on its historically preferred (and more viable) commerce-raiding strategy that ran to the detriment of both. In this sense, the Dreadnought Race further

Figure 5. Ratio of British torpedo-craft fleet to German torpedo-craft fleet, 1894–1914

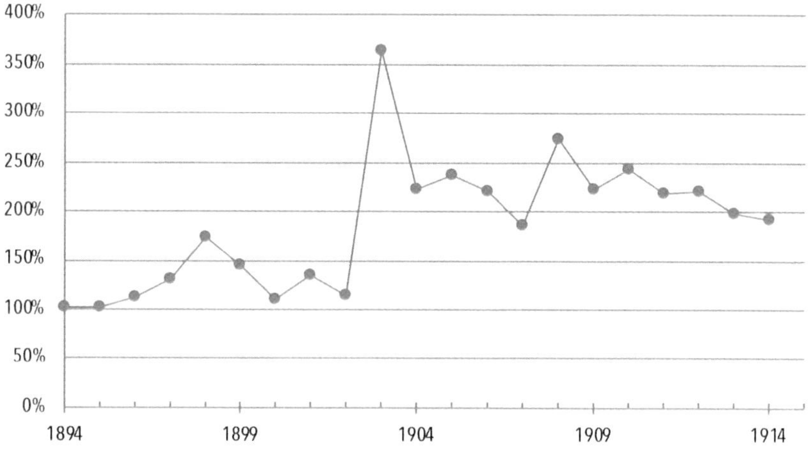

Source: compiled by the author, adapted by MCUP. See appendix for sources and methodology.

contributed to a military balance that was advantageous for Britain but disadvantageous for Germany.

In the political realm, Britain, paradoxically, both succeeded and failed. It succeeded in containing its competitor and bringing to bear the maximum amount of international effort against Germany, ensuring that Britain faced as little force as possible and Germany faced as much. If one ends the narrative in 1912, when Germany effectively abandoned *Weltpolitik* as a policy, scaled back its rhetoric, engaged in arms control talks, and initiated a process of diplomatic conciliation that included some substantive overtures, the Dreadnought Race had been won outright by Britain there and then. It is only if one links the developments in the Dreadnought Race to British involvement in World War I—arguing either that they precipitated the war or at the least ensured that Britain would be dragged into it—that the Dreadnought Race's outcome appears tragic.

It goes without saying that World War I, though it did not threaten British home soil, nevertheless weakened Britain greatly and ruined its international position, quite apart from its human costs.[26] In the lead-up to the war, Britain's policies of conciliation toward France and Russia, undertaken at least in part to contain Germany—however debatable the exact motives of each decision maker—served to accomplish precisely that. The practical effect of the Triple Entente was not merely to contain Germany, but to ensure, in a zero-sum world, that the maximum number of guns was pointed at Germany and the minimum number pointed at Britain. As former German chancellor Otto von Bismarck had understood, but as Kaiser Wilhelm and his ministers apparently had not, in a multipolar world there were obvious benefits to belonging to the larger coalition. Because Germany, post-Bismarck, declined to achieve this, Britain did so, with the result that Britain, being allied with France and Russia, did not have to fear them and could end military competition in favor of open cooperation, while Germany lost all diplomatic and military flexibility.

It is even possible that the Dreadnought Race compromised Germany militarily in the opening phase of the war. As Peter Padfield notes, the German Army was as vital to Germany's basic territorial security as the Royal Navy was to Britain's, and the funds appropriated for the horrendously expensive High Seas Fleet could have been more useful paying for a few extra infantry divisions—which may, just possibly, have robbed Germany of a necessary quick victory in the western theater in the war's opening weeks.[27] It is impossible to say for certain, but the famous "Miracle on the Marne" in September 1914, in which British and French troops turned back the German advance just east of Paris, may well have been made possible almost exclusively by Germany's antagonism of an offshore power that historically had acted as an occasional ally. Competing with Britain at sea diverted resources that Germany needed to win on land while simultaneously making the deployment of the British Expeditionary Force (BEF) into the fight on land more likely—a double blow to Germany's national security that was acutely felt. The Dreadnought Race had therefore affected the military balance on land at the start of World War I as assuredly as it had failed to alter the military balance at sea against Britain.

However, whether the Dreadnought Race alone caused the dispatch of the BEF is a questionable proposition given Britain's historical attachment to securing Belgium and the Netherlands, which lay in the path of most reasonably anticipatable German war plans from the 1890s onward, and also given the other factors involved. The political fundamentals—Britain's aversion to involvement on the continent pitted against its need to preserve a balance of power there—were not entirely altered by the Dreadnought Race, which merely added some additional weight.[28] All that is assumed here is that there was a tenuous but observable connection between the Dreadnought Race, the military conversations, Britain's ultimate decision during the July Crisis to go to war (at least insofar as Britain had only one meaningful potential enemy to consider by this point), and the dispatch of the BEF to the continent once war was underway—and that, to the extent that the former two made the last likely, British involvement in World War I must be counted among their costs.[29]

Conversely, as noted, even if the war occurred for other reasons, and even if the dispatch of the BEF was a tragic inevitability, one can at least say that the Dreadnought Race did nothing for Germany where it mattered most, and that it may have made the BEF's task ever so slightly easier. Britain had done what it could to make the German naval challenge futile; the war was out of any single state's hands. In that sense, Britain won the Dreadnought Race. If, as noted in chapter 3, war cannot be prevented, then the best one can do along the security metric is to ensure that one is properly prepared. Britain had, if not maximized its security on the eve of war, at least increased it relative to what it might have been. The opposite was true for Germany.

Given all this, the way to understand Britain's performance in the security metric is twofold: whether it neutralized the German threat to the naval balance that its security depended on; and whether doing so affected its readiness for a war that it at least arguably could not prevent. It achieved the former and at least arguably the latter. Where the former was concerned, the German abandonment of *Weltpolitik* as anything more than a halfhearted slogan, its shift in focus back to continental matters, and its conciliatory gestures after 1912 all argue that Britain had decisively outcompeted Germany and restored an adequate military balance. There is little evidence that the Dreadnought Race contributed to British unreadiness for war. To the contrary, the upshot of the Dreadnought Race was a reinvigorated Royal Navy.

Winning the race with no finish line, as the quotation that inspired the title of this book reminds us, requires convincing the challenger to quit. As discussed with regard to Gary P. Hamel and C. K. Prahalad's arguments above, and in more detail regarding this case below, Britain did ultimately succeed in making further competition futile, even if the Dreadnought Race did not entirely end. Britain's security was not assured—the war, after all, was ruinous—but Britain was as secure as it could reasonably be expected to be, and it survived.

Finance and Welfare

It was in the financial realm that Britain suffered most as a result of the Dreadnought Race. Although it was able to use resources more efficiently than its competitor, it suffered a net loss to its people's welfare and treasury's assets, albeit one minimized by frugality.

The most obvious measure of the cost of the Dreadnought Race to Britain's finances was the increase in defense spending to finance the necessary naval expansion, an increase not offset by any obvious financial gain. In a different universe, an increase in defense spending might have amounted to an investment in the riches of empire that, however illegitimate by modern standards, might have added to Britain's coffers. Deterring German expansion and confronting its naval buildup offered no such offset. Where the finances of the British Empire were concerned, it amounted to a deadweight loss.

The Dreadnought Race in Strategic Perspective

The gradual increases in defense expenditures that occurred during the timeframe of the Dreadnought Race appear small only by the standards of the era, in which government spending was much lower than today. In relative terms, they were both sudden and monumental. From 1875 to 1898 (the year before the Boer War, which took its own toll on Britain's finances), Britain's overall defense expenditure (figure 6) averaged about 2 percent of its gross national product (GNP).[30] From 1903 (after the end of the Boer War, and therefore not counting increases for it) to 1913, Britain's overall defense expenditure averaged 3.1 percent of GNP—notwithstanding the fact that GNP was rising. In net terms, defense expenditure as a percentage of GNP rose in Britain from 2.16 percent in 1898 to 3.22 percent in 1913, a net increase of 1.1 percent.

The increase was attributable in the main, if not entirely, to naval spending, and although Britain had other concerns during this time, this increase in naval spending in turn is attributable in aggregate terms to Britain's naval competition with Germany. Although historical analyses may differ as to the degree to which Britain's naval policies were driven by the Dreadnought Race as opposed to other factors, as well as on the degree to which Britain could be said to have specifically concerned itself with the German challenge at the start of the decade, it can nevertheless be said with some accuracy that Britain's increases in naval expenditure served the purpose of confronting Germany and would have had to do so whether or not its decision makers had that specific purpose in mind. Such being the case, Royal Navy estimates (figure 7) ran from £22,338,000 in 1898 to £51,550,000 in 1914, an increase of 131 percent. As a percentage of GNP (figure 8), Royal Navy estimates increased from 1.18 percent in 1898 to more than 2 percent in 1913. The average naval expenditure as a share of GNP between 1875 and 1898 was 0.86 percent; after 1903 (and the end of the Boer War) until 1913, naval expenditure averaged 1.75 percent of GNP.

The overall effect can be seen in naval estimates as a share of overall defense spending (figure 9). In 1898, estimates for

Figure 6. British total defense expenditures as percentage of gross national product, 1875–1913

Source: compiled by the author, adapted by MCUP. See appendix for sources and methodology.

the Royal Navy amounted to 54.9 percent of British estimated overall defense expenditure; as of 1914, the figure was 64.1 percent. Historically, the 1898 figure was high—the average share of overall expenditure allotted to the Royal Navy between 1875 and 1898 had been 43.5 percent. After the Boer War, from 1903 to 1914, the Royal Navy's share of the overall estimates averaged 57 percent.

The Royal Navy estimates increased by an average of 3.6 percent annually between 1875 and 1898. Between 1899 and 1913, the estimates increased by an annual average of 5.6 percent. Not only was the naval budget increasing, but it was also increasing at a faster rate (figure 10).

All these data can serve to illustrate, in very tangible terms, the price of naval primacy. Although in any particular instance

The Dreadnought Race in Strategic Perspective

Figure 7. British naval estimates in pounds, 1875–1914

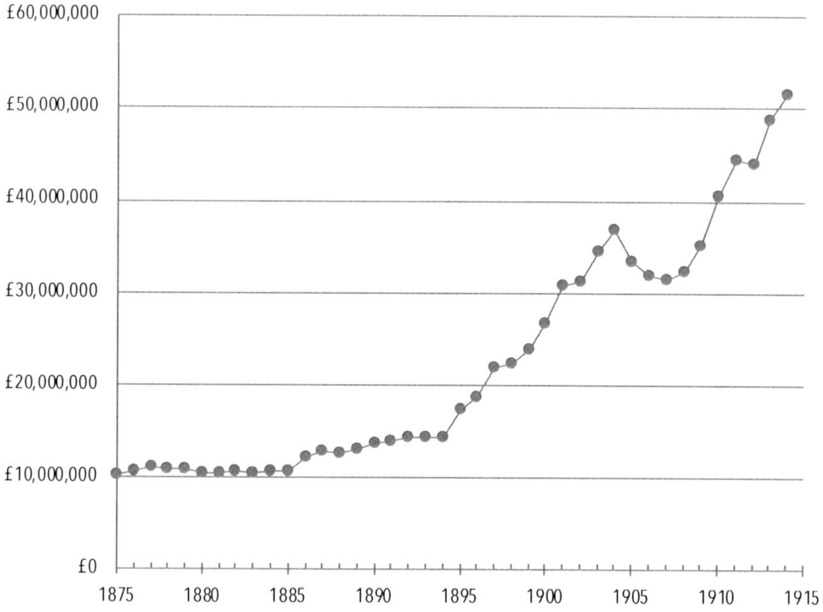

Source: compiled by the author, adapted by MCUP. See appendix for sources and methodology.

other factors besides the Dreadnought Race can be cited as underlying Britain's naval expenditure, it is indisputable that Britain's naval competition with Germany became the main driver of its naval policy during the course of the first decade of the twentieth century, and that this is reflected in its naval expenditures.

By contrast, the British Army showed negligible increases in personnel during this time, as well as comparatively small increases in absolute and relative spending (figure 11). Once Boer War mobilization and demobilization are accounted for, Britain's Regular Army was about the same size at the height of the Dreadnought Race as it had been before: 184,853 in 1898, versus 181,100 in 1913. If anything, as these figures show, it had shrunk slightly.

Once various part-time forces (e.g., territorials, yeomen, etc.) are included, the figures show only a small increase: 677,314

Figure 8. British naval estimates as percentage of gross national product, 1875–1914

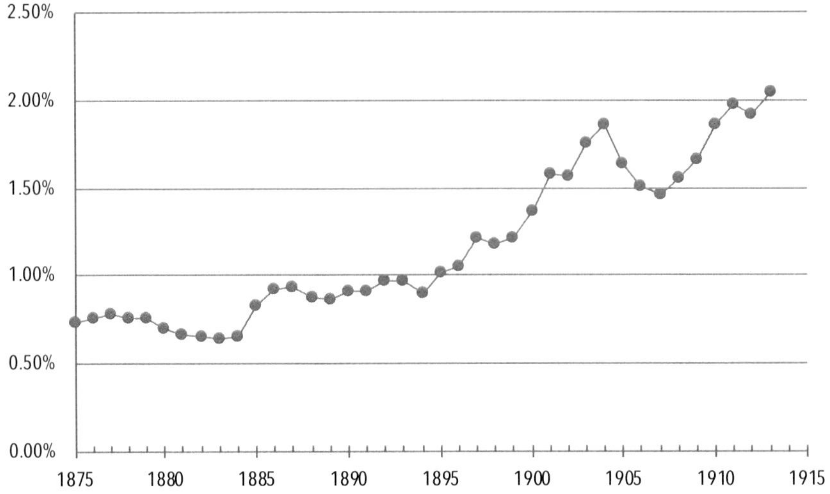

Source: compiled by the author, adapted by MCUP. See appendix for sources and methodology.

in 1900 (the first year for which such figures were available), and 711,575 in 1914 (figure 12).

Expenditure per professional soldier (the army estimates divided by the strength of the regular army) rose from £99 in 1898 to £156 in 1913, and in general, the army estimates also increased, from £18,340,500 in 1898 to £28,845,000 in 1914 (figures 13 and 14).

It is clear, however, that the British Army's funding growth was less dramatic than the Royal Navy's, showing a net increase of 57.3 percent as opposed to the navy's 131 percent (figure 15). In general, army estimates increased annually by an average of 1.2 percent from 1875 to 1898. From 1904 (by which time the army had completed its post-Boer War drawdown) to 1914, army estimates were flat (with 0.04 percent average annual growth).

Unlike the Royal Navy, the British Army not only grew at slower rates generally, but it also had its rate of funding growth slowed by the Dreadnought Race, during which the navy's funding growth had accelerated. This might be fully expected

The Dreadnought Race in Strategic Perspective

Figure 9. British naval estimates as percentage of total defense budget, 1875–1914

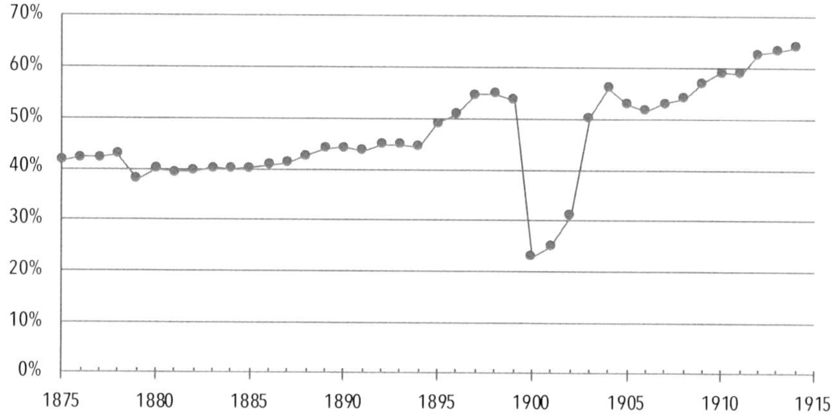

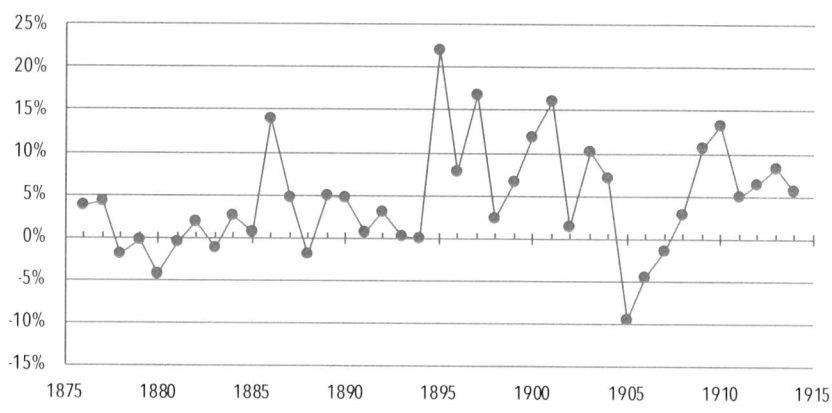

Source: compiled by the author, adapted by MCUP. See appendix for sources and methodology.

Figure 10. British naval estimates' growth rate from previous year, 1876–1914

Source: compiled by the author, adapted by MCUP. See appendix for sources and methodology.

under the circumstances, but in any event the statistics tell the tale quite clearly.

All in all, the share of defense budget growth for Britain that could be attributed to the Royal Navy as opposed to the British Army had long been higher, but the rate of growth increased in a serious fashion during the Dreadnought Race. The average annual rate of increase in overall British defense estimates from 1875 to 1898 was 2.3 percent, with increases in naval expenditure accounting for an average of 77.4 percent of these overall increases (figures 16 and 17).[31]

From 1905 to 1914 (setting aside the Boer War increases and drawdowns), overall British defense estimates increased at the slightly slower average rate of 2.2 percent, but the Royal Navy accounted for an average of more than 100 percent of annual increases in defense estimates where increases occurred (figure 17). The British Army's share was actually negative, as army estimates were reduced to make room for naval spending. In net terms, increases in naval expenditure accounted for 63.3 percent of increased defense spending in 1898 and 81.4 percent in 1914—a pronounced upward trend. By any account, the Royal Navy was the senior service by more than honorifics during this period. Again, one does not have to accept that all British naval policy makers were at all times concerned solely with Germany during this period to see the general trend at work and to note that the Dreadnought Race was an obvious driver of it.

The cost to the British taxpayer was in proportion to this financial reality. British defense spending grew relative to GNP during this time. Where the latter was concerned, Britain's GNP grew at an average rate of 1.9 percent per year from 1875 to 1913, but its growth rate slowed during the Dreadnought Race. From 1875 to 1898, British GNP grew an average of 2.1 percent per year; from 1899 to 1913 the annual average was 1.6 percent (figure 18).

Though one cannot know for sure, it is possible that the diversion of significant funds from the private sector to the government impacted investment, and therefore GNP growth, during this period. Even if this was not the case, the Dreadnought Race was a drain on British taxpayers' pocketbooks, increasing taxation and deficit spending, and in this way consuming resources that could have been used for government social spending, investment, or private consumption. Moreover, the least that can be said is that this occurred at a time

The Dreadnought Race in Strategic Perspective

Figure 11. British Regular Army total personnel, 1875–1913

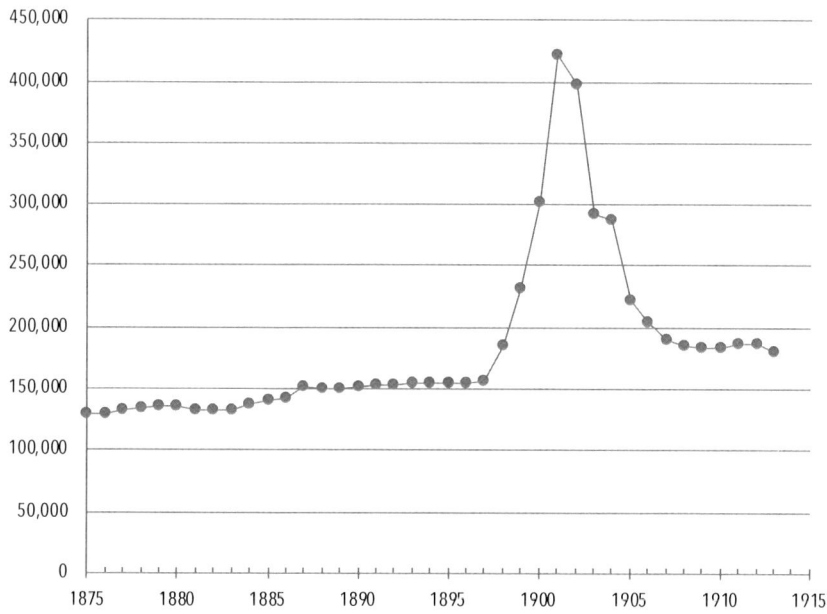

Source: compiled by the author, adapted by MCUP. See appendix for sources and methodology.

Figure 12. British Army total effective personnel, including reserves, militias, and others, 1900–14

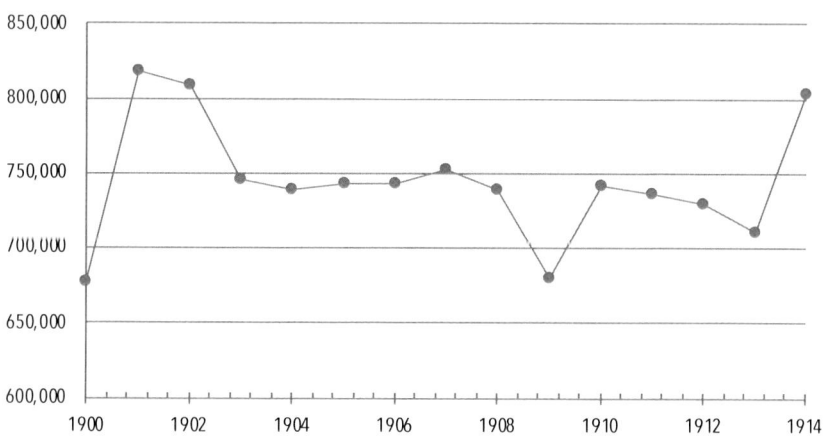

Source: compiled by the author, adapted by MCUP. See appendix for sources and methodology.

Figure 13. British Army estimates in pounds, 1875–1914

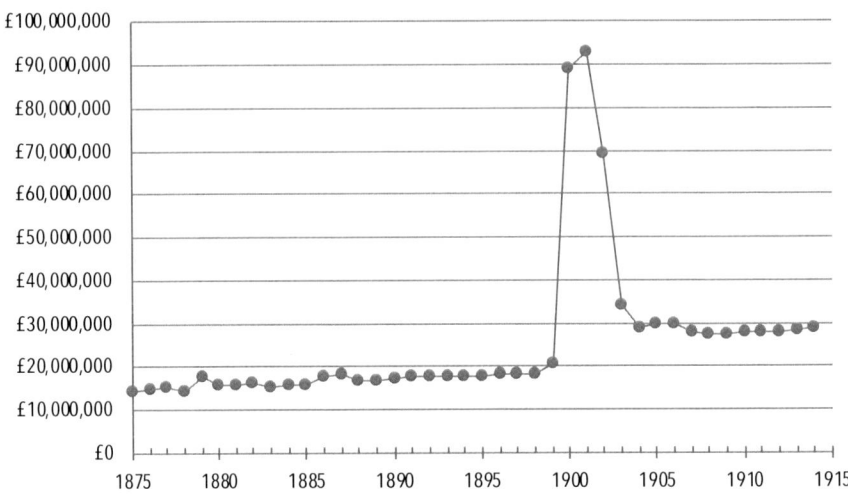

Source: compiled by the author, adapted by MCUP. See appendix for sources and methodology.

Figure 14. British expenditures per soldier in pounds, 1875–1913

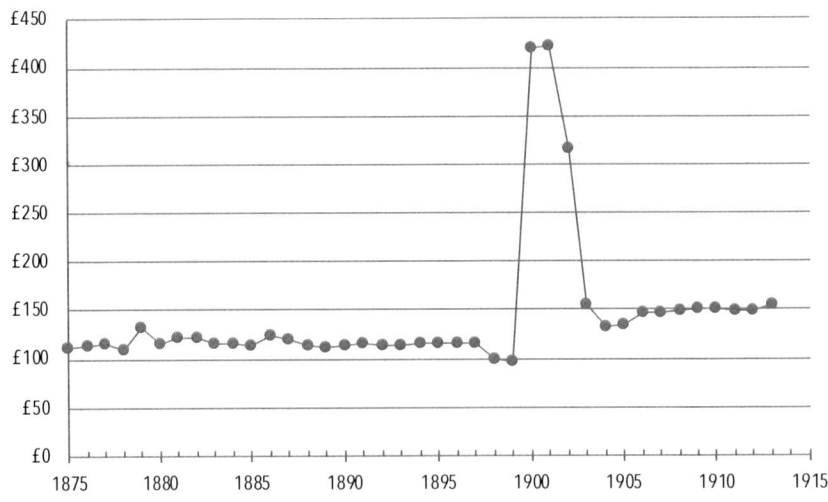

Source: compiled by the author, adapted by MCUP. See appendix for sources and methodology.

The Dreadnought Race in Strategic Perspective

Figure 15. British Army estimates' growth rate from previous year, 1876–1914

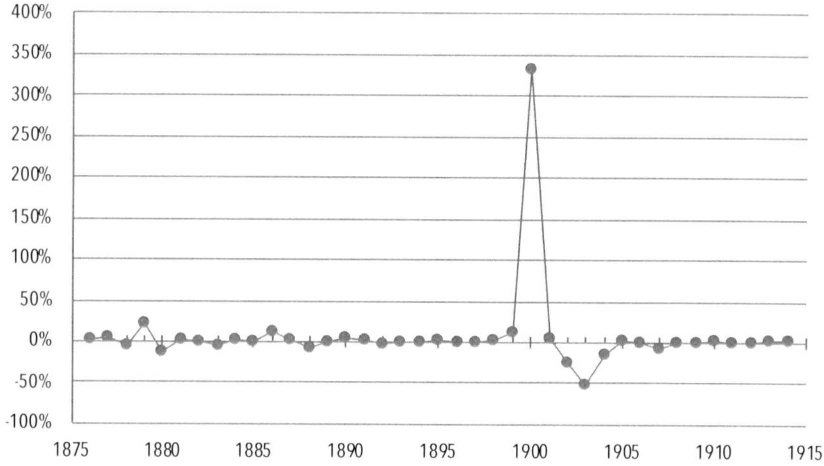

Source: compiled by the author, adapted by MCUP. See appendix for sources and methodology.

of slowing growth, when private citizens could justly be said to need the funds being requisitioned all the more.

Britain's public spending grew during the Dreadnought Race for reasons of both foreign and domestic policy (figures 19 and 20). It is true that this period also saw the beginnings of the modern British welfare state, including old age pensions and the national insurance program. Funding for these programs is partly reflected in increases in British revenues relative to GNP at this time, and in the aggregate these represent funds diverted for legitimate domestic purposes. But competition with Germany also played a role, which can be quantified. From 1875 to 1898, British revenues as a percentage of GNP averaged 5.8 percent. From 1898 to 1913, they averaged 7.2 percent—effectively a 25-percent increase in taxation relative to production and a net increase of 1.4 percent. Likewise, from 1875 to 1898, British overall expenditure as a percentage of GNP averaged 5.7 percent; from 1898 to 1913, it averaged 7.4 percent, a 30-percent increase in expenditure relative to production and a net increase of 1.7 percent.

Chapter Six

Figure 16. British defense budget growth rate from previous year, 1876–1914

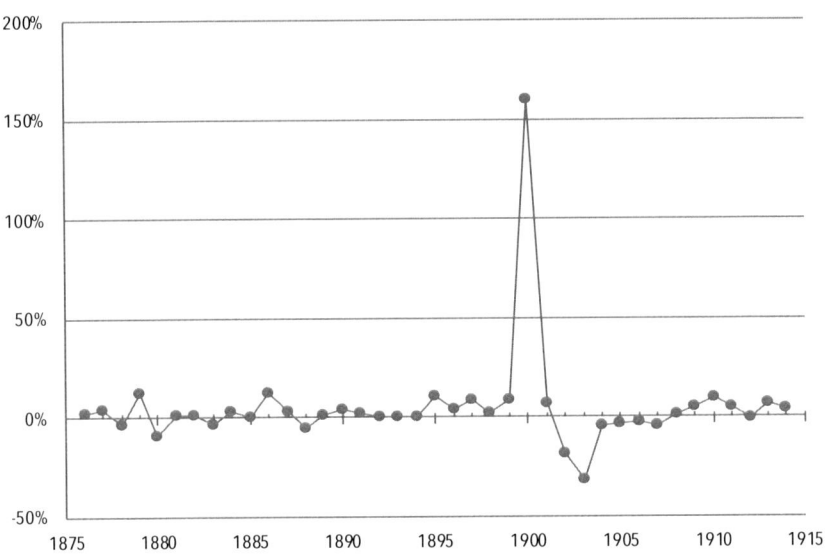

Source: compiled by the author, adapted by MCUP. See appendix for sources and methodology.

Figure 17. Share of British defense budget growth attributable to Royal Navy estimates increase, 1876–1914

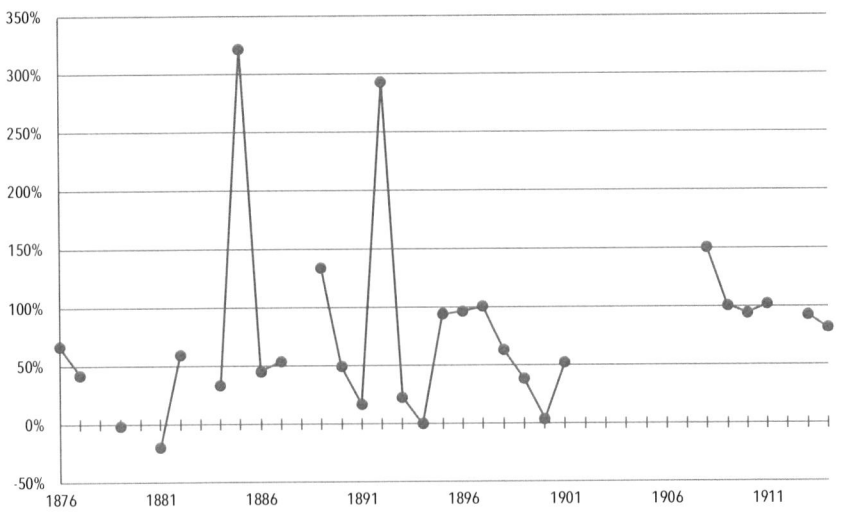

Source: compiled by the author, adapted by MCUP. Gaps indicate years where no documentable positive defense budget increase occurred. See appendix for sources and methodology.

The Dreadnought Race in Strategic Perspective

Figure 18. British gross national product growth rate from previous year, 1876-1913

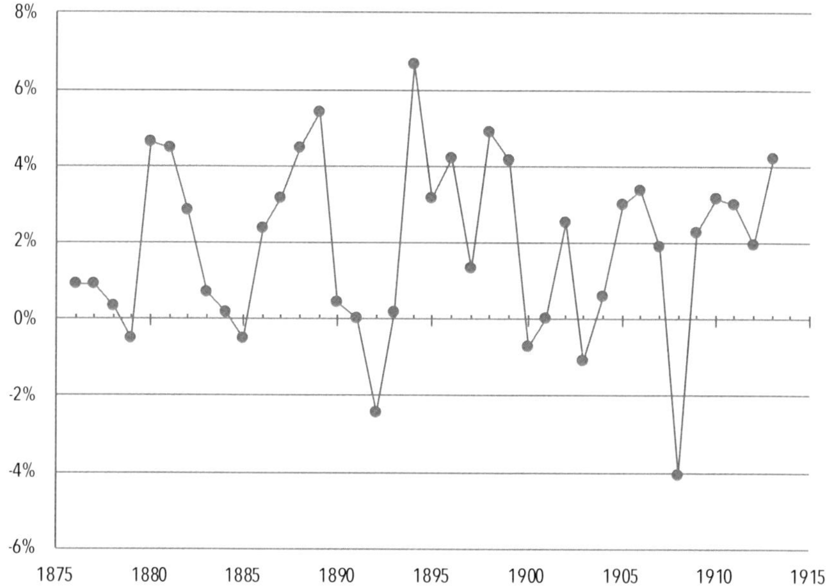

Source: compiled by the author, adapted by MCUP. See appendix for sources and methodology.

As noted, Britain's average defense expenditure was higher by an amount equal to 1.1 percent of GNP in this period, almost 80 percent of the total increase in tax revenue and 65 percent of the total increase in expenditure. And, as seen above, effectively all of that increase occurred as naval expenditure. Once more, one does not have to accept that these increases were intended solely to confront Germany, or that they ultimately had that effect, to accept that a significant part of them did. The Dreadnought Race, quite simply, ate revenue.

Furthermore, Britain's defense expenditure began to slowly crowd out nondefense spending during this period (figure 21). Prior to the Dreadnought Race, from 1875 to 1898, defense expenditures accounted for an average of 34.1 percent of overall expenditure. From 1899 to 1913, defense expenditure averaged 46.5 percent of overall expenditure. Even discounting the dramatic uptick in defense spending during the Boer War, defense spending increased relative to overall expenditure:

Figure 19. British revenue as percentage of gross national product, 1875–1913

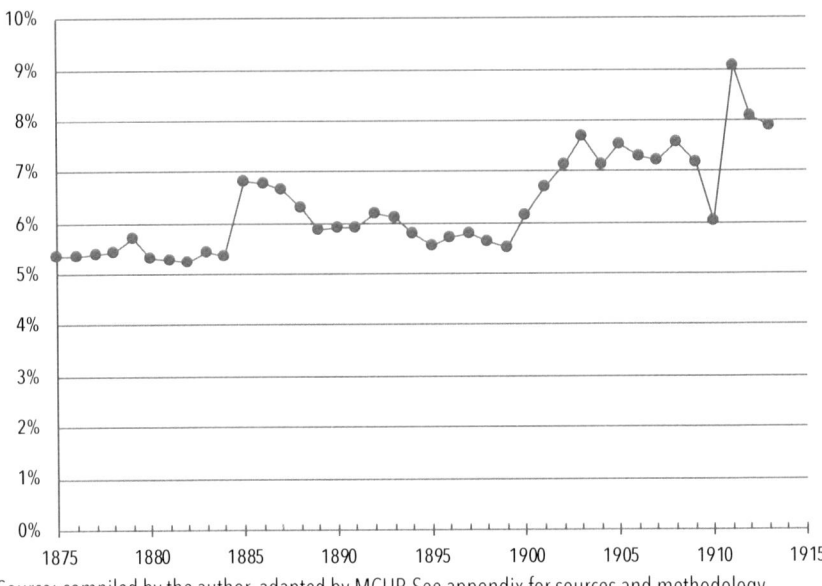

Source: compiled by the author, adapted by MCUP. See appendix for sources and methodology.

Figure 20. British overall expenditures as percentage of gross national product, 1875–1913

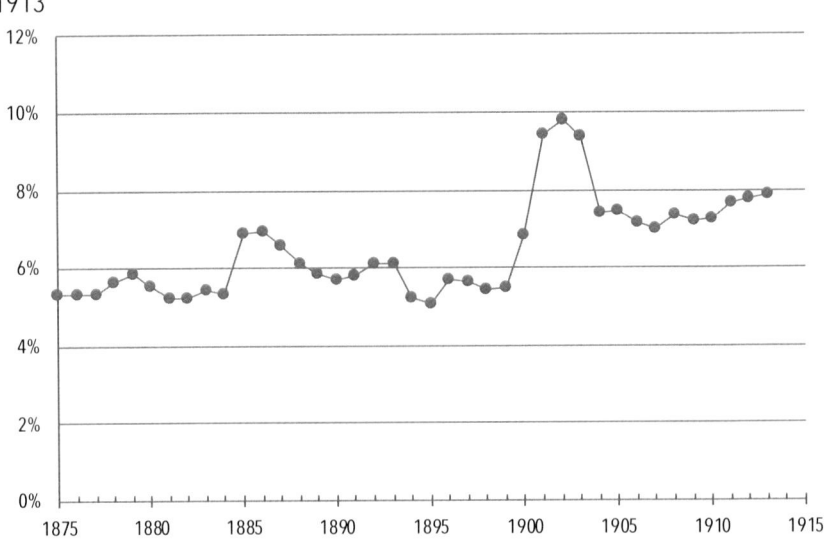

Source: compiled by the author, adapted by MCUP. See appendix for sources and methodology.

defense expenditure from 1903 to 1913 averaged 41 percent of overall expenditure. As of 1913, it was 40.8 percent of the overall total. At least in relative terms, Britain was trading butter for guns, almost literally, to stay afloat. Yet, once more, one may note that although the Dreadnought Race was not solely responsible for this trend, it was most definitely a major part of it, particularly given that the Royal Navy took the lion's share of the increases.

Allowing for all the appropriate historical caveats, the Dreadnought Race, all told, and just in terms of the increases in spending on the Royal Navy, can be said to have accounted for an increase of anything up to 20 percent in British annual spending relative to GNP for each of the years it took place. Effectively, Britain poured more than 1 percent of its annual income down the drain—or, if one prefers, into the sea.

But this is not quite the end of the matter. The increased spending could, of course, be offset both by taxation or debt issuance. Britain was much more heavily burdened by debt than Germany. In 1898, Britain carried a national debt equal to 33.6 percent of GNP (figure 22). What is amazing, however, and what offers some insight into the lack of urgency felt by British policy makers at the time, is that this ratio declined. By 1914, Britain's national debt amounted to a mere 27.7 percent of GNP, a fall of 24 percent in the space of 16 years. In absolute terms, the national debt rose only 4 percent, from £634,435,704 in 1898 to £661,473,765 in 1914 (figure 23). In the interim, the debt rose to a peak of £770,778,762 in 1903 before declining. Incredibly, Britain had actually managed to pay off some debt while competing with Germany for naval supremacy, a fiscal stiff upper lip if ever there were one. Whatever else may be said, the competition did not sap British resources to the point where fiscal prudence had to be sacrificed, and indeed this tracks with the revisionist argument that Fisher's reforms were as much to do with cost-cutting as with confronting Germany, and that a true sense of urgency developed late.

Figure 21. British defense expenditures as percentage of overall expenditure, 1875–1913

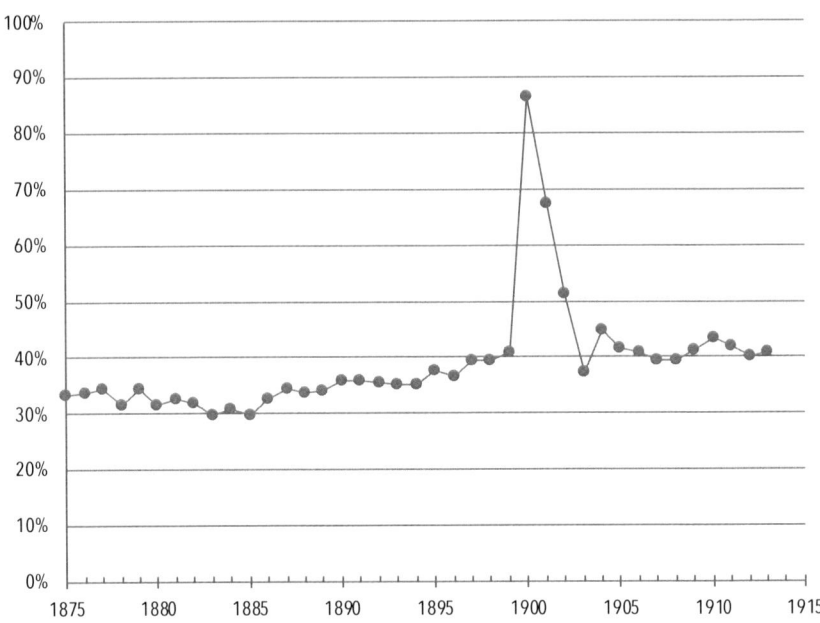

Source: compiled by the author, adapted by MCUP. See appendix for sources and methodology.

Although this may be read as the final word on the matter, there is a little more yet. Where the welfare metric is concerned, this study has yet to reckon with the concern noted earlier regarding the intervention of World War I in the best-laid plans of Britain's policy makers. If one holds that Britain's entry into World War I had somewhat to do with the Dreadnought Race, then Britain's astounding fiscal restraint, which left room for debt reduction but failed to intimidate Germany enough to prevent a threat to the Triple Entente's members, seems not prudent but reckless. Where the million or more British and British Imperial dead of World War I were concerned, they were better indebted than dead.[32] If, however, one hews to the quite reasonable view that, although the Anglo-German rivalry increased the likelihood of Britain's involvement in the war but that the war had a multitude of contributory factors, and that Britain might well have gotten involved regardless, then

The Dreadnought Race in Strategic Perspective

Figure 22. British national debt as percentage of gross national product, 1875–1913

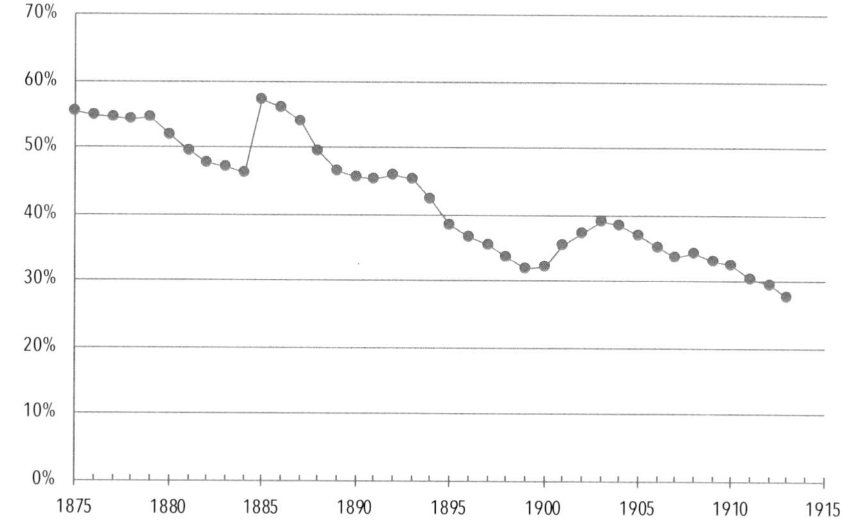

Source: compiled by the author, adapted by MCUP. See appendix for sources and methodology.

Britain's fiscal restraint assumes less significance. This study is prepared to hold that the Dreadnought Race contributed to a less than salubrious international atmosphere and locked Britain into a policy of supporting France against Germany, even on land and at great cost—but it is not prepared to argue that the war could necessarily have been prevented simply by a greater British military deterrent or diplomatic initiative. The international environment in which the war ultimately occurred was far too complex for that.

If anything, therefore, Britain did about as well as it could from the standpoint of its citizens' and treasury's finances, at least in the short run. It is only if one adopts the assumption that a more aggressive British armament policy would somehow have prevented World War I that this policy looks ineffective. Assuming that this was not the case (or that it cannot be proven), Britain actually managed to maintain its position along the first metric (security) at an acceptable cost to the second (welfare). But there is no question that welfare was impacted.

Figure 23. British national debt in pounds, 1875–1913

Source: compiled by the author, adapted by MCUP. See appendix for sources and methodology.

Intangibles

Although one could argue that Britain was not damaged in the area of intangibles by the Dreadnought Race, one must remember that in certain intangible areas, Germany's gain was Britain's loss even if no harm was done.

The British Empire—including the African colonies and the unimpeded access to them that Kaiser Wilhelm had coveted—was probably never in better shape than in the decade preceding World War I. Britain's international prestige had taken somewhat of a blow as a result of the Boer War, which showed not only that ragtag militias could humble a world power but also that a world power that claimed to be liberal could stoop to considerable savagery. In broad terms, however, it is difficult to find any tangible effects of this in the diplomatic developments that occurred in the subsequent decade—except, of course, for the turn that Anglo-German relations took. What is true, however, is that Britain's overall prestige was damaged by the need to confront a rising great power threat for the first time since Napoléon.

What one observes in viewing the deployments of the Royal Navy during the Dreadnought Race are the beginnings of a siege mentality. Initially, as noted, the Royal Navy was a prestige instrument, its captains chosen and promoted for their ability to keep their ships spotless and beautiful. The very fact that gunnery was an afterthought and that, in Peter Padfield's almost poetic characterization, "tennis was played on the Admiralty lawn" bespeaks fearlessness and self-confidence.[33] The Royal Navy was ubiquitous, based not solely in European waters, but worldwide. But during the course of the Dreadnought Race, this posture changed, with almost tragic results. Under Fisher, bases were consolidated, with an eye toward maintaining strength in the North Sea and Channel, effectively localizing Britain's naval hegemony. In a way, the formation of the Triple Entente minimized the harm from this, in that it prevented France, historically a more formidable competitor with Britain for African colonies than Germany (as the 1898 Fashoda Incident demonstrates), from taking advantage of the situation, particularly as it effectively put paid to the longstanding (and slow-moving) Anglo-French naval rivalry, in the process unintentionally rendering the *Jeune École* in its original form a mere hypothetical.[34] The entente also reduced the possibility of general hostilities, and it amounted to an agreement that France would focus its military growth on the landward side. After Fisher's time as First Sea Lord, Britain's naval posture shrunk still further, with the bulk of the Royal Navy now concentrated exclusively in home waters, a buildup to the naval events of World War I, when the Royal Navy concerned almost entirely with blockading German ports. As noted earlier, this even extended to the Mediterranean, where the Royal Navy, under the guidance of Winston Churchill as First Lord of the Admiralty, took the unprecedented step of turning to a flotilla defense utilizing small craft, even as it eschewed this option

Chapter Six

in the much more critical defense of the North Sea, where it adopted the more conservative option of concentrating its capital ships.[35] All of this may have been economical, but the image it presented was scarcely one of strength.

Although it is difficult to quantify or even qualify, Britain's prestige had therefore been dampened. It had been reduced from a world power, to a regional power, to a regional competitor. It will never be known how long this might have continued had World War I not intervened. What is certain, however, is that, to the extent that Germany had sought the humiliation of Britain at any cost (and really to that extent alone), Germany had succeeded in its undertakings. Britain endured a blow to its prestige and therefore suffered a loss in terms of intangibles. British imperial pride, which had tremendous sentimental value (as the Rudyard Kipling quotations that began the preceding chapters can readily attest), had been chastened. Given a choice, British policy makers would no doubt have preferred it not end this way, even before World War I began its terrible course. Although they maintained a kind of stiff upper lip, and although, as Paul Kennedy in particular has noted, they were wont to prefer conciliation and appeasement (before the word acquired its modern connotation) over open confrontation, British policy makers were forced, inch by inch, to cede British prestige in exchange for the more tangible outcome of basic security (the first metric) at acceptable cost (the second).[36] In that particular sense, Germany's foolish confrontation with Britain about North Sea naval hegemony had a peculiarly vengeful quality to it. It may, in fact, have provided some tiny satisfaction to Tirpitz and Wilhelm that no matter how badly their country suffered, they had succeeded for a time in taking Britain down a peg. To a more cold-hearted (or clear-headed) strategist, the blow to Britain's prestige may well have been insubstantial and therefore not worth considering. Indeed, a little humiliation was very much an acceptable price to pay when threats to basic national security and fiscal livelihood were at stake. Objectively, and in material terms, the cost was not worth counting. But it was there all the same, and it stung.

Germany

Analysis of Germany's performance along the three metrics is in many ways redundant, not only because some of that performance has been discussed above, but also because Germany's loss, even before World War I removed any ambiguity, was near-total. For thoroughness sake, however, an accounting follows.

Security

Everything that can be said regarding Britain's performance in this area can be said in reverse about Germany's, in that security can be regarded in a competitive scenario as being essentially zero-sum. To challenge Britain's North Sea hegemony, Germany had to take measures that amounted to seeking a loss of security for Britain; Britain's losses and gains in this area were therefore Germany's gains and losses, respectively. As above, the security aspect must be understood through both raw strength and political maneuvering. Although Germany's fevered efforts to build a fleet able to confront Britain's came to some success and in a narrow sense diminished Britain's military position, in the aggregate it served to box Germany in, leaving it many adversaries, comparatively few military resources, and fewer options. Quantitatively, by a certain narrow measurement, Germany improved its position, but qualitatively and holistically its situation was grim well before the guns began to fire.

In purely naval terms, by one measure, Germany came out somewhat ahead. As noted above, Germany did succeed, by certain measures, in improving the ratio of its own naval strength to Britain's, at least in totality (figure 24). For the first time in history, Germany had put to sea a fleet of modern battleships that could challenge Britain on an approximately 2:3 or 1:2 ratio, depending on the year. As noted above, Germany was closing the capital ship gap as of 1914, with a fleet of capital ships 67.74 percent the size of Britain's. This may have satisfied

Chapter Six

Tirpitz's minimum requirements, but not the more expansive dreams that accompanied them.

Germany's High Seas Fleet never achieved parity—much less dominance—and was in an inferior and vulnerable position. It spent World War I rusting in port. It was not even able to act as a Corbettian "fleet in being," since this would have required mobility and a high seas presence, which the inaptly named High Seas Fleet did not have.[37] Apart from this, as discussed, Britain actually succeeded in increasing the overall size of its fleet relative to Germany's, both in terms of the overall number of ships and in terms of major combatant vessels. Britain also maintained a 2:1 or (at the end) 3:2 edge in small attack craft (figures 25 and 26).

Germany's overall fleet, as noted, never approached the size of Britain's, ending the Dreadnought Race in a somewhat worse position to where it began (figure 27).

Even if only large combatant vessels (battleships, battlecruisers, and large surface ships) were counted, the picture scarcely improved. As noted above regarding the Royal Navy, the German Imperial Navy improved its overall ratio of major surface combatants to those of the Royal Navy, but not enough to gain parity (figure 28). As of 1914, Germany had a fleet of large surface combatant vessels only 44 percent the size of Britain's—an improvement from the 12.7-percent ratio with which it began the Dreadnought Race, but not enough to close the gap. Given the vital importance that the North Sea had for British security, to compete with Britain there was to compete with the Royal Navy in its totality, and Germany could not accomplish this.

For all this, as has been said, Germany did succeed in building a modern navy for perhaps the first time and narrowing the gap in capital ships with Britain. If one wishes, one may award it marks for effort. Otherwise, one must conclude that the whole exercise was a dismal failure. By the end of the Dreadnought Race—indeed, about the time HMS *Dreadnought* was launched—the walls were closing in. Whereas Germany

The Dreadnought Race in Strategic Perspective

Figure 24. Ratio of German capital ships to British capital ships, 1875–1915

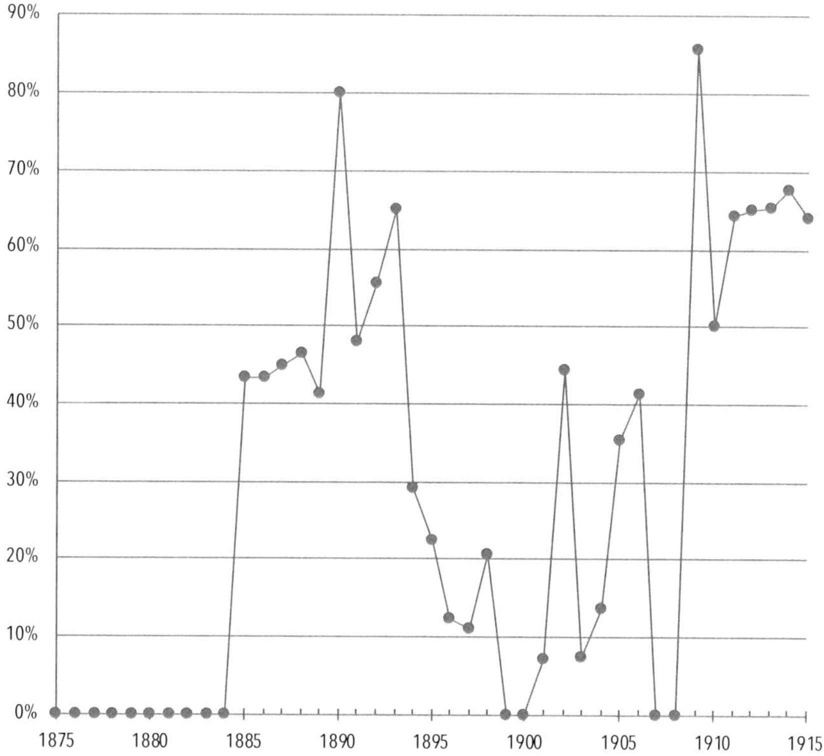

Source: compiled by the author, adapted by MCUP. See appendix for sources and methodology.

had had some strategic options prior to the Dreadnought Race, the formation of the Entente Cordiale isolated it politically and militarily, even as the High Seas Fleet diverted scarce resources away from both commerce raiding and the army.

This, however, is not necessarily the most significant result of Germany's political isolation. More damaging was the loss of options with regard to dealing with various combinations of powers. As Paul Kennedy notes, the one thing that was guaranteed where the British response to the July Crisis was concerned was that Britain would be considering war with Germany and no other power: the formation of the entente had calcified British policy and put it in knee-jerk opposition to Germany.[38]

Figure 25. Ratio of German torpedo-craft fleet to British torpedo-craft fleet, 1875–1915

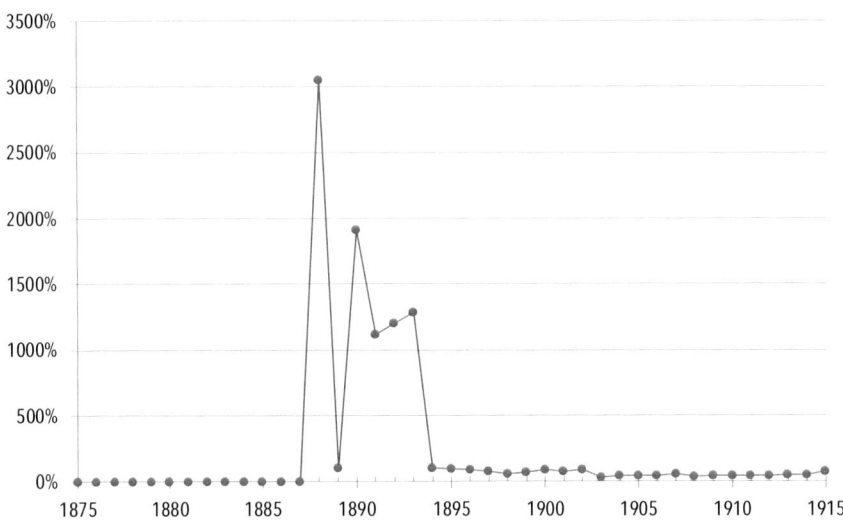

Source: compiled by the author, adapted by MCUP. Points that show 0 percent are a divide-by-zero error. See appendix for sources and methodology.

Figure 26. Ratio of German torpedo-craft fleet to British torpedo-craft fleet during (and immediately before) the Dreadnought Race, 1895–1915

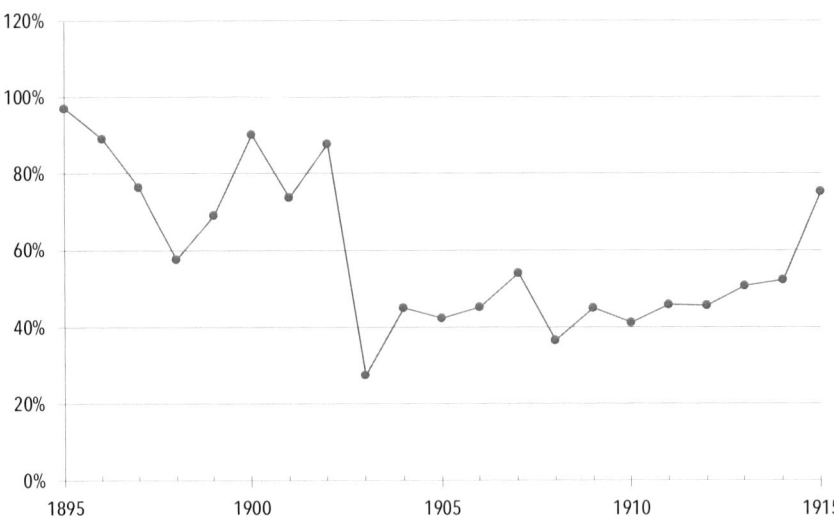

Source: compiled by the author, adapted by MCUP. Points that show 0 percent are a divide-by-zero error. See appendix for sources and methodology.

The Dreadnought Race in Strategic Perspective

Figure 27. Ratio of German fleet size to British fleet size, 1875–1915

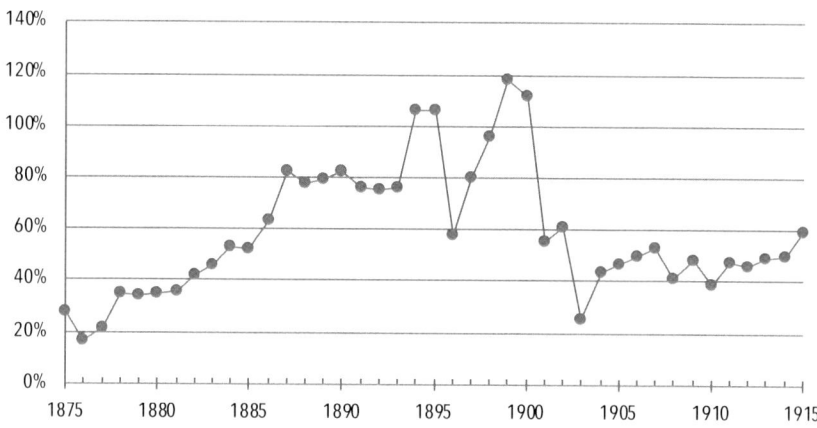

Source: compiled by the author, adapted by MCUP. See appendix for sources and methodology.

Figure 28. Ratio of German major surface combatant fleet size to British major surface combatant fleet size, 1875–1915

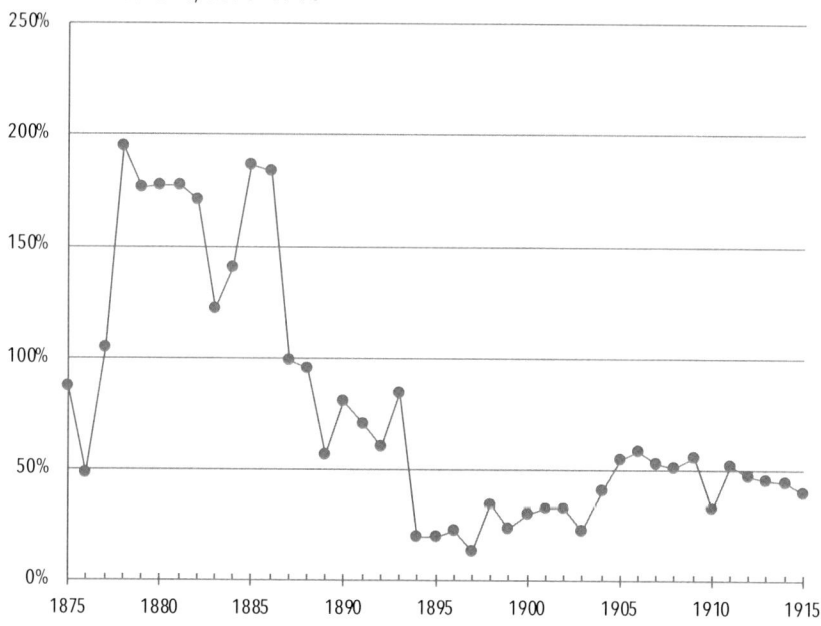

Source: compiled by the author, adapted by MCUP. See appendix for sources and methodology.

Chapter Six

Likewise, as noted by Arne Røksund and others, the entente put an end to any real or potential Anglo-French rivalry, ending the *Jeune École* in its original form in the process, and in so doing ensured that Germany would not have the option of playing potential adversaries off one another that it had enjoyed in the past.[39] By this, then, Germany had not only made it somewhat more likely that it would face the BEF on land at the commencement of hostilities, but it had also eliminated its ability to turn to diplomacy for a solution.

Here again, the numbers tell a story. Unlike Britain, which apart from its temporary mobilization for the Boer War kept the strength of its land forces essentially constant, Germany steadily increased the size of its active army during the Dreadnought Race, as in fact it had done previously, from 585,453 in 1898 to 790,985 in 1913 (figure 29).

Throughout this period, Germany relied, famously, on its military reserve system to swell the ranks in the event of war. In theory, as many as 3 million troops could be called to the colors in wartime throughout this period. The theoretical wartime figure, as of the eve of war, stood at 3.25 million (figure 30).

Throughout this period, about 1 percent of Germany's population were on active duty in the army at any given time (figure 31).

In net terms, this was a loss. Germany was forced to use essentially the same army it had relied on in the past to deter attack in a changed threat environment—one in which diplomacy was less useful as a tool for mitigating crises because adversaries could not be so easily divided or played off against one another; in which a full-scale British response to military action in Belgium and France was marginally more likely; and in which Britain and France had more closely coordinated their plans for land and naval war. The difference may have been slight. Nevertheless, Germany's reduced diplomatic flexibility and the tighter coordination of potential adversaries were unmitigated by any additional available forces except for an ineffective battleship fleet.

The Dreadnought Race in Strategic Perspective

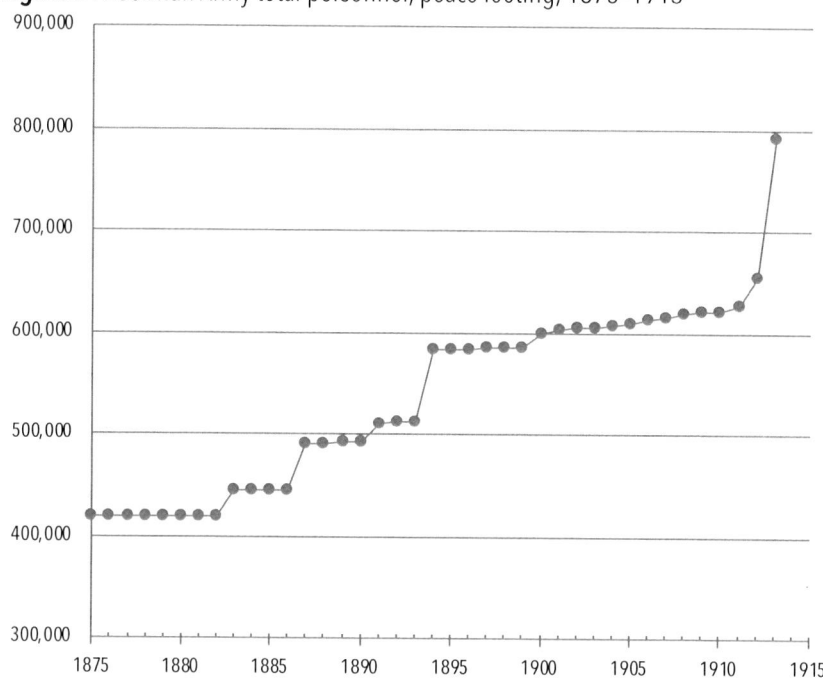

Figure 29. German Army total personnel, peace footing, 1875–1913

Source: compiled by the author, adapted by MCUP. See appendix for sources and methodology.

War is the ultimate laboratory, and in the final analysis the opening moves of World War I—even before the Battle of Jutland put paid to the notion that the High Seas Fleet might be militarily relevant—revealed the essential flaw in Germany's defense posture. As Peter Padfield notes, although one can never know for certain, the funds poured into the High Seas Fleet probably denuded the German Army of much-needed personnel, training, and equipment.[40] In the end, the initial German offensive bogged down before Paris, having failed to destroy the French Army. Though certainty is elusive, one can only assume that had the resources that had gone into the (useless) High Seas Fleet gone into the German Army instead, the extra divisions needed to prevent the Miracle on the Marne might have been there.[41] In essence, Germany had not only needlessly made enemies, lost political options, and boxed itself

Figure 30. German Army estimated total personnel, war footing, 1875–1913

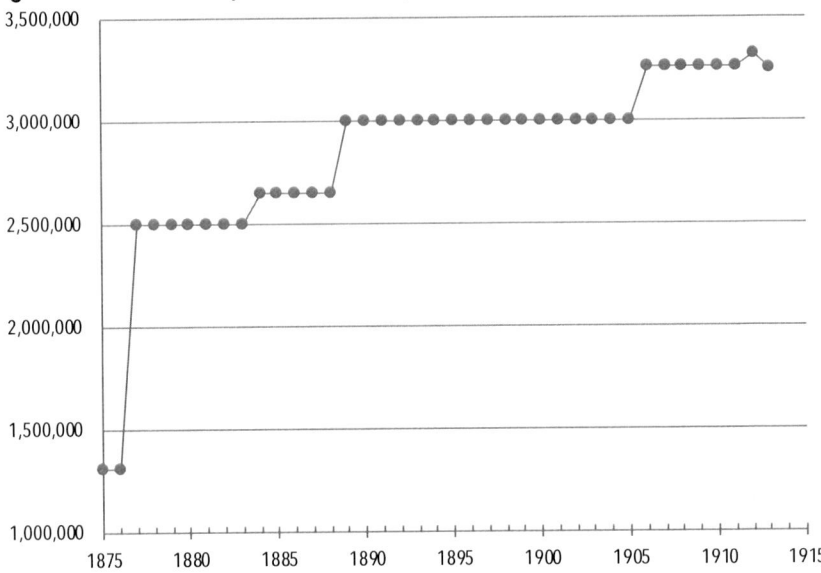

Source: compiled by the author, adapted by MCUP. See appendix for sources and methodology.

Figure 31. German peacetime military participation, 1875–1913

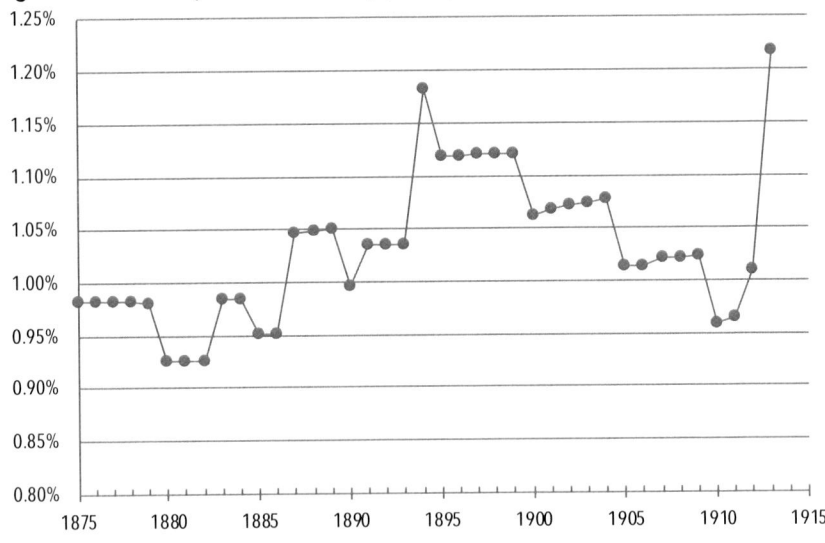

Source: compiled by the author, adapted by MCUP. See appendix for sources and methodology.

in with adversaries on all sides, but it had forgone an opportunity to play to its strengths and perhaps even repeat its triumph in the Franco-Prussian War a generation earlier.

It did so, in particular, in the face of better options. Insofar as the Royal Navy was deterred from a close blockade for reasons having nothing to do with the High Seas Fleet and everything to do with advances in technology and tactics, the High Seas Fleet constituted an almost criminal waste of resources.[42] By contrast, as the squabbles between the German High Command and Tirpitz's own office regarding resources for commerce raiding amply demonstrate, the High Seas Fleet starved the Imperial German Navy of resources for a more effective wartime naval strategy—one which would have played to Germany's existing core competencies, particularly given that Germany had thoroughly absorbed the doctrines of the *Jeune École*. Even if the money that had gone into the High Seas Fleet had not been used to augment Germany's capabilities on land, there were better uses for it at sea that were widely understood at the time. The seemingly unimaginative concerns of the lesser mortals in the Imperial German Navy's other departments may have been a distraction for Tirpitz from implementing his great vision, but they were born of a great deal of military good sense.[43]

To sum up, whether to connect the British entry into World War I specifically to the Dreadnought Race is an open question, although the Dreadnought Race undoubtedly made that entry more likely, and in this sense it can only be called a detriment to Germany's overall security even if events had transpired differently. Regardless, however, it is certainly the case that, on the eve of World War I, the High Seas Fleet was nowhere near the capability Tirpitz had sought for it, and the ultimate German aim—to alter the North Sea naval balance and thereby move the political situation in Germany's favor, either by war or by suasion—was nowhere close to fulfillment. This would not have changed even had World War I not intervened.

On the other hand, if Germany had to go to war with Britain—if it had in truth set on this course from the beginning of their rivalry, and if World War I was the culmination of that course, a thesis this study does not necessarily credit but can consider as a hypothetical—then it went about preparing for it in the worst possible way: it built ships when it needed soldiers, and it built useless battleships when it could have used smaller cruisers and submarines. It neither achieved its original goals nor a favorable political-military outcome, and it was far more unprepared for World War I than it might have been. Though much more may yet be said, the matter may be comfortably left there.

Finance and Welfare

Germany's finances were strained in a similar manner to Britain's, but not to the same degree. Although both Britain and Germany were forced to expend financial resources on the Dreadnought Race, the remarkable fact is that Britain exerted itself more heavily in this regard than did Germany. Compared to Britain's dramatic increases in public expenditure to meet the crisis, Germany's increases were more muted. This is attributable in the main to the fact that, at least initially, Germany retained some initiative—it could control the degree to which it competed with Britain, whereas Britain's naval planners had to reason that they were better safe than sorry. The lackadaisical attitude of Germany's senior decision makers toward what should have been a top priority once the decision to compete with Britain's naval superiority was taken go a long way toward explaining Germany's failure in the Dreadnought Race. On the other hand, on paper, Germany did not experience the same level of financial strain that Britain experienced.

Germany's main problem was actually not increases in its overall defense expenditure or even the Imperial Germany Navy's share of it—it was that its competition in the security realm was limited by its people's, and especially its aristocracy's, willingness to put up with the necessary taxation and expenditure. As noted in the preceding chapter, this led to constant friction with the Reichstag whenever fleet funding came to the floor. Faced with this, Germany was forced to take on debt, which it actually had greater capacity to do than did Britain, on account of the fact that, since the 1870s, it had been essentially starting from scratch in this regard. For all this, the Dreadnought Race was indeed costly for Germany, and the effects can be seen in a comparison of its spending patterns before and after the race started.

Germany's official defense estimates (naval and army estimates added together) did not change a great deal before and after the First Naval Law was passed in 1898 (figure 32). Combined, Germany's Army and Navy estimates averaged 1.52 percent of GNP between 1875 and 1898. From 1899 to 1914, the average was 1.8 percent. As of 1914, total defense estimates had risen to slightly above 2 percent of GNP.

The trend was up, but in the official estimates Germany was actually well behind Britain's transition from 2 to more than 3 percent of GNP at this juncture, and as the averages show, the transition was not great. Indeed, as the chart illustrates, the averages were partly skewed: overall, there was very little fluctuation. Germany's official estimates, however, did not reflect actual total expenditure, which could be and often was higher.[44] From 1892 (the first year for which numbers are available) to 1898, Germany's actual defense expenditure averaged 2.35 percent of GNP; from 1899 to 1913, total actual defense expenditure averaged 2.57 percent (figure 33). The averages obscure a rather dramatic upward trend. As of 1913, total actual defense expenditure was 3.69 percent of GNP, up from 2.8 percent the year before, and up from 2.26 percent in 1897 and 2.27 percent in 1898, when the race began. Still, until 1913, defense expenditure, all told, had accounted for a smaller percentage of Germany's GNP than was the case for Britain, which consistently spent more than 2.7 percent of GNP on defense for every year that followed the Boer War (and more than 2.8 percent in every year except 1907).[45]

For all this, the increase is there: Germany averaged an increase in defense spending equal to 0.2 percent of GNP

Figure 32. German total official defense estimates as percentage of gross national product, 1875–1914

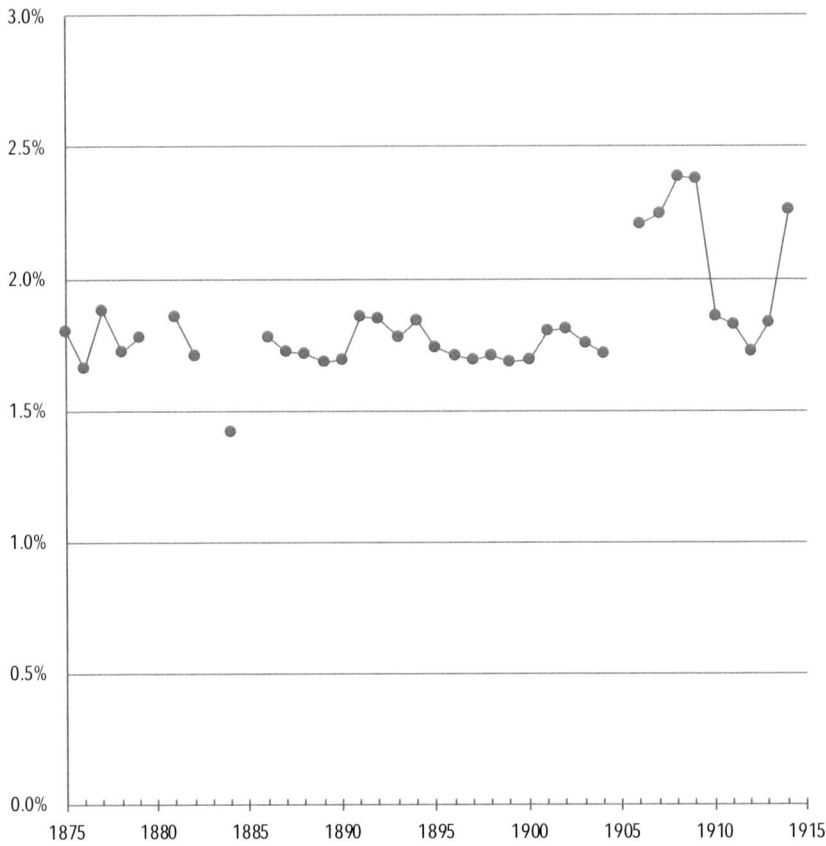

Source: compiled by the author, adapted by MCUP. Gaps indicate lack of data for the relevant years. See appendix for sources and methodology.

during the Dreadnought Race and, by the trend, registered a defense spending increase of 1.42 percent of GNP from 1898 to 1913. It had taken longer to get there, but in the end Germany was spending somewhat more on defense relative to its starting point as Britain was.

Some of the increase was indeed due to higher naval expenditures, although this was less obviously the case with Germany than with Britain. By the official estimates, the Imperial German Navy's share of the German defense budget averaged only 9.5

Figure 33. German total defense expenditures as percentage of gross national product, 1893–1913

Source: compiled by the author, adapted by MCUP. See appendix for sources and methodology.

percent between 1875 and 1898, before almost doubling to an 18.4 percent average share of the official estimates for the years between 1899 and 1914 inclusive (figures 34 and 35). In terms of actual expenditure, between 1892 and 1898, the navy's average share of overall defense expenditure was 13.8 percent. This average shot up to 27.2 percent for the period from 1899 to 1913.[46]

A fuller picture is given by Germany's increases in defense spending (figure 36). By the estimates, Germany was averaging an annual increase of 2.1 percent in its defense budget from 1875 to 1898. The average increased to 2.74 percent for the years 1899 to 1914.

Figure 34. German naval estimates as percentage of total army and navy estimates, 1875–1914

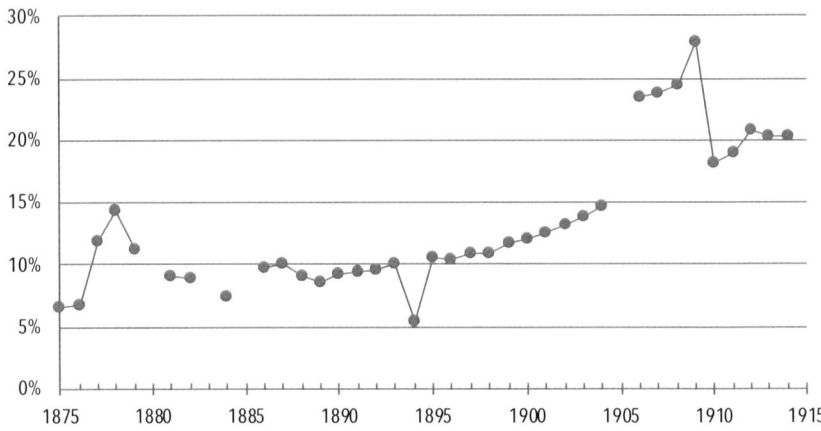

Source: compiled by the author, adapted by MCUP. Gaps indicate lack of data for the relevant years. See appendix for sources and methodology.

Data on actual expenditures show a more pronounced upward trend (figure 37). Between 1893 and 1898, total defense expenditures increased by an average of 2.72 percent. From 1899 to 1912, the annual average increase was 7 percent. The last few years showed dramatic increases: 11.1 percent in 1911 and 38.23 percent in 1912.

However, the navy's share of these increases was not as large as with Britain. From 1893 to 1898, naval spending increases accounted for an average of only 52.9 percent of Germany's defense expenditures increases; from 1899 to 1913, it accounted for 73.5 percent (figure 38).[47] In other words, although naval spending increases accounted for a proportionately higher share of Germany's defense spending increases after the Dreadnought Race began than before, at the height of the Dreadnought Race, they were only averaging significantly less than Britain's average of more than 100 percent. The share in any given year fluctuated wildly. In 1911, the navy got 66 percent of the increase in expenditure; in 1912 and 1913, it accounted for only 8.8 and 1.8 percent, respectively.[48]

The Dreadnought Race in Strategic Perspective

Figure 35. German naval expenditures as percentage of defense expenditures, 1892–1913

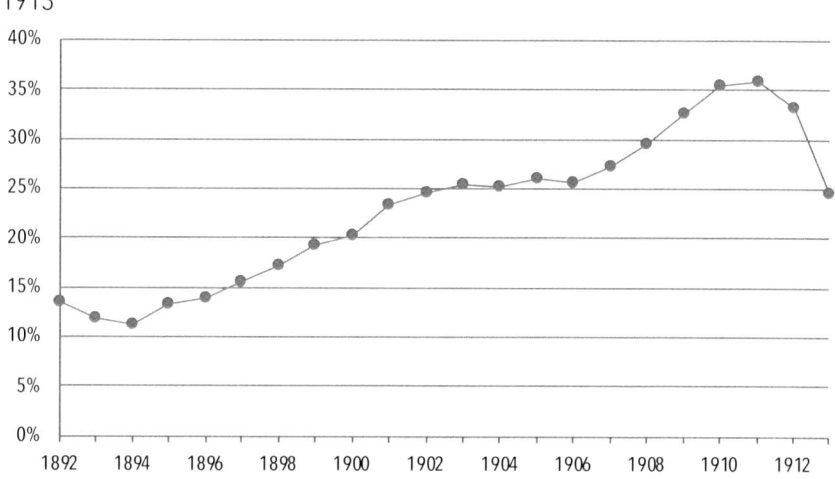

Source: compiled by the author, adapted by MCUP. See appendix for sources and methodology.

Figure 36. Annual percent increase in total German defense estimates, 1876–1914

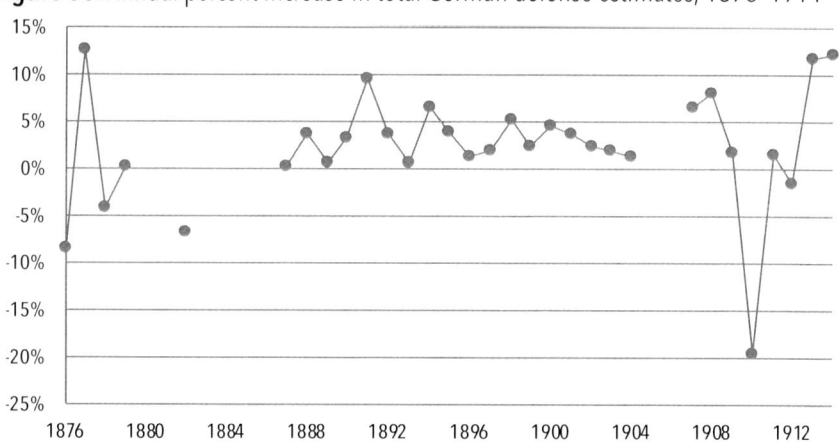

Source: compiled by the author, adapted by MCUP. Gaps indicate lack of data for the relevant years. See appendix for sources and methodology.

For what they are worth, the official estimates tell a more pronounced story (figure 39). From 1875 to 1898 (excluding 1879, an anomalous year in which naval estimates were massively and atypically curtailed), the navy's average share of increases in the official defense estimates was 28.6 percent, while from 1899

Figure 37. Percent increase in German total defense expenditures, 1893–1913

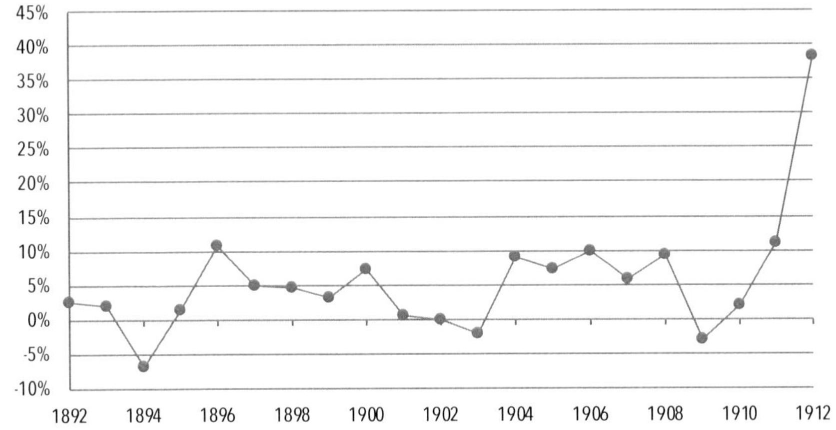

Source: compiled by the author, adapted by MCUP. See appendix for sources and methodology.

to 1914 it accounted for an average of 53.3 percent of annual increases, with 1914 showing a 19.8 percent share.

In short, while effectively 100 percent of Britain's annual increases in defense spending during the Dreadnought Race went to the Royal Navy, the Imperial German Navy had to make do with an average of at most three quarters of each year's spending increase, with the army getting the rest. In fact, the Imperial German Navy's average share of the increases in defense spending at the height of the Dreadnought Race was (whether one uses estimates or expenditures) less than what the Royal Navy's share of defense spending increases was before the race began. This is not to say that the Imperial German Navy was not crowding out the army. As noted, Germany could readily have used the extra funds to pay for more divisions rather than more ships, and the German national security debate—from the high command down to the Junker-bourgeois clash in the Reichstag about shipbuilding—most definitely reflected this. It is, however, to say that, particularly when compared to Britain, Germany's increases in defense expenditure only partly reflected the naval race as opposed to land-based concerns. If ever there existed a numerical illustration of Germany's permanent and irrevocable status as a land power, this was surely it.

Unlike Britain's, Germany's national debt increased during the Dreadnought Race. In their efforts to appease the public, Kaiser Wilhelm and Tirpitz had financed much of the additional defense expenditures with debt rather than through taxation. German tax revenues did increase relative to GNP during this timeframe. In 1898, tax revenues amounted to 5.5 percent of GNP; in 1914, they amounted to 7.66 percent (figure 40). The 1875–98 average for tax revenues was 3.7 percent of GNP; from 1899 to 1914, revenues averaged 6.2 percent of GNP. Similarly, overall expenditures relative to GNP increased (figure 41). In 1898, overall expenditures was 5.5 percent of GNP, while in 1913 it was 7 percent and in 1914 it was 7.7 percent. The 1875–98 average was 3.67 percent, and the 1899–1914 average was 6.23 percent. Although these official estimates suggest a balanced budget, in practice, debt financing prevailed.

A significant percentage of the increased spending was offset through borrowing. Germany's national debt, measured at contemporary exchange rates in pounds, was £112,673,626 in 1898; as of 1914, it was £224,662,467 (figure 42). Debt peaked in 1911 at £245,052,383 before decreasing; at no point prior to that had it decreased between fiscal years. In effect, Germany's national debt doubled during the Dreadnought Race.

Relative to GNP, a similar situation prevailed: Germany's national debt stood at 6.84 percent of GNP in 1898; as of 1914, it was 10.2 percent of GNP (figure 43). These increases were partially attributable to increased defense spending, in proportion to the degree to which defense spending accounted for overall spending increases. As noted, these were within what might be described as comfortable margins.[49]

Interestingly, whereas British defense expenditure accounted for an increasing percentage of overall expenditure during the Dreadnought Race, the picture was more nuanced on the German side. German actual defense expenditures averaged 47.5 percent of the total estimated expenditures from 1892

Figure 38. Share of German total defense expenditure growth attributable to navy expenditure increase, 1893–1914

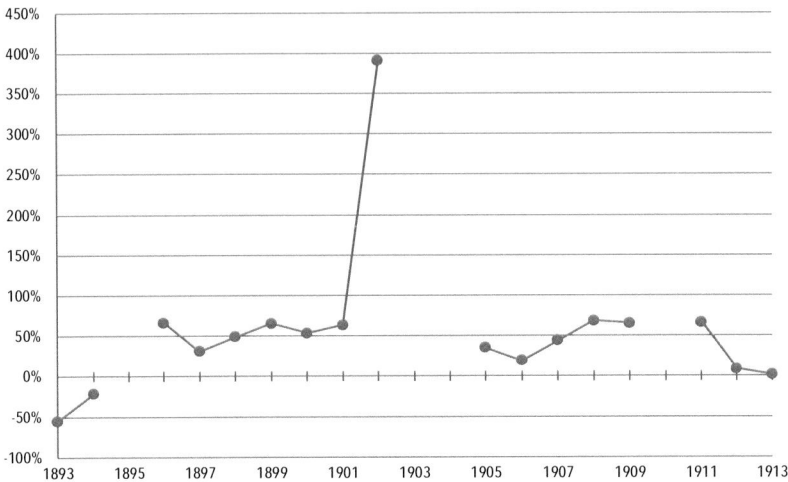

Source: compiled by the author, adapted by MCUP. Gaps indicate lack of data for the relevant years. See appendix for sources and methodology.

Figure 39. Share of German defense estimates growth attributable to increase in navy estimates, 1876–1914

Source: compiled by the author, adapted by MCUP. Gaps indicate lack of data for the relevant years. See appendix for sources and methodology.

(the first year for which these data are available) to 1898 (figure 44). From 1899 to 1913, it averaged 42 percent.[50] Conversely, this was due to a marked decline in defense spending relative to overall estimates prior to the Dreadnought Race; from 1899 onward, German defense spending did once again begin to eat into overall spending.

The official defense estimates averaged 51.45 percent of overall expenditures from 1875 to 1898; from 1899 to 1914, they averaged only 31.1 percent (figure 45). In fact, the trend was slightly upward; actual defense expenditures were 53 percent of Germany's overall expenditures in 1913, but the wild fluctuations that produced the aforementioned average serve to illustrate that Germany was not consistently increasing defense expenditures relative to overall expenditures during the Dreadnought Race.

It has already been remarked that the bulk of what defense spending increases did take place went to the German Army, not the High Seas Fleet, even if it is true that the fleet was consuming resources that the army could have used. The picture that emerges here is of a lackadaisical approach to defense that is at odds not only with the stereotype of Germany as a militarist state but also with the pretensions of Kaiser Wilhelm and others to a strong military and an aggressive foreign policy. Unlike Britain, whose increases in expenditure were primarily taken up by defense spending, Germany had other uses for its spending. Put together with Germany's comparatively low average defense expenditure relative to GNP (compared to Britain's), these numbers tell a story of a society on a peace footing even as it menaced its neighbors and competed for North Sea hegemony. Even—indeed, especially—allowing for the fact that Germany's federal system allocated most of the responsibility for social programs to its individual states, it becomes clear that, by its actions instead of its words, Germany during this period was scarcely concerned with winning its competition with Britain, so little did it spend on the Dreadnought Race relative to what it could have spent.

Chapter Six

Figure 40. German revenue as percentage of gross national product, 1875–1914

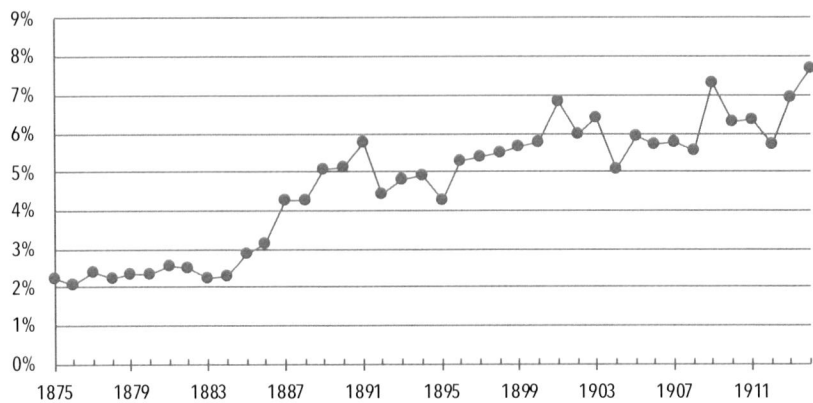

Source: compiled by the author, adapted by MCUP. See appendix for sources and methodology.

Figure 41. German overall expenditures as percentage of gross national product, 1875–1914

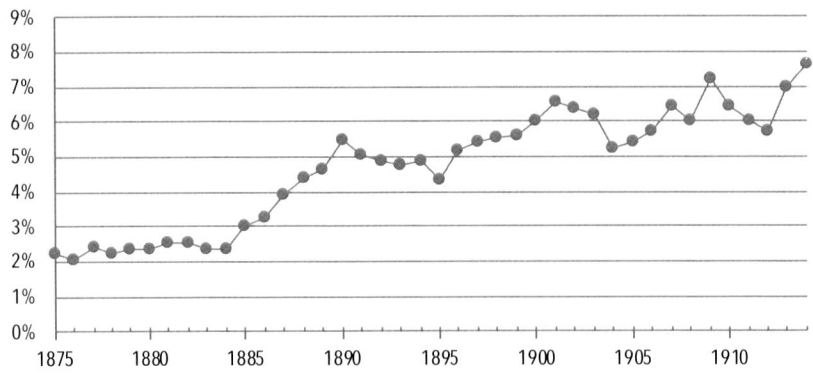

Source: compiled by the author, adapted by MCUP. See appendix for sources and methodology.

The Dreadnought Race in Strategic Perspective

Figure 42. German national debt in pounds, 1875–1914

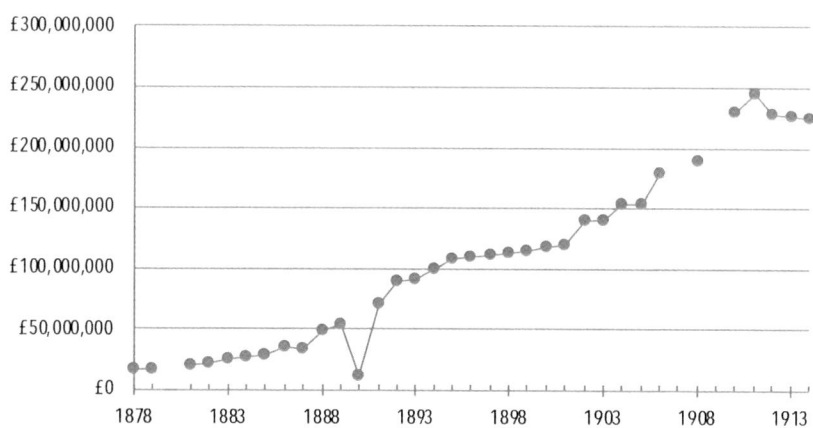

Source: compiled by the author, adapted by MCUP. Gaps indicate lack of data for the relevant years. See appendix for sources and methodology.

Figure 43. German national debt as percentage of gross national product, 1875–1914

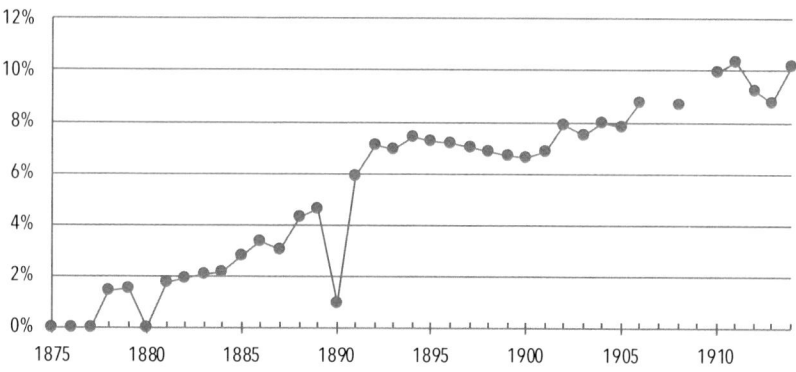

Source: compiled by the author, adapted by MCUP. Gaps indicate lack of data for the relevant years. See appendix for sources and methodology.

Chapter Six

There is a powerful case to be made, therefore, that in the short run, the German taxpayer, as an abstract concept, was only minimally hurt by Germany's competition with Britain, precisely because Germany did so little to compete. It is difficult to gauge political will at this distance in time: there is no way to know how much Germany could have spent given the political attitudes prevalent at the time within its own society. Tirpitz and German chancellor Bernhard von Bülow's difficulties with the Reichstag, however, suggest a society more complacent than warlike, and suggest that political will for this foreign policy project was in short supply. What is indisputable based on the raw data, in any event, is that Germany could, at least in theory, have sunk considerably more of its income into the Dreadnought Race than it did, without even approaching the level of financial constraints that Britain found itself in. The decision not to do so was a choice, both made collectively (and perhaps unconsciously) by the German public, which set the limits on what was politically acceptable, or individually by Wilhelm, Tirpitz, and all the others involved in plotting Germany's "new course."

Where actual public attitudes in Germany are concerned, however, there is no need to indulge in hypotheticals where facts may be cited. The Dreadnought Race pitted Germany's industrial middle class—both in the abstract and in specific terms where its members could in any way tie their interests to shipbuilding—against its nobility, with the industrialists broadly favoring the High Seas Fleet and the Junkers favoring the army. To be sure, some significant part of the country benefited, at least in relative terms, from the shipbuilding boom, even though the benefit was often abstract. Tirpitz was able to browbeat shipyards into building ships at a loss, whether out of patriotism or as a loss-leader that allowed for bigger enterprises later. But the cost of the whole project had constantly to be shifted: what part of it was not financed by debt (to be paid off later) was offset by regressive indirect taxation that hit ordinary workers the hardest and accounts in part for the Social

Figure 44. German actual defense expenditures as percentage of overall estimates, 1892–1913

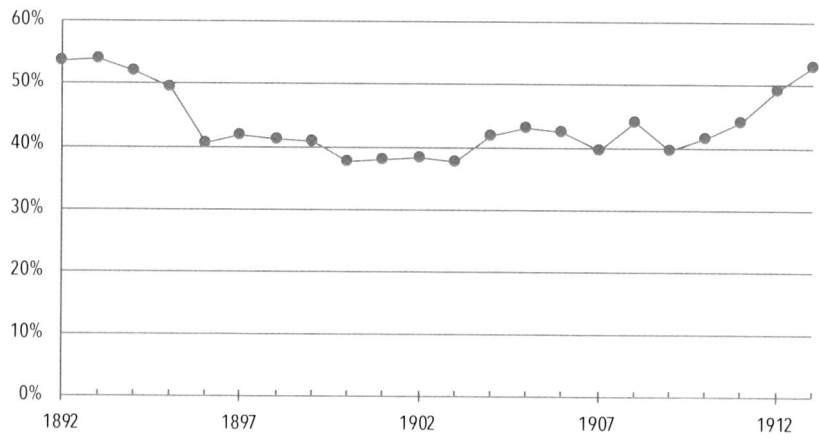

Source: compiled by the author, adapted by MCUP. See appendix for sources and methodology.

Figure 45. German defense estimates as percentage of overall estimates, 1875–1914

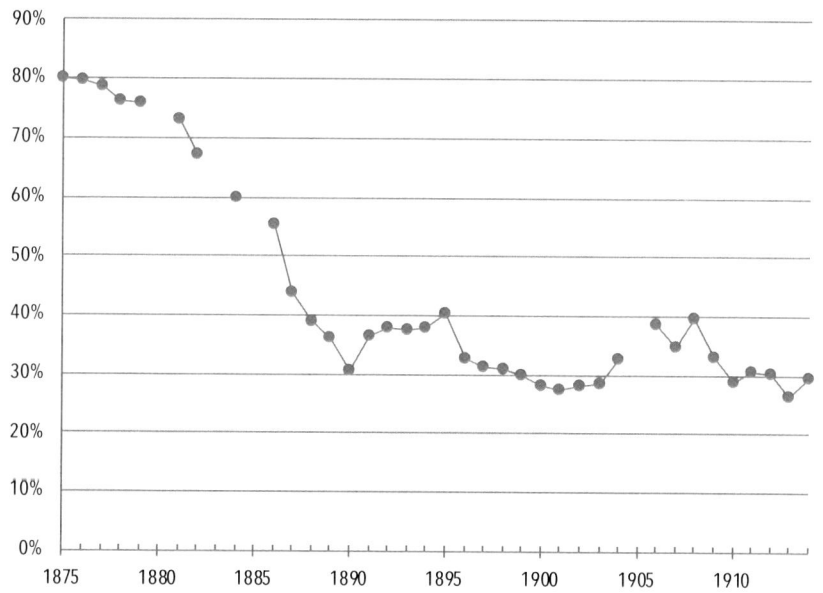

Source: compiled by the author, adapted by MCUP. Gaps indicate lack of data for the relevant years. See appendix for sources and methodology.

Democrats' opposition to the Dreadnought Race, though other ideological reasons also existed. Attempts to make the nobility, who were opposed to the very idea, pay for the naval race with direct taxation (inheritance taxes) not only failed but exposed the limits of the whole project, for once regressive indirect taxes had reached their politically acceptable maximum, relatively progressive direct taxes were a nonstarter.[51] It is therefore accurate to state that the Dreadnought Race left the average German taxpayer poorer to such a degree as to doom the project, even if some resources were yet available to devote to it.

In consideration of this, one can only say that Germany fell victim to the same short-term thinking that Britain did, but from an opposite perspective. Whereas Britain, financially on the ropes, managed to avoid going further into debt and in so doing missed its last, best opportunity to outspend Germany and decisively outmatch it in naval power in home waters, Germany failed to compete to the extent that it could and therefore put itself in the worst of all possible positions: weak and beset by antagonists. It did so for the most understandable of reasons: even from their high position, Wilhelm and Tirpitz had to account for public opinion, and public opinion did not share their grand vision. The absence of serious strategic intent in Germany is illustrated better than anywhere else by Germany's inability to put financial resources into the Dreadnought Race even to the same extent as its old, battered, tired hegemonic competitor. If Britain was exhausted, Germany had yet to really run.

There remains, though, the thorny question of whether Germany failed its citizens more fundamentally by starting a competition with another great power that is widely seen as having contributed to one of history's most destructive wars, which quite obviously left the average German worse off and Germany's wealth and welfare depleted. This question is less easily answered. As noted above, questions of German war guilt are complicated, as is the question of to what extent the Dreadnought Race precipitated World War I, and it is important not to claim too much. But in strict human security terms, if it is true that the Dreadnought Race made British involvement in the war more likely, and if this, in turn, worsened the lot of the average German in tangible terms, then it must at least be said that, in terms of the risks, if not the costs, that Germany's citizens were expected to take on, the Dreadnought Race added nothing to the positive side of the ledger, and that this would have been the case practically regardless of whether or how it influenced the coming of the war. Regardless of exactly the degree to which the Dreadnought Race influenced the lead-up to World War I, Germany was playing with fire by engaging in it, and it is difficult to say this was not at least partly by choice, or without consequence.

Intangibles

In an important sense, however, the Dreadnought Race, from Germany's perspective, was not about either security or financial welfare. Allowing for some intelligence on the part of Kaiser Wilhelm, Tirpitz, and others, one can say that had such been the case, the Dreadnought Race never would have happened. This requires some qualification. It is certainly true that Tirpitz sold his idea as a "risk fleet" and that the nominal purpose of the High Seas Fleet was deterrence—albeit, as noted, of a very ephemeral type—in that the threat it addressed was scarcely pressing at the time and that it was probably the wrong means of addressing it, as it deterred a close blockade that was not necessary and, in the process, did nothing to address the prospect of a more distant blockade.

Yet, as is readily clear and has been discussed, the High Seas Fleet was inseparably tied to Wilhelm's *Weltpolitik* and served as the means toward it. As such, it served several purposes, some direct, some indirect. Altering the naval balance in the North Sea in Germany's favor—however that was ultimately to be accomplished—was supposed to pave the way for a breakout, in which at minimum Germany could prevent Britain from exercising its naval leverage over Germany as it competed for distant colonies, and at maximum Germany could exercise the kind of sea control that Britain had so far exercised without fear of military reprisal. At minimum, the reason for this was psychological. *Weltpolitik*, the High Seas Fleet, and the overturning of the North Sea naval balance were first and foremost intended to address the insecurities and salve the pride of Germany's head of state and his ideological fellow travelers, who had chafed under the patronizing condescension of his British royal relatives for far too long. Although other motives existed, this most personal of motives was both necessary and sufficient for the Dreadnought Race to happen. Wilhelm wanted a fleet; a fleet therefore had to be procured.

The High Seas Fleet was not merely an end in itself but an instrument to be used in the pursuit of more nebulous goals, the most obvious among them being greater influence in the colonial arena and the national pride that came with it. To what degree the High Seas Fleet was meant to serve German colonial policy and aspirations is debatable. Certainly Wilhelm and Bülow envisioned a more expansive German world role that would allow Germany to take and keep colonies in a more British fashion, moving as it wished rather than as it must, and asking no one's permission. But to what degree this was the main goal, rather than a subsidiary of it, is another question, the more so since, as Paul Kennedy and others have noted, and as Eyre Crowe implied in his own memorandum, differences between Britain and Germany about colonies were far more bridgeable than those about the question of North Sea naval hegemony per se.[52] Although colonial aspirations sometimes served as a vehicle for the pride of Wilhelm and his officials (e.g., in the Boer War, in passing references to Spain's demise as a colonial power, and in repeated attempts to intervene in Africa), the actual question was who would wield naval power, and possess the political influence that came with it, much closer to home. Be this as it may, the question of colonial successes is an important one in assessing Germany's net gains or losses along the intangibles metric, insofar as they represented the international prestige

Germany under Wilhelm clearly craved and had somewhat to do with the acquisition of greater naval power.

The Dreadnought Race was also tied to concerns at home. *Sammlungspolitik*—binding together naval policy with attempts to resolve differences and keep harmony within Germany—cannot be ignored as a major German goal in the Dreadnought Race, and it must be understood as an intangible one, since it did not directly impact German power and security or its people's welfare (though it surely had a secondary effect on all of these). It, too, represented a royal vision—in this case, a vision of how Germany would be at home instead of abroad, namely, free from socialism and with at least reasonable political amity between a dying but still potent aristocracy and the industrial middle class that was replacing it. However vague this goal might have been, it can be understood qualitatively.

Where each of these was concerned, Germany failed to achieve what it set out to do. To proceed in reverse order, *Sammlungspolitik*, taken literally, was an overly ambitious goal that probably could not be achieved in any remotely democratic system, even one with as powerful an executive as the German system possessed. Real divisions existed within German society, and they were exposed and exacerbated rather than submerged and ameliorated by the Dreadnought Race during the long term. As Rolf Hobson has noted, it is far from clear from the available evidence even that the High Seas Fleet was ever envisioned as achieving such harmony.[53] The basic failure of Wilhelm and Tirpitz to think the whole matter through from the start is obvious in retrospect. While the aristocrats might be persuaded to indulge Tirpitz and the navalists in their naval project as long as it did not too badly affect their interests, they were never fully on board, and the instant that resources became scarce and hard decisions had to be made—as was almost inevitable in a competitive environment in which serious dedication was required to overthrow a naval adversary determined to sell its supremacy dearly or not at all—Germany's social divisions came to the forefront. The Dreadnought Race touched off a nasty altercation between the navalists and the aristocrats about inheritance taxes that can scarcely be said to have created the political unity the Kaiser had hoped for. It also exercised the Social Democrats, who turned to dialogue with Britain (no less) as a course of opposition. At minimum, one can say that Germany ended the Dreadnought Race with essentially the same set of social problems with which it began it; at the outside, the High Seas Fleet had rubbed seawater in existing wounds.

Where colonies were concerned, to the extent that colonial prestige can be considered an indirect goal of German policy in confronting Britain, Germany left the Dreadnought Race with essentially the same colonies with which it began it, and it was no closer to being able to administer them independently and without British consent. Apart from some worthless swampland that Germany extorted during the "Panthersprung" affair, Germany made no new acquisitions. Without entering into the moral questions of colonialism, even from the simple point of view of what Wilhelm and others may have hoped to accomplish by a naval-spearheaded *Weltpolitik*, it was also a pathetic failure. It is indeed ironic that Bismarck, whom Wilhelm had fired, was a more effective African colonizer, with less of a navy and more ambiguous motivations, than Wilhelm and his lackeys came to be. It is difficult to know how matters might have played out differently. Certainly there is some evidence that Germany had some sort of leverage over Britain in this regard, in that it might have traded a favorable end to the Dreadnought Race for some sort of colonial favors. Crowe's recommendations in his famous memorandum suggest that the idea must have crossed at least certain officials' minds, and, as Kennedy has famously noted, Britain was often willing to offer small favors (what would later be termed *appeasement*) to would-be adversaries as a cheap way to placate them. The fact that Germany did not seriously pursue any such settlement can be said to indicate where Germany's real priorities lay. North Sea naval supremacy was more valuable in its own right, in terms of what it represented and what it allowed, than any colonial settlement it might have brought in.

The Dreadnought Race in Strategic Perspective

As to that, the real prestige that Germany sought lay in its ability to copy Britain's dominance of north European waters, achieving the security and freedom that Britain enjoyed and the admiration that came with it. In this, it failed. If the success of the High Seas Fleet was an end in itself, then its failure speaks volumes. Moreover, insofar as the real intangible goals of the project, from Wilhelm's standpoint, were prestige not merely among nations but among royal relatives, then the loss of face he received as a result of having failed to alter the naval balance in his favor can only be counted as a loss of an intangible benefit. As noted above, Germany did succeed in scaring Britain and forcing it to consolidate its naval bases. To the extent that the High Seas Fleet was merely intended to diminish Britain's prestige, it may be said to have had some qualified success. But that success was entirely negative: Britain lost; Germany did not gain.

Moreover, the diplomatic costs of the whole move imposed costs in terms of prestige. The formation of the Triple Entente should be sufficient evidence of Germany's diminished, rather than enhanced, international position. Although one can debate to what degree this impacted the international course of events—once again, as noted, Britain never claimed the entente to be binding, and it had its reasons for wanting to intervene in the event of war regardless—it is certainly the case that Germany lost the international political initiative. If British prestige was damaged by its consolidation of its naval bases, Germany's prestige was damaged at least as badly by its loss of the diplomatic initiative internationally. If it can be said to have dreamed of influencing great events by acquiring sea power, in the end it was reduced to waiting on events.

Germany's newly acquired seapower was never demonstrated. Probably the closest Germany ever came during the Dreadnought Race to actually employing seapower for the political purposes for which it intended it was the Panthersprung incident and the follow-on Agadir crisis. However, the High Seas Fleet played no part in the Agadir crisis; the ship Wilhelm dispatched for the occasion was a humble gunboat. In the end, the High Seas Fleet spent World War I docked miserably in port, and prior to that time—admittedly allowing for the fact that it existed for so short a time—it never had the opportunity to prove its usefulness or its mettle. Given the constraints on its usage, it is difficult even to see how it could have had such an opportunity. At best, one can say that Wilhelm, Tirpitz, and their circle, in their more feverish dreams, seem genuinely to have believed that a day would come when Britain would be the less powerful state in North Sea waters and that it would therefore have no choice but to deal with Germany as Germany had once dealt with it. Resource constraints, and British determination, ensured that this dream would remain just that. The end result was not what Wilhelm and Tirpitz had hoped for, and that simple phrase is the epitaph for a failed strategy.

The Three Metrics: General Remarks

The Dreadnought Race demonstrates quite clearly that the three-part division of metrics on the basis of security, finance, and intangible goals can provide for thorough and meaningful analysis of states' competitive behavior. Perhaps most relevant is the implication, in the case of both Britain and Germany, that a refusal to choose between such metrics can lead to strategic failure. Both Britain and Germany sought to limit their financial expenditures, but far from saving them money, this merely ensured defeat along all three metrics. Britain managed to keep its expenditures within bounds, and although it did not definitively end the Dreadnought Race, by the end it had probably put Germany in a position where future competition would have been eventually deemed futile even if war had not intervened. Britain also suffered a (comparatively minor) black eye in the intangible realm of international prestige. Germany, meanwhile, avoided even the level of expenditures that Britain pursued (irrelevant success on the welfare metric), but completely failed to build a fleet to rival the Royal Navy and succeeded only in surrounding itself with adversaries (failure on the security metric), all the while failing to achieve the international respect and deference it sought (abject failure on the intangibles metric). The lesson here is quite simply that a state engaged in competition for regional hegemony must be willing to accept a loss on one of the three metrics, preferably the one on which it can afford to lose and with which it can afford to play. As an old engineering saying would have it, "You can have good, you can have fast, or you can have cheap—pick any two."[54] In a way, failure to recognize these tradeoffs led both parties to pursue the competition down a road that led to war, failing either to deter or to bow out.

Although one could say that it is always the financial metric that is doomed to suffer in a competition for regional hegemony of the type under study here, a closer look at the Dreadnought Race may suggest a qualified rebuttal. In one sense, Britain very nearly won the Dreadnought Race, and it did so less at financial cost that, though onerous, was supportable and left room to spare than at some cost in intangibles—the loss of prestige associated with bringing the Royal Navy closer to home waters rather than lay the keels for yet more ships in a (perhaps futile) effort to have both a far-called fleet and protection at home. As things stood, even without a final arms control agreement, Britain had effectively forced Germany to change its policy and might have forced it to abandon its naval program. Britain could afford to save money if it could be parsimonious in its prestige.

The only metric that seems not to be easily "tradeable" where regional hegemony is concerned is the security metric, for obvious reasons: if regional hegemony is based in part on hard power, then hard power must be maintained. In this sense, the decision to compete is a decision to prioritize the security metric as it pertains to regional hegemony (and not in other ways). As noted, this particularly applies if one is expecting to someday go to war. Even here, however, a state is left with options. Where the Dreadnought Race was concerned, Germany's primary goal—prestige and freedom within the international system— fell in the intangibles metric; naval power was a means to this end, albeit closely intertwined with it. In Germany's case, therefore, it could probably have traded some security in one area for more power in another: the obvious tradeoffs between army and navy, and between competition on land and completion at sea, that Germany refused to make can be seen as illustrative. Germany could compete with Britain or its neighbors, but not both; if it wanted to compete with Britain, it would have to accept some other risks and shape the political environment so as to minimize them. At least in the abstract, this is possible, though the difficulties in Germany's case are readily apparent. In a sense, therefore, Germany could have traded some security and power via limiting its continental ambitions in exchange for seeking dominance over Britain, which was less about security than intangibles.

What is certain, in any case, is that the willingness to make tradeoffs among metrics amounts to a conscious decision to allocate resources in pursuit of an objective—in short, a sound strategy. As the foregoing analysis shows, this was indeed crucial to the success and failure of the competitors in the Dreadnought Race. It is to the broader questions of the general formulation of strategy—or the lack thereof—that the analysis here must now turn.

Strategic Lessons: The Applicability of Strategic Intent, Core Competence and Competitive Advantage, and Decision Cycle Analysis

All of the strategic concepts outlined in the preceding chapters may be seen to have relevance in this case, and they go a long way toward explaining what went so terribly wrong for Germany and what went both right and wrong for Britain. In serving as an empirical test for these concepts, the Dreadnought Race case does appear to validate them, as well as show how they can be appropriately and insightfully applied. Accordingly, the strategic intent, core competence and competitive advantage,

Strategic Intent and the Creation of a Finish Line

4. Assess more generally the ability of each competitor to formulate strategic intent—in particular its leadership, ability to leverage resources and core competencies, and overall understanding of the nature of the project it is facing. Assess in particular which competitor is more able to intelligently and ruthlessly make tradeoffs among the three metrics in pursuit of its competitive goals.
5. Assess whether either player has an endgame in mind—either to outrun the adversary (in alpha's case) or to force a finish (beta), and whether that endgame is achievable within the context of the player's strategic intent—the decisions it knows to make and is able and willing to make.

The basic thrust of strategy as a concept, discussed in the preceding chapter—adopting a plan (even a malleable one) to allocate resources in pursuit of an organizational objective—was honored in the breach by Germany and in practice by Britain, although in the latter case a critique can be applied. Where Germany was concerned, its twin (and twinned) goals of altering the North Sea naval balance and international prestige resulting from the implications of this foundered on the rocks of resource constraints on the one hand and bad policy on the other. Both had a role in bringing down first Germany's ambitions and then its national security. By contrast, although Britain's democracy and the ambivalence of its leaders never allowed for a truly well-developed plan for allocating resources in competition, and although, as revisionists have noted, British policy was driven by a host of other concerns, Britain nevertheless acted with some degree of coherence in allocating its resources to compete with Germany, even if it took some time before true strategic intent emerged.

Chapter Six

Where Germany was concerned, strategic intent never fully materialized, a more crucial factor with Germany acting as beta in this competition. The reasons were manifold. First, however, there was the simple question of framing the competition. Since ends must be decided before one can speak of means, it pays to investigate to what degree Germany's ends were even clear. This study has treated the Dreadnought Race as a competition for regional hegemony—specifically, for the relative share of hard power in the North Sea and the political influence that that carried. This is an accurate enough characterization of the competition, but as noted in preceding chapters, even this implies a greater level of clarity than Germany actually achieved. Briefly, if it is possible even at this late date for historians to dispute whether Tirpitz was really concerned merely with deterrence rather than North Sea dominance, this lack of clarity was amplified at the time, to the point at which it was not even understood at the highest levels in Germany what, exactly, the High Seas Fleet was supposed to be and do.

Nor were the surrounding implications (and they were considerable) ever adequately addressed. As noted in the discussion of Tirpitz's intentions at the beginning of the Dreadnought Race, the very idea posed problems that were never discussed, much less solved. While Tirpitz had every reason to be confident in his ability to stretch his limited resources and at least some plausible reasons to be confident in his organization's learning curve, if not any initial advantages in ship production, the capacity of his dockyards relative to what he would need to exceed the Royal Navy's building tempo seems not to have bothered him. Neither does he seem to have thought through the relationship between the likely funding he would have at his disposal and the task he had undertaken.

At the political level, while Bülow and the Kaiser clearly drew a connection—*Weltpolitik*—between the High Seas Fleet and the ability to gain greater policy freedom with regard to Britain—and just possibly with regard to overseas colonization—exactly how this connection was to be realized was left hopelessly vague. It is not an exaggeration to say that Wilhelm and Bülow actually seem to have supposed that they merely needed to build ships and the British Empire would bow before them. The essential ludicrousness of this proposal was not obvious to them at the time.

It is accurate, therefore, to say that Germany could not have had the necessary strategic intent because its goals were vague, even before discussing whether and how it could have pulled together its resources in an effort to stretch its capabilities and oust the British alpha per Gary P. Hamel and C. K. Prahalad's formulation. But to the extent that the basic goal existed—supplant Britain as the dominant naval power in North Sea waters for political profit—Germany failed to organize its resources as well.

First, German naval planners failed in truly epic fashion to get all the relevant players on board. Although in theory strategic direction was set by Kaiser Wilhelm, the kaiser was famously unstable and feckless, unable to coordinate policy on the one hand and prone to extreme foolishness on the other.[55] He was prone to gaffes that needlessly antagonized Britain when a more subtle approach would have bought time, and he was tragically inept in his personal diplomacy, mismanaging the Algeçiras crisis even as his advisors utterly failed to produce a workable policy. Where the German Imperial Navy was concerned, Tirpitz's point of view prevailed only with difficulty, and he was embedded in a perennial bureaucratic feud (made possible by the kaiser's ridiculous reorganization of his government) with the naval high command about whether to continue building the fleet or move scarce resources toward actual war preparation, a feud that constantly threatened to scuttle the whole enterprise. Nor, in the end, did Tirpitz cooperate with Wilhelm's foreign policy, or even understand it—whether that policy took the form of attempting to avert the formation of the Triple Entente or achieving an arms control settlement—all this despite the stated goal of using a tilt in the North Sea naval balance for political purposes. In purely bureaucratic terms, Germany failed to pull its resources together.

Moreover, Wilhelm's desired budgets were subject to quasi-democratic audit via the Reichstag, and the Reichstag was at best a reluctant participant. The results of this situation were all too predictable: Germany could not focus or mobilize all its resources in pursuit of its strategic objective. The budgetary figures analyzed above speak for themselves. More fundamentally, however, the disputes about funding, which exposed latent societal tensions once the issue of inheritance taxes was raised, illustrated the limits of German strategic focus: only a limited contribution could be expected from a limited segment of German society before serious disputes over costs and benefits arose.

What was true militarily and financially was true politically as well. For practical purposes, given the budgetary constraints that Germany faced, without a serious effort to mend Germany's international image and its relations with its neighbors, the Dreadnought Race was unwinnable. The problem was the number of missions that the German Army and Navy both had to accomplish, which were imposed by political realities. Too much of Germany's military budget—effectively three-quarters, as the above fiscal analysis shows—was tied up in its army, which in turn had to be constantly strengthened in anticipation of a land war. This, moreover, does not even take into account the Imperial German Navy's own day-to-day (or year-to-year) financial requirements to fulfill its mission as a coastal defense force and commerce raiding force, which necessitated submarines and the aforementioned shortcuts with regard to outfitting ocean liners as cruisers. Going beyond this to build a blue water fleet of capital ships was an exercise in "stretch," defined by Hamel and Prahalad as reaching for an additional capability not yet present.[56] It had to be treated as such. This could only be accomplished by shortchanging the army and the navy's guerre de course capabilities, and this, in turn, would have left Germany exposed to even more immediate national security threats unless the political environment could be radically reshaped.

Herein lay the problem. As should be clear by now, Germany could only have hoped for success in the Dreadnought Race—and, indeed, would have had only a fighting chance in any event—if it could throw all of its resources behind a navalist and Atlanticist policy. To become a naval power and acquire all that came with that status, Germany had to accomplish the already difficult task of transitioning away from a land-based foreign policy. To accomplish this at all would have required playing France and Russia off Britain, as Bülow and Wilhelm attempted without success and over Tirpitz's obstruction. Exploiting these gaps was far from a guaranteed success—but only by doing so could Germany reduce the diplomatic tensions that required it to be prepared for war on all fronts all the time and thereby free up resources for the High Seas Fleet.

Germany had failed, therefore, to make its grand strategy match its competitive strategy. In such circumstances, goals must themselves be treated as resources: the more goals one pursues with fewer resources, the more one's resources are stretched. In simpler terms, trying to do too much with too little results in failure. While, as noted in the preliminary chapters to this study, grand strategy is beyond this study's scope, its handling has a direct impact on competitive strategy. One cannot compete if one does not allocate the resources to do so.

Ultimately, Germany, as beta, could not create a finish line. Karen Rasler, William R. Thompson, and Sumit Ganguly's argument that rivalries—a close relation to competition for regional hegemony as understood here—terminate when both sides reciprocate the reduction of tensions offers an insight into how Germany might have produced a finish line, as well as the inherent difficulties of doing so.[57] Because German diplomacy was so ham-fisted, on the one hand, and because control of the region in question was so important to Britain, on the other, there was little chance of this kind of deescalation occurring. The determination of Germany to "shake our mailed fist in his face" left few options other than war. But if that was the objective—or if a showdown in which Britain realized it could not win was—then Germany needed a concept of how to get there, and its leaders never posited one.

In sum, then, Germany failed to clearly articulate its aims. It failed to focus its policy on the competition it sought to undertake, thereby making its grand strategy an impediment to its competitive strategy rather than an asset to it. It had failed internally to allocate sufficient fiscal resources, and it failed to bring its chaotic and ponderous national security bureaucracy into line behind its goal. It had, in short, failed to develop strategic intent.

Where Britain was concerned, the beauty of its strategic position was that, despite its initial lethargy, its course was already charted for it. On no account could Germany (or anyone else) be allowed to challenge Britain's North Sea naval preeminence. Other areas of the globe offered some flexibility. Once again, as Kennedy notes, Britain preferred to appease small powers over small issues rather than wage costly wars (the Boer War being a great exception to this otherwise sensible policy), and as Eyre Crowe suggested in his memorandum, a few colonies here and there were a less serious matter than the fundamental national security problem of securing the North Sea.[58] In view of this, Britain eventually came to do what Germany could not: adapt its grand strategy to its competitive strategy. As alpha, Britain initially had other options than to compete with Germany, but as the German challenge became more urgent, it was forced to adopt a strategic focus of its own.

How and to what extent it did this deserves some explication. As the narrative in the foregoing chapters shows, Britain, as both a parliamentary democracy and a comfortable imperial power, took some time to acquire a consensus in favor of confronting Germany as its primary adversary, and in fact it is debatable to what extent it ever did so.

In that sense, Britain did not develop true strategic intent. However, coordination of its own resources was good enough

The Dreadnought Race in Strategic Perspective

in the end. Once the Royal Navy and its civilian ministers overcame their organizational and mental inertia and addressed the German challenge directly, there was only one thing left for them to do, a policy that required very little thought or subtlety: build ships and make friends.

These twin tasks were two sides of the same strategic coin; they amounted to the basic requisites of being a seapower. Building ships needs little explanation, but it is important to qualify Britain's behavior in this regard in light of the major controversies that have swirled around the interpretation of Britain's naval policy during this time. Suffice it to say that, although Britain most certainly possessed other priorities besides Germany that drove its naval policy (as was likewise true with its foreign policy), and although Fisher was assuredly a genius (and perhaps a mad one) with a vision of his own that went beyond reacting to a challenge, Britain nevertheless proceeded during the Fisher era to focus its naval policy on a set of reforms—in ship design, fleet deployment, manning, personnel policy, and war preparation—that gave it a decided edge over Germany in the Dreadnought Race. Although it came at the cost of other priorities—this is a sure sign of strategic focus properly understood—Britain's naval policy allowed it to fortify the North Sea against a breakout by the High Seas Fleet and any threat that might pose and to gain a decisive edge in shipbuilding that rendered the outcome of the competition a foregone conclusion. It did so by a combination of careful resource usage (the design of HMS *Dreadnought* and HMS *Invincible* and their successors allowing for added capabilities at little extra cost), dipping into resource reserves (the added naval expenditure from 1908 onward), and incurring acceptable opportunity costs (notably the redeployment of capital ships from outlying regions to the North Sea). In short, once it became clear that competition was going on in earnest, and that resources could no longer be considered infinite, Britain focused its limited resources on the task at hand. The fact that other motives also drove British policy—and meddled within it—notwithstanding, Britain achieved a kind of strategic focus in at least this aspect of its policy.

Something similar prevailed in the diplomatic realm. As it had done historically, Britain reacted diplomatically to a developing threat from the continent in a manner calculated to maximize its chances against that threat, even at some cost to its other interests (again, a sure sign of focusing resources on the competition at hand). The effects on Germany's political position have already been discussed. What is noteworthy is that, from Britain's perspective, the formation of the Triple Entente and the subsequent military conversations allowed Britain to focus its military resources on deterring Germany and, in turn, to maximize the resources it could devote to the naval arms race. This occurred at the cost of other diplomatic objectives and at some (acceptable) risk. In short, Britain succeeded in making its grand strategy fit its competitive strategy, and in so doing it achieved strategic focus in its foreign policy as well as in its fiscal and military policy. As noted, this came at the cost primarily of Britain's diplomatic flexibility and may have somewhat increased the likelihood of British involvement in World War I, allowing for appropriate caveats regarding the near-inevitability of a British response to an invasion of Belgium, on the one hand, and Britain's policy ambivalence all the way up to the end of the July Crisis, on the other. In general terms, Britain solved its competitive problems to about the extent it could have been expected to. It focused its resources and its resource usages appropriately.

None of this should obscure Britain's difficulties in formulating strategy, or even the fact that it never consciously had one. As noted several times above, one does not need to be perfectly consistent to have a strategy—even, and particularly, if it is a relatively bad one. It is indeed true that the ambivalence of British political leaders (especially before 1908), the inevitable tradeoffs and shortsightedness imposed by democratic governance and coalition-building, the multiple imperatives of naval policy, and simple bureaucratic squabbling (most famously

The Dreadnought Race in Strategic Perspective

between Fisher and Lord Charles Beresford, but also elsewhere) all created difficulties. It was only in the essentials that Britain got it right. While Britain's internal policy process may not have been pretty to look at, policy was adopted at the top and coordinated downward, and there was never an instance in which Britain's diplomacy completely contradicted its military policy, as was so obviously the case with Germany. Britain may not have played the game brilliantly, but the beauty of its long-term strategy is that it did not have to. Its policy makers knew what needed to be done and did it well enough.

Britain did have at least two major problems in formulating strategic intent, both stemming from its policy process and ultimately from its democracy. The first was Britain's initial inability to recognize the German challenge for what it was. The second problem was related to the first: certain inconsistencies in British strategy were never actually worked out.

As to British recognition of the seriousness of the German threat, plausible dates for such recognition would include 1904, when Fisher and others first suggested preventive war and the Entente Cordiale was signed; 1905, when the keel was laid for HMS *Dreadnought*, even allowing for its multiple potential uses; 1908–9, when the Naval Scare finally jolted the British government out of its complacency and led to an accelerated building program; or even 1911, when the Agadir crisis decisively ended Prime Minister David Lloyd George's opposition to naval rearmament.

The degree to which Fisher in particular made countering Germany his major project, as opposed to naval innovation more generally, is a matter of considerable debate, as are the actual intentions behind his numerous reforms and their scope. In the traditional interpretation, first put forward in detail by Arthur J. Marder and broadly adhered to in historical narratives by Peter Padfield and Robert K. Massie, Fisher overturned a somnolent and apathetic Royal Navy bureaucracy, forcing it to adapt to technological change (most famously in ship construction) and putting it on a footing to confront Germany, which he quite early on perceived to be the Royal Navy's greatest challenge.[59] In the revisionist interpretation, whose major proponents have been Jon T. Sumida and (especially) Nicholas A. Lambert, all of these assertions are up for review: the Royal Navy's officer corps was hardly apathetic or averse to new ideas, Germany featured as but one of many threats the Royal Navy might face, and Fisher's genius lay in manipulating the political system (and the fears of his political masters) almost as his opposite number Tirpitz had done, while focusing on modernizing the Royal Navy for his own purposes to the extent his political masters allowed.[60]

Where Fisher's focus or lack thereof on Germany—and his overall intentions—are concerned, some synthesis is possible. Fisher's mind, and British policy, contained multitudes, and it is by no means unreasonable to note that at the start of Fisher's tenure of office as First Sea Lord opinion had not yet completely congealed either in his own mind, in the minds of his political superiors, in the Royal Navy, or among the British public regarding the German challenge to British naval primacy—although, as the aforementioned sourness of the public mood and among some of the cabinet demonstrated all too well, the trend was in that direction. Nor is it unreasonable to hold that Fisher and his colleagues came into office holding ideas about naval technology and tactics that did not depend on a particular adversary to exist and propagate themselves. Still less is it unreasonable to note that policy, particularly when made by committee, seldom reflects a single person's agenda or a single set of aims, and that such was certainly the case with the series of policy decisions—the decisions, all discussed in detail below, to cut the naval estimates, to build *Dreadnought* and *Invincible* and to build squadrons of battleships, to overhaul the Royal Navy's manning policies, and especially to overhaul of the fleet and redeploy it to home waters—made by the Royal Navy's leaders during this time. All of this can be acknowledged.

But the trend was decidedly in the direction of a British awakening to the German challenge. As discussed extensively by Matthew Seligmann, Frank Nagler, Michael Epkenhans, and

Christopher M. Bell, the Royal Navy did indeed move toward a wide-ranging set of plans for dealing with a major naval war with Germany at this time, even if some of these plans were more outré than the classical historiography (Marder in particular) could have imagined.[61] Moreover, as noted in particular by Bell, if the Royal Navy's intentions are discerned primarily by its actions rather than merely by documentary evidence, then it cannot be denied that, during the course of the decade preceding World War I, the Royal Navy did indeed move toward a defensive posture toward Germany that relied on battleships for its substance.[62]

Where Fisher was concerned, his ability to harness fears of Germany for political purposes requires, at the very least, an understanding of the arguments for that position, while a repeated argument for "Copenhagening" the German fleet—which Fisher unquestionably made, and which went well to the right wing of what political consensus existed at the time, and therefore cannot entirely be dismissed as mere time-serving—unquestionably suggests that Fisher was at least aware of the concerns.[63] Likewise, while he was occasionally cagey, Fisher's almost single defining feature was brashness in conversation and correspondence, suggesting that, when he suggested aggressive measures to counter Germany, he at least meant what he said.[64] This does not diminish the likelihood that the revisionists are also correct that Fisher held *avant-garde* ideas about naval warfare which he sought to implement not only to counter Germany, but for other purposes and for their own sake. Nor does it mean that all of the ideas that the revisionists have imputed to Fisher and to his successors (particularly Churchill as the civilian First Lord)—notably flotilla defense of British home waters—were necessarily held or implemented as the revisionists suggest. It does mean, on the other hand, that the intentionality of Royal Navy decision making, and that of Britain more generally, relative to countering the German challenge increased over time.

Chapter Six

There is a sense in which Britain never fully zeroed in on Germany as a singular threat—as noted, up through the end of the July Crisis, Britain was ambivalent about everything from its alliance commitments to its ultimate decision to go to war. None of this can obscure the fact that Britain did, nevertheless, make key resource allocation decisions as a state in an increasingly focused manner, as necessary to meet the German competitive challenge. To the extent that it did so imperfectly, it goes without saying that Britain performed best when its civilian ministers and military leadership were in full agreement, an intuitively obvious statement that nevertheless reinforces the notion that strategic intent matters in such cases. It also goes to show how difficult it was to manufacture strategic intent for an arms race and regional security competition in a democracy with inborn incentives to avoid excessive or even necessary defense spending and a natural aversion to war.

As to the second problem, strategic inconsistencies continued throughout. Though this was inevitable to some extent, certain major instances of this hampered Britain's strategic focus. Fisher's reforms to this day (as the controversy about revisionism shows) are seen to have had multiple purposes and to have been imperfectly implemented. Bureaucratic infighting serially prevented the Royal Navy from organizing efficiently. And the cavalier behavior of Sir Edward Grey's Foreign Office ensured that the French Army knew more about how Britain and France would cooperate in the event of war than the Royal Navy and British Army knew about how they would cooperate in such a situation, even as Grey delayed in notifying the British government and insisted publicly that the entente was nonbinding. Although it is legitimate to question how much difference all of this made in the aggregate, it is nonetheless true that thoroughgoing coordination of resources was lacking in places.

This, however, still demonstrates the importance of generating, sustaining, and implementing strategic intent. If it also shows that Britain, like Hamel and Prahalad's hypothetical alpha, was initially quite lax in its approach to the problem because of the inertia associated with being at the top, it merely confirms that their hypothesis is relevant to cases such as this. In sum, and allowing for the suboptimal performance of both players in the Dreadnought Race, all of this tends to vindicate both the broad understanding that strategic intent is vital to competitive success. The state that could coordinate and focus its efforts won, albeit not completely; the state that failed to do so lost miserably, precisely because of this lack of coordination and focus.

Competitive Advantage and Core Competence

6. Determine what core competencies each player has—what abilities it possesses that will enable it to obtain an advantage in the competition. Assess in particular each competitor's astuteness in leveraging these competencies behind strategic intent. Assess more specifically whether it has an advantage in performing the actions it will have to perform to win the competition, relative to its competitor.

Where competitive advantage and core competencies are concerned, the Dreadnought Race offers a similar validation. It can readily be shown that Britain possessed and used a significant competitive advantage rooted in certain core competencies essential to the nature of the competition.

Briefly, Britain's competitive advantage lay in its status as alpha, for it possessed all the capabilities necessary to do what Germany wanted to do, as Germany initially did not. Britain's obvious core competence was in ship construction. As noted in the narrative, the Royal Navy built ships both more quickly and more cheaply than its Imperial German counterpart, a basic fact of the competition that Tirpitz did his utmost to negate but never succeeded in negating. It also had a technological edge, although this was not obvious at the outset. Britain revolutionized ship design in the dreadnought era as well, repeatedly rendering existing German ship classes obsolete (first with

Chapter Six

HMS *Dreadnought*, then with the innovation of all-in-line-turret battleships, then with the adoption of oil-fired vessels). Even allowing for disputes over to what extent Fisher's "naval revolution" was ruined by hesitation and conservatism, the Royal Navy maintained its technological advantage, an outgrowth of innate competencies that a landlubber navy such as Germany's simply could not imitate—the essence of core competence.[65]

Intriguingly, Britain can also be said to have possessed a core competence in something so basic to statecraft that it might go unrecognized, namely, diplomacy. Britain's reputation for carefully managing international relationships was well-known and well-deserved at the start of the Dreadnought Race, and it was on display during it. The entente was a masterpiece of careful ground-laying and compromise that German diplomats in the post-Bismarck era probably could not have imitated in similar circumstances; more to the point, it amounted to a victory in a zero-sum game to gain and retain the initiative in European international politics. Because Britain was able to patch up its relationships with France and Russia—in the face of considerable disincentives to do so, no less—it was able to avoid a situation in which it could be politically neutralized by diplomatic maneuvering, and in so doing to force Germany to maintain (and heighten) the garrison mentality that prevented it from devoting more than token resources to its navy. In the end, these two core competencies—in naval administration and technology and in political acumen—were enough to force Germany to abandon *Weltpolitik* and at least begin to draw down its naval ambitions. Had war not intervened, this would still have been enough.

Germany, by contrast, was not playing to any readily identifiable strengths. The fact that Britain and Germany had historically been compatible and complementary allies—the one blockading Napoléon's France while the other sent army after army after him—should suggest that for the one to try to become like the other would have been a difficult proposition. Germany was not a naval power; to try to become one required it to master skills it did not have. Tirpitz's battleships were inferior in speed and weaponry to Fisher's, and yet they were superior in cost and took longer to build. As the foregoing analysis shows, moreover, Germany had no extensive experience in building the type of political coalitions abroad that were necessary to build and maintain seapower. In fact, it did not even have a consensus in favor of seapower at home. It is not much of an exaggeration to say that not only did Germany act like a land power while trying to be a seapower, but it thought like one too, repeatedly concerning itself with being able to win wars rather than avoid them. Germany simply did not have a core competence in the area in which it chose to compete with Britain. Acquiring the needed competencies would have required that much more investment of resources and, above all, official attention; Kaiser Wilhelm and his advisors never supplied it.

Hamel and Prahalad's theory that the challenger (beta) must be all the more focused in its efforts, and leverage the advantages that it does have, if it is to overthrow and succeed the leader (alpha), is therefore applicable here. While Hamel and Prahalad envisioned energetic and clever betas overthrowing lumbering and stupid alphas, there is nothing inconsistent with their theory about imagining events happening the other way around. A foolish beta can fail to unseat a reasonably competent alpha; such is, in fact, to be expected normally, and it happened here.

As a matter of fact, Hamel and Prahalad's more profound argument—that competing successfully is as much about defining the terms of the competition as it is about actually winning it—is quite applicable here. Germany was not trying to do anything new; it was trying to copy the old. Rather than seek its own path, it chose to compete head to head with Britain in an area where Britain already held all the relevant advantages. Germany was destroyed, well before World War I finally removed any doubt, as much by a failure of imagination as one of execution. Hamel and Prahalad's insights into the strategic requirements of competition by asymmetric players are therefore applicable in the case of the Dreadnought Race and are validated by it.

Decision Cycles and Decision Making

7. Assess the ability of each competitor to read its adversary and make decisions that will correspondingly limit its competitor's options and situational understanding—the ability to operate "within" the adversary's decision cycle. Reassess as the competition unfolds.

It remains to apply the decision cycle analysis portion of the framework. As discussed in chapter 2, the concept of the OODA loop, as described by U.S. Air Force colonel John R. Boyd, posits that an individual actor's decision cycle can be broken into a series of four sequential and overlapping stages: observation, orientation, decision, and action. All else equal, the ability to complete the cycle more quickly than an opponent is thought to confer a strategic advantage. Likewise, the loop allows for analysis of a decision cycle by breaking it into its component parts; if a wrong turn is made, it is possible to see which aspect of the cycle led to it and to correct the problem in the future. Ultimately, a strong competitor will use their own decision cycle to attack their opponent's, disrupting the opponent's ability to get anything done while maintaining their own. To see how the concept applies here, it is useful to address each stage in its turn, bearing in mind nevertheless that the competition between Britain and Germany in fact involved innumerable cycles of the loop.

Where observation is concerned, Britain appears to have enjoyed a very slight advantage in this area. There was never a point at which Britain was completely blind-sided by Germany. Individually, naval officers such as Fisher and politicians as early as Lord Salisbury's government were quite aware of both Germany's intentions and growing naval capabilities, as well as of developments in naval technology in general. And although a consensus in favor of action on the part of the civilian leadership was slow in coming, information traveled quite easily.

The Dreadnought Race in Strategic Perspective

Britain also possessed the political acuity to see the (probably perishable) opportunity for the formation of the Triple Entente in time to seize it, even if its motives in doing so were ambiguous. All in all, this may not represent extraordinary acuity, but it is difficult to argue that the relevant information was not gathered when it was needed.

By contrast, Germany's strategy, to its detriment, did not place great emphasis on accurate observation. From the point of view of Tirpitz and Wilhelm, once the decision to construct a fleet to challenge Britain had been taken, paying attention to what Britain was doing in response was at best secondary. This manifested in curious ways. At the level of naval technology, Tirpitz seems not to have been aware of some of the relevant technological developments. Not only did HMS *Dreadnought* render his fleet obsolete, but Tirpitz was slow to pick up on certain technological advantages of the British ships that could be observed with the naked eye, most notably in-line gun turrets, which were lacking even on his later designs. At the higher levels of strategic analysis, Tirpitz clung to his vision of crossing the "danger zone" well after Britain had made it abundantly clear that the danger zone was impassable. He and Wilhelm were also willfully blind—it is difficult to see how else to describe the matter—to political realities within their own country, most notably the intractable problem of extracting funding from the fleet from unwilling Junkers. This assessment is necessarily qualitative and somewhat counterfactual, but the problems of successfully running the Dreadnought Race that have been repeatedly remarked on by historians seem to have escaped the notice of the people who initiated it. To an extent, this analysis necessarily bleeds over into the question of orientation, vice observation, but it is notable that Wilhelm and Tirpitz seem not to have paid conscious attention to problems as they arose, whether or not they might have analyzed them.

Where orientation was concerned, Britain had a notable edge, in that its leaders interpreted data correctly and Germany's did not. This requires some qualification. It is certainly true that at a number of points Britain lagged in truly understanding what was going on. The 1908 Naval Scare is perhaps the most notable, in that the relevant information was available but a consensus that Britain was in danger of falling behind Germany in capital ships took months to emerge. One can likewise point to the rather chaotic approach of the Salisbury government to hostile German rhetoric as another instance of political leadership failing to interpret available data in quite the necessary way. In general, however, Britain made no notable errors of interpretation in understanding the German challenge.

Germany's difficulties interpreting available information, which have already been discussed, were voluminous. Right from the start, plenty of information was available to Tirpitz—and theoretically to Wilhelm and the other politicians to whom he sold his scheme—that, correctly interpreted, would have suggested all the difficulties he eventually faced. Such difficulties included the conceptual problems associated with defining the size and scope of the fleet, whether it was merely to achieve a certain force ratio or continue until parity was reached; the fundamental, *a priori* unwillingness of the Royal Navy to ever part with control of an area of serious national security import; the inevitable funding problems and the limits of the German political process; and the likely diplomatic costs of making yet another enemy. All of these and more went unnoticed not because of a lack of information but because of an almost pigheaded determination not to consider it. When it came to more mundane matters, the seemingly small-minded bureaucratic pessimism of Germany's less illustrious and less remembered military and political leaders seems not limiting but prescient: those with day-to-day jobs to do seemed to understand the military and diplomatic situation better than their political masters.[66] For this reason, the German Foreign Office was more unnerved by the entente than Wilhelm, and the Naval High Command was less enthusiastic about Tirpitz's battleship construction project than Tirpitz was. What more pedestrian officials saw quite clearly the putatively visionary senior leadership saw imperfectly or not at all.

Where decision is considered, the picture becomes less clear. Britain's decision making depended on democratic consensus, maintained over time. The Admiralty, under Fisher, was engaged in an embarrassing public row involving both substantive technological debate and personality conflicts. Fisher's idiosyncratic decision-making style, his abrasive and argumentative personality, his abrupt statements of opinion on policy, and his lack of a formalized staff all contributed to what must have appeared to be utter organizational chaos. At the political level, the Dreadnought Race chanced to occur at a time when Britain was engaged in a contentious national debate about social spending and the creation of the modern welfare state, and finessing the necessary increases in naval spending, across multiple governments of opposite parties, required considerable political effort and skill and took time. Because of this, Britain often lagged Germany in the time it took to make a major decision, at least at the highest levels. Fisher's ability to ram new ship designs through to construction ameliorated this on the purely military side.

But Germany was in no better position. The Naval Laws may have given Wilhelm and Tirpitz carte blanche to build a certain number of ships, but getting authorization to build over that limit—the point of the successive novelles—required a political consensus in favor of building that simply did not exist. Consensus failed in particular once the question was raised of funding shipbuilding by direct as opposed to indirect taxation. In the end, Tirpitz ran out of money and was anticipating a building pause. At the diplomatic level, bureaucratic infighting, particularly between Tirpitz and Bethmann Hollweg during the arms control talks, effectively paralyzed German decision making. Wilhelm's incoherent decision making—from the very beginning, he was confused about how to move forward with the naval scheme—and inability to coordinate policy merely added to the chaos.

Chapter Six

Both states, therefore, had problems with decision making, if not for the exact same reasons. Britain's process was moderately functional but slow; Germany could move quickly at times but suffered from its infamous "polycratic chaos." For these purposes, it is enough here to note that, qualitatively, neither state was truly superior to the other in this regard.

But in the realm of action, Britain again appeared to have the advantage. Partly just because of its relative competence in this regard, Britain was able to compete militarily with what can only be called agility. HMS *Dreadnought* was built in the space of a calendar year. The newer oil-fired ships were likewise built at a record pace. By contrast, Tirpitz's own inferior ships took at least twice as long on average to build and were more expensive to boot. Politically, Britain was able to lock down a diplomatic coalition against Germany in the space of about two years' time and hold it together. On the German side, for all its insistence on competing in an arms race, Germany had terrible difficulty executing its strategy once adopted, repeatedly coming up short in terms both of quality of ships produced and of speed of production. In the international political arena, it is difficult to assess events that never took place: Germany had made precious few decisions and did not therefore need to act on them. Nevertheless, Bethmann Hollweg's last, desperate attempts to put the situation right should speak for themselves. In political matters, Germany had an implementation problem.

All in all, Britain appeared to enjoy a slight advantage in most aspects of the cycle. *In toto*, however, what is really noteworthy is not that Britain appeared to have a slight qualitative advantage in decision making but that it actually did operate within Germany's decision cycle. The military side of the equation shows an obvious British advantage. On repeated occasions, Fisher ruined Tirpitz's plans, putting more advanced ships to sea and forcing Tirpitz almost literally back to the drawing board. The first instance was the launching of HMS *Dreadnought*; the second was the launching of HMS *Invincible*. The new super-dreadnoughts of the next decade frustrated Tirpitz further: at

The Dreadnought Race in Strategic Perspective

literally no point was Tirpitz or the High Seas Fleet ahead of the design curve. Nor did the Imperial German Navy ever solve problems of money and time as successfully as the Royal Navy (which, after all, had years of practice): given the cost-inefficient and slow building processes for German ships that turned out to be inferior to their British counterparts (right up to the last, when the High Seas Fleet deployed ships with obsolete gun emplacements), Fisher (and later Churchill) were inside Tirpitz's decision cycle for almost the entire length of the race.

Politically, it was the same story. It is not clear, when all is said and done, that Germany even had a decision cycle on the diplomatic front. Its leaders were highly militarist and regarded political compromise as weakness and war as an inevitability. Precisely for that reason, in Peter Padfield's apt phrasing, British diplomats "ran rings around" them.[67] Germany's few attempts to force a favorable diplomatic outcome—whether in the Algeçiras crisis, in which it attempted to disrupt the nascent entente; in the Agadir crisis, in which it attempted to force colonial concessions from a position of weakness; or in the arms control talks, in which it attempted to consolidate what relative gains it had made—came to little, as in all such cases Britain and the rest of the entente acted more swiftly and decisively than Germany did and with greater resolve. Insofar as flinching in the face of one's adversary counts as having one's decision cycle disrupted, Germany suffered this serially. Whatever one may say about the British Foreign Office's arguably reckless conduct under Grey, it did succeed in isolating Germany without hindrance, even as it maintained correspondence with the domestic faction in Germany (the Social Democrats) who were opposed to the Dreadnought Race. Britain was acting against and frustrating Germany; Germany was not doing the same to Britain.

Did Britain actively target Germany's decision cycle, as Boyd exhorts his followers to do, and as his predecessor Sun Tzu obliquely urges when directing his disciples to target the adversary's strategy first?[68] This cannot be said for most of the Dreadnought Race—in fact, Britain was mostly reactive throughout; it merely happened to be the case that when Britain did engage it had an edge in initiative and directed effort. Ironically, by attempting to deceive Britain about his country's intentions, Tirpitz attempted this, only to fail to do so when various aspects of his scheme—notably, the diplomatic exigencies of avoiding attention—clashed with the foreign policy necessities. As noted, it would have been better if Germany's diplomats had been free to attempt to exploit divisions between Russia and Britain over the Dogger Bank incident and elsewhere. Neither side played this aspect of the game well. But by the end, a different picture emerges: Churchill, by pulling out all of the stops in ship construction (even going so far as to seek additional ships from the dominions) and driving hard diplomatically at the same time, seems to have forced the kind of paralysis and collapse that Boyd suggests can be achieved by entering an adversary's cycle. Bethmann-Hollweg's ultimate lack of initiative in the face of both British intransigence and the collapse of Germany's diplomatic position—including his inability to bring the reckless and feckless Alfred von Kiderlen-Wächter to heel—are evidence of this kind of collapse. By the end, Germany was not working internally any more than externally, and it was the pressure Britain had brought to bear that explained a good deal of it. If Germany was forced to use gunboats tactically instead of battleships strategically to achieve minor aims, and if this in turn caused no end of internal chaos, then some kind of decision cycle collapse can indeed be said to have occurred.

Although this analysis is admittedly qualitative and intuitive, in the end it must admit that Germany, having first failed relative to Britain in its ability to formulate strategic intent and put together a competent plan of action, failed subsequently relative to Britain in implementing its own design. Germany was weak first on paper and then in the fluid, iterative process of confronting its adversary. If this speaks ill of Wilhelmine Germany, it nevertheless vindicates the usefulness of classical strategic thought as a guide to interstate competition.

General Conclusions

The one lesson from Sun Tzu that is indubitably applicable to this case is never to start a fight that one is unprepared to win. To pull these disparate threads together, if a state competes for regional hegemony, it must decide what that entails in terms of the three metrics and choose from which of them it is willing to sacrifice and which of them it is trying to enhance. Ideally, the one it gives up should be the one it can most readily part with. Doing so involves playing to a strength; doing otherwise amounts to relying on one's weaker areas. Neither Britain nor Germany fared well when it came to making these choices, although Britain performed better.

Once one is committed to the competition, however, one must win. As shown, Hamel and Prahalad's insights regarding what each competitor must do are valid: one must maintain its position and wear down its adversary; the other must force a favorable decision. In either case, it is vital for someone to take charge and pull together all the available resources, making best use of their side's strengths and competencies and using them to best advantage to achieve the strategic objective. Likewise, one must be the actor, not the acted on: one must stay inside the adversary's decision cycle. Britain did this somewhat better than Germany did, and it was somewhat better positioned at the end of the Dreadnought Race in spite of the intervention of war.

As stated before, strategy is a process as well as a plan; it is certainly not a static plan. Those who are clear in their intentions and willing and able to allocate resources behind them are more likely to succeed than those than cannot decide. Once competition is joined, those who make decisions better and more quickly have a better chance of coming out ahead and of disrupting their adversaries' efforts to do the same. All told, the Dreadnought Race serves as a very illustrative example of how strategic concepts can be put to use in the analysis of international politics as well as other fields.

Chapter Six

Endnotes

1. Rudyard Kipling, "Recessional," *Spectator*, 24 July 1897.
2. Sun Tzu, *The Art of War*, tr. Samuel B. Griffith (New York: Oxford University Press, 1963), 73.
3. Sun Tzu, *The Art of War*, 142-43.
4. Henry Mintzberg, *The Rise and Fall of Strategic Planning* (New York: Free Press, 1994), 23. This concept is generally captured in the idea of strategy being an ongoing process of decision making, but it can be accounted for in Mintzberg's original formulation here: there are inherent difficulties associated with speaking of a corporate organization of any type as having a "strategy" or an "official mind," even as it is often necessary to do so. As noted above, this study contends with those difficulties simply by accounting for the decisions made by key individuals as well as the thought process behind them to the extent it is known.
5. See Peter Padfield, *The Great Naval Race: The Anglo-German Naval Rivalry, 1900-1914* (Edinburgh, Scotland: Birlinn, 2005). This study, in particular, takes issue with Padfield's statement that "there was no British failure—save from those who encouraged German ambition by preaching peace as a higher value than integrity or self-preservation. Both in diplomacy and naval strategy the Liberals ran rings around Wilhelm's second eleven of *Real* politicians. One can only marvel at a golden age which could harness such . . . giants." Padfield, *The Great Naval Race*, 346. There is no doubt that Britain repeatedly outmaneuvered Germany and that its strategy was sounder. As will be shown, however, considering the totality of what took place and the opportunities that were missed, it is difficult to say that Britain succeeded in all its stated and implicit objectives.
6. See Arthur J. Marder, *The Anatomy of British Sea Power: A History of British Naval Policy in the Pre-Dreadnought Era, 1880-1905* (Hamden, CT: Archon Books, 1940), 543-44; and Robert K. Massie, *Dreadnought: Britain, Germany, and the Coming of the Great War* (London: Vintage Books, 2004), 184.
7. Paul M. Kennedy, *Strategy and Diplomacy, 1870-1945: Eight Studies* (London: Allen and Unwin, 1983), 20-25.
8. The argument is made forcefully in Christopher Clark, *The Sleepwalkers: How Europe Went to War in 1914* (New York: HarperCollins, 2014), xxvii, 560-62.
9. The argument, now a classic one, can be found in Paul M. Kennedy, *The Rise of the Anglo-German Antagonism, 1860-1914* (London: Ashfield Press, 1987). It can survive the counterpoint, made by Zara S. Steiner and Keith Neilson that, oddly, "There was talk of

war on both sides but there was nothing concrete to fight about." Zara S. Steiner and Keith Neilson, *Britain and the Origins of the First World War*, 2d ed. (Hampshire, UK: Palgrave Macmillan, 2003), 82. As will be discussed, although there may well have been no specific reason for Britain and Germany to come to blows, and although Britain may indeed have had options, the relationship between the two had been framed by the Dreadnought Race and the general deterioration of relations to such an extent that the dominant question in Britain was whether to go to war with Germany, not which side to come down on if pressed. For the role of the Anglo-French naval cooperation, see Samuel R. Williamson Jr., *The Politics of Grand Strategy: Britain and France Prepare for War, 1904-1914* (London: Ashfield Press, 1990), 370.

10. See, for example, Hew Strachan, *The First World War*, vol. 1, *To Arms* (Oxford, UK: Oxford University Press, 2001), 93.
11. This discussion hinges in particular on whether the British decision to enter the war was a foregone conclusion or a happenstance decision made at the last minute in the midst of the political and bureaucratic chaos of the July Crisis. For a treatment of this issue, see J. Paul Harris, "Great Britain," in *The Origins of World War I*, ed. Richard F. Hamilton and Holger H. Herwig (Cambridge, UK: Cambridge University Press, 2003), Kindle ed., loc. 6701-7629, https://doi.org/10.1017/CBO9780511550171.008. Harris points out that Britain's decision to go to war, which was taken at the last possible minute, was anything but predetermined and that preexisting war planning for an expeditionary force—along with the generals—were ignored until the decision was made, two days after war was declared, to send a land force to France. Harris, "Great Britain," loc. 7246-63. This is a point partially conceded by Samuel R. Williamson Jr., who argues that "the entente and the staff talks played influential but not decisive roles in the British decision to go to war." But once "that step was taken, the impact of the entente and the conversation upon the related question of how the [British Expeditionary Force (BEF)] would be employed was crucial and unambiguous: for the staff talks, with an assist from the Admiralty, were responsible for British presence on the western front." Williamson, *The Politics of Grand Strategy*, 361. Williamson particularly notes that the military staff talks created an organizational lobby for war whose weight was indeed felt in the final days of the July Crisis. Williamson, *The Politics of Grand Strategy*, 370-71. In sum, as Harris notes, the military conversations may not have predetermined the dispatch of the BEF to the continent, but they assuredly laid the ground-

work for it and made it more thinkable, even as they made war itself more thinkable. On this point, this study is again prepared to accept Paul M. Kennedy's argument that Britain's decision to go to war may have hinged on any number of immediate factors, and Germany in 1914 may have welcomed British neutrality had such been declared, but that the fact that war was being debated at all in British Parliament, and that it concerned Germany and no other power, bespeaks a hostility born of long and ongoing experience. See Kennedy, *The Rise of the Anglo-German Antagonism*, 422–23, 456–58.

12. Fritz Fischer's now-classic argument that the German decision to go to war was effectively made in the famous war council of December 1912, and that subsequent developments merely tracked with this decision, is refuted or qualified by most of the commentators on whom this study relies, and this study does not subscribe to it. See, for example, Strachan, *To Arms*, 52–54; Kennedy, *The Rise of the Anglo-German Antagonism*, 457, 467; and Ivo Nikolai Lambi, *The Navy and German Power Politics, 1862–1914* (Boston, MA: Allen and Unwin, 1984), 383–84. Strachan is quite correct in noting that Fischer's arguments, as they relate specifically to German war guilt, are not easy to summarize and are perhaps best understood as a school of thought, and not an entirely coherent one. Strachan, *To Arms*, 3. A notable exception of a more popular nature is Peter Padfield, whose narrative of the Dreadnought Race explicitly begins with the broad premise of German war guilt, although he does not see the war council as the deciding factor. Padfield, *The Great Naval Race*, xv–xvi. It is not necessary, however, to believe that Germany made the decision to go to war early and with aggressive intent to subscribe to the more modest premise—classically argued by Kennedy and repeated here—that the Dreadnought Race helped to drive the unraveling of the Anglo-German diplomatic relationship and with it made war more and more thinkable.

13. Matthew S. Seligmann, Frank Nägler, and Michael Epkenhans, *The Naval Race to the Abyss: The Anglo-German Naval Race, 1895–1914* (London: Routledge, 2016), 402–4; Michael Epkenhans, *Tirpitz: Architect of the German High Seas Fleet* (Washington, DC: Potomac Books, 2008), 55; Kennedy, *The Rise of the Anglo-German Antagonism*, 423–24, 447, 448–52, 458–63; and Lambi, *The Navy and German Power Politics*, 322, 427.

14. Steiner and Neilson, *Britain and the Origins of the First World War*, 275.

The Dreadnought Race in Strategic Perspective

15. Or, put differently, that the naval race had *something* to do with the Anglo-German enmity, even if it was not the only source of it. This commonsense proposition is affirmed by Matthew S. Seligmann, Frank Nägler, and Michael Epkenhans. See Seligmann, Nägler, and Epkenhans, *The Naval Race to the Abyss*, 402-4; and Epkenhans, *Tirpitz*, 55.
16. See Carl von Clausewitz, *On War*, trans. Michael Howard and Peter Paret (New York: Alfred A. Knopf, 1993), 154, 156, 161; and Martin van Creveld, *The Transformation of War* (New York: Free Press, 1991), 171.
17. Clausewitz, *On War*, 86, 726.
18. See Kennedy, *Strategy and Diplomacy*, 141. As noted in the preceding chapter, it is clear that at least some of the ideas imputed to Fisher and others regarding naval strategy—notably flotilla defense of British home waters—were never seriously under consideration. In what follows, this study will mostly concern itself with what options were seriously available, but some analysis is offered regarding what might have been possible.
19. Quotation found in Jon Tetsuro Sumida, *In Defence of Naval Supremacy: Finance, Technology, and British Naval Policy, 1889-1914* (London: Unwin Hyman, 1989), 58.
20. As noted above, this is due as much to agnosticism as anything else. Doctrinal disputes about the role and use of the light-armored battlecruisers were never fully resolved, and historical disputes about the causes of the battlecruisers' poor performance in the Battle of Jutland—and the degree of poverty of that performance—still remain, and they complicate the task of assessing the naval balance. On the theory that the battlecruisers could have served as adequate substitutes for the heavier battleships—that is, that Fisher was right—they are counted together here.
21. See Nicholas A. Lambert, *Sir John Fisher's Naval Revolution* (Columbia: University of South Carolina Press, 1999), 277-80. As noted in the preceding chapter and above, although Fisher and others were aware of the growing capabilities of submarines, and although Winston Churchill decided on flotilla defense as a stopgap measure for the Mediterranean, surface battleships remained the capital ship standard and the preferred method for defending British home waters; the emphasis here reflects that decision mentality.
22. Since, for purposes of this study, after the launching of HMS *Dreadnought* in 1906 all older capital ships are treated as obsolete, one could say that between 1907 and 1908 the ratio was infinite, as Germany had not yet launched a truly modern battle-

ship. Figure 2 accounts for these instances of infinite ratios as zeroes. These brief instances of infinite naval supremacy were extremely short-lived, however, as is noted in the chart.

23. Readers of this work in its earlier form as a PhD dissertation will note that this chart has been extensively revised. In the course of developing this work, it was noted that the previous chart contained an anomaly, in which German fleet size was shown to be larger than that of Britain for a substantial period before the Dreadnought Race began. The error lay in the coding of major surface combatants, which required significant judgment calls regarding the relative fighting value of ships during a time of extensive technological change and fleet modernization. The appendix offers a discussion of the methodology used here. This author freely admits that significant latitude is possible in terms of how to characterize naval strengths throughout this period — this chart represents a best guess but should be understood as an approximate picture.

24. As the chart shows, there are limitations to this analysis. In terms of total quantity of ships (but not of quality or usefulness), Britain did have a low ratio of total ships to Germany for the last two decades of the nineteenth century, which in turn underlay the sense of urgency that led to the Two-Power Standard from the late 1880s onward. The fact that Britain was not able to regain its former preeminence in sheer fleet size, however, must be considered.

25. See Frederick Martin, ed., *The Statesman's Yearbook* (London: Macmillan, 1878), 106; Martin, *The Statesman's Yearbook* (1879), 106; Lawrence Sondhaus, *Preparing for Weltpolitik: German Sea Power before the Tirpitz Era* (Annapolis, MD: Naval Institute Press, 1997), 141; and Massie, *Dreadnought*, 167.

26. The human cost of the war, per se, is best understood under the rubric of finance and welfare, discussed below; this section concerns Britain's national security rather than its people's human security. The sometimes antagonistic relationship of the two concepts is noteworthy, notwithstanding the fact that citizens are traditionally seen to have their welfare maximized when the security of their community is most assured. To quote John Keegan: "[H]ow self-defeating is the effort to run in harness in the same society two mutually contradictory public codes: that of 'inalienable rights,' including life, liberty, and the pursuit of happiness, and that of total self-abnegation when strategic necessity demands it." John Keegan, *A History of Warfare* (New York: Vintage Books, 1994), 50. The strategic dilemma of balancing the

The Dreadnought Race in Strategic Perspective

three metrics discussed here (security, welfare, and intangible goals) has to be resolved by choosing among them.

27. Padfield, *The Great Naval Race*, 343–44. The Schlieffen Plan's—so named for Alfred, Graf (count) von Schlieffen, chief of the German General Staff, to wage a successful two-front war—very existence as a single meaningful concept has been debated of late, with Terence Zuber arguing that there never was a single plan, but rather a series of wargames and studies that acquired their principal originator's name and were elided with the actual military decisions made in the opening moves of the war. Terence Zuber, *The Real German War Plan: 1904–1914* (Gloucestershire, UK: History Press, 2011), 180–86. For a discussion, see Hans Ehlert, Michael Epkenhans, and Gerhard P. Gross, eds., *The Schlieffen Plan: International Perspectives on the German Strategy for World War I*, ed. MajGen David T. Zabecki, USA (Ret) (English translation) (Lexington: University Press of Kentucky, 2014). Gross' argument that Helmuth von Moltke's decisions at the war's outbreak strongly resembled previous planning, whether or not there was a singular plan, is perhaps the heart of the matter. See Gerhard P. Gross, "There Was a Schlieffen Plan: New Sources on the History of German Military Planning," in *The Schlieffen Plan*, 119. Such may be the case, but the general notion that British involvement on the continent added extra enemies for Germany while the partial cause of that involvement—the High Seas Fleet—sapped its strength is accurate enough. In addition, it is not too farfetched to argue, as Epkenhans does, that although many factors may have driven the decision by Britain to intervene, the Dreadnought Race was at least a contributor to the enmity that preceded that decision. See Epkenhans, *Tirpitz*, 55.

28. Zara S. Steiner and Keith Neilson's argument that the continental balance of power, and not any specific German policy, was the crucial factor behind British involvement in World War I can be conceded without rejecting Paul M. Kennedy's point that the construction of France and Russia as allies and Germany as a potential adversary was driven by a general breakdown of Anglo-German relations in which the Dreadnought Race played a crucial role. See Steiner and Neilson, *Britain and the Origins of the First World War*, 275; and Kennedy, *The Rise of the Anglo-German Antagonism*, 458. Conversely, as Samuel R. Williamson Jr.'s summary of the Salisbury government's attitude toward Belgium shows, Britain had long been ambivalent about the matter. See Williamson, *The Politics of Grand Strategy*, 21.

Chapter Six

29. As the previous chapter noted, there is plenty of evidence, cited by Hew Strachan, among others, that Britain was ambivalent about the need to go to war throughout most of the July Crisis, and that in fact the final decision for war was made only at the last minute and amid substantial cabinet dysfunction. See Strachan, *To Arms*, 93–98. Whatever the case, as Paul M. Kennedy notes, the question during the July Crisis was whether to go to war with Germany or merely wait on events, not any broader policy question—and the Dreadnought Race was a contributing factor to this. See Kennedy, *The Rise of the Anglo-German Antagonism*, 422–23, 456–58.
30. Precisely 1.97 percent, per figures derived from *The Statesman's Yearbook* and Angus Maddison's calculations (see the appendix). Unless otherwise noted, all data in this chapter are derived from these sources, as described in the appendix.
31. Readers of this work in its dissertation form will note a small change here, as well as with figures 38 and 39. Previously, these charts simply divided the change in naval estimates or expenditures by the total defense estimates or expenditures for the relevant country. This method, however, produced anomalies in which total defense budget changes were negative; consequently, only instances in which overall defense estimates or expenditures increased in a given year are counted here. This picture is also somewhat incomplete, in that it does not account for the overall net change in naval versus total defense spending, though this is accounted for in figures 9, 34, and 35. In general, both Great Britain and Germany were increasing their naval spending at the expense of their armies, but not to the same degree or with the same result.
31. The oft-repeated figure is given in Keegan, *A History of Warfare*, 365. The Commonwealth War Graves Commission records a total of more than 1.1 million. See *Commonwealth War Graves Commission Annual Report, 2013–2014* (Maidenhead, UK: Commonwealth War Graves Commission, 2014).
32. Padfield, *The Great Naval Race*, 47.
33. See Arne Røksund, *The Jeune École: The Strategy of the Weak* (Boston, MA: Brill, 2007), 221–22.
34. Christopher M. Bell, "Sir John Fisher's Naval Revolution Reconsidered: Winston Churchill at the Admiralty, 1911–1914," *War in History* 18, no. 3 (July 2011): 333–37, 355–56, https://doi.org/10.11177/0968344511401489.
35. Kennedy, *Strategy and Diplomacy*, 20–25.

36. Julian S. Corbett, *Some Principles of Maritime Strategy* (London: Longmans, Green, 1911; ed. and repr. Project Gutenberg, 2005), Kindle ed., loc. 2520–30; and Padfield, *The Great Naval Race*, 343.
37. Kennedy, *The Rise of the Anglo-German Antagonism*, 458.
38. Røksund, *The Jeune École*, 221–22.
39. Padfield, *The Great Naval Race*, 343, 345.
40. As Peter Padfield notes, Tirpitz was rebuffed in budgetary discussions by Imperial German Army generals who wanted to prioritize extra divisions over new ships. Padfield particularly notes that whereas the Imperial German Navy only ever inherited between 19 and 26 percent of overall German defense spending, the Royal Navy could count on as much as 60 percent of Britain's overall defense expenditure, figures that are broadly confirmed by the quantitative analysis presented here. See Padfield, *The Great Naval Race*, 210, 298–89.
41. See Kennedy, *Strategy and Diplomacy*, 146–48; and Patrick J. Kelly, *Tirpitz and the Imperial German Navy* (Bloomington: Indiana University Press, 2011), 313–14.
42. Matthew S. Seligmann, *The Royal Navy and the German Threat, 1901–1914: Admiralty Plans to Protect British Trade in a War against Germany* (Oxford, UK: Oxford University Press, 2012), 10.
43. From 1893 onward, *The Statesman's Yearbook* gives two different statements of Germany's naval budget, one referred to as "estimates" and the other as "expenditure." The latter is likely the more accurate figure. As a sanity check, Germany's total naval expenditure (vice estimates) from 1901 to 1909 is given by *The Statesman's Yearbook* as approximately £119,400,000, using current exchange rates (see the appendix), which is quite close to Peter Padfield's estimate of German total naval expenditure during that timeframe of £125 million. See Padfield, *The Great Naval Race*, 234.
44. The year 1913 was the last year for which *The Statesman's Yearbook* gives peacetime defense expenditures.
45. This dovetails well with Peter Padfield's estimation of naval expenditures accounting for between 19 and 26 percent of overall military spending. See Padfield, *The Great Naval Race*, 210.
47. As noted with regard to figure 17, this has been amended from the original dissertation version to include only those years in which overall defense spending increased. Years in which naval spending declined but overall defense spending increased are thereby counted as negative naval shares of the total.
48. This sentence has been corrected from the original dissertation version, which erroneously put the 1.8 percent figure in 1912 and

which cited an estimates figure for 1914 instead of an expenditures figure (which is not available).

46. As noted in the appendix, debt figures are derived from *The Statesman's Yearbook* for the relevant years. They are in keeping with the figures used by Peter Padfield, who gives a German debt figure for 1909 of £200 million. See Padfield, *The Great Naval Race*, 175. This figure is very close to the figure of £188,799,623 given by *The Statesman's Yearbook*, after converting at current rates, for 1908 and £228,762,295 for 1910 (no figure is given for 1909). Although different debt accounting methods exist, it can be fairly said that the ones used here correspond to those cited by the historical secondary sources used. The most noteworthy alternate source for debt relative to output is historical debt/GDP (gross domestic, vice national, product) estimates provided by the International Monetary Fund (IMF). See S. Ali Abbas et al., *A Historical Public Debt Database*, IMF Working Paper no. 10/245 (Washington, DC: International Monetary Fund, 2010); and S. Ali Abbas et al., *Debt Database Fall 2013 Vintage* (Washington, DC: International Monetary Fund, 2013). These estimates provide a different picture of a much higher, but also unchanging, German debt/output ratio, averaging 39.1 percent of GDP for 1880-98, and 39.5 percent for 1899-1913, ending at 38.5 percent in 1913. Out of confidence in Angus Maddison's methods and the raw debt data collected here (which, as noted, is also trusted by the historians), and out of consistency, the aforementioned method, rather than that of the IMF, is preferred here.

47. Again, these figures, taken from *The Statesman's Yearbook*, appear to reflect actual defense spending, while only the official estimates are available as a guide to total expenditures.

48. See Strachan, *To Arms*, 21-22, 31, 407.

49. Kennedy, *The Rise of the Anglo-German Antagonism*, 410-11; Kennedy, *Strategy and Diplomacy*, 139-41; and Eyre Crowe, "Memorandum on the Present State of British Relations with France and Germany," in *The Hidden Perspective: The Military Conversations 1906-1914*, ed. David Owen (London: Haus Publishing, 2014), 230-50.

50. Rolf Hobson, *Imperialism At Sea: Naval Strategic Thought, the Ideology of Sea Power, and the Tirpitz Plan, 1875-1914* (Boston: Brill Academic Publishers, 2002), 316. This point is echoed in Kelly, *Tirpitz and the Imperial German Navy*, 153.

51. A civil engineering version of this maxim is "You can have it on time, on budget, or up to code—but not all three."

52. Padfield, *The Great Naval Race*, 344–45. Padfield writes, "Knowing only what he wanted, and reaching out for it greedily like a child without thought to others' reactions, then blaming them for the consequences of his thoughtlessness, he tried to lead Germany towards the world power and equal recognition with England's world Empire which he craved above all else, along several diverging paths at once. His Chancellors took one road, his Foreign Office others, his Army another, and his Navy the most dangerous of all while he strutted and postured, now before this column, now before that, proclaiming peaceable intentions one moment, showing his 'mailed fist' the next."
53. See Gary Hamel and C. K. Prahalad, "Strategy as Stretch and Leverage," *Harvard Business Review* 71, no. 2 (March–April 1993): 75–84.
54. See Karen Rasler, William R. Thompson, and Sumit Ganguly, *How Rivalries End* (Philadelphia: University of Pennsylvania Press, 2013).
55. Kennedy, *Strategy and Diplomacy*, 20–25, 141; and Crowe, "Memorandum on the Present State of British Relations with France and Germany," 240–41, 255–56. The idea that the ultimate aims of the competition involved more than colonies appears to have become a standard talking point among British policy makers; thereby the behind-the-scenes liberal political fixer Viscount Esher, in a memo to Winston Churchill in the latter's capacity as First Lord of the Admiralty, wrote (and then deleted) a sentence in 1912 saying, "What is mere imperialism to them is life and death to us." The statement may have been too blunt, but not inaccurate. See Randolph S. Churchill, ed., *The Churchill Documents*, vol. 5, *At the Admiralty, 1911–1914* (Hillsdale, MI: Hillsdale College Press, 2019), 1492–93.
56. See Arthur J. Marder, *From the Dreadnought to Scapa Flow*, vol. 1, *The Road to War, 1904–1914* (Annapolis, MD: Naval Institute Press, 2013), 3–45; Kennedy, *Strategy and Diplomacy*, 111–26; Seligmann, Nägler, and Epkenhans, *The Naval Race to the Abyss*, Kindle ed., loc. 299–327; Christopher M. Bell, "Contested Waters: The Royal Navy in the Fisher Era," *War in History* 23, no. 1 (January 2016): 2–6, https://doi.org/10.1177/0968344515595330; Bell, "Sir John Fisher's Naval Revolution Reconsidered," 333–37; Massie, *Dreadnought*, 373–543; and Padfield, *The Great Naval Race*, 94–98, 115–56.
57. Lambert, *Sir John Fisher's Naval Revolution*, 38–40, 172, 177–82, 239; Sumida, *In Defence of Naval Supremacy*; and Nicholas A. Lambert, *Planning Armageddon: British Economic Warfare and the First World War* (Cambridge, MA: Harvard University Press, 2012), 76–77.

Chapter Six

For a discussion, see Matthew S. Seligmann, "Naval History by Conspiracy Theory: The British Admiralty before the First World War and the Methodology of Revisionism," *Journal of Strategic Studies* 38, no. 7 (July 2015): 966–84, https://doi.org/10.1080/01402390.2015.1005443; and Bell, "Contested Waters," 6–7, 11–12. Lambert is criticized by Seligmann for reading too much into Fisher's attempts to manipulate politicians and consequently into Fisher's extant writings themselves. See Seligmann, "Naval History by Conspiracy Theory," 971. Suffice it to say, there is vast room for interpretation without final resolution in this area.

58. Seligmann, *The Royal Navy and the German Threat*, 25–45, 65–88, 173–75; Seligmann, Nägler, and Epkenhans, *The Naval Race to the Abyss*, 106; and Bell, "Sir John Fisher's Naval Revolution Reconsidered," 355–56.
59. Bell, "Sir John Fisher's Naval Revolution Reconsidered," 333–35, 355–56.
60. See, for example, Marder, *The Road to War*, 26, 112–13.
61. Marder, *The Road to War*, 26, 112–13. The "Copenhagen" threat is perhaps merely one of the better-known examples of this. It was not a carefully considered remark, but an exclamation.
62. It was suggested to this author by a reviewer that British seamanship was a core competence in that British personnel were better trained and more suited to sea service than their German counterparts. This is undoubtedly true, but one should be wary here of attributing strategic significance to this where the Dreadnought Race is concerned. If anything, the actual performance of British ships at sea when used was, as Churchill's judgment noted above shows, disappointing.
63. Seligmann, *The Royal Navy and the German Threat*, 10.
64. Padfield, *The Great Naval Race*, 346.
65. See, for example, Frans P. B. Osinga, *Science, Strategy, and War: The Strategic Theory of John Boyd* (London: Routledge, 2007), 235–37; and Col John R. Boyd, USAF (Ret), "Patterns of Conflict" (lecture, U.S. Marine Corps Command and Staff College, Marine Corps University, Quantico, VA, 25 April and 2–3 May 1989; transcribed by Maj Ian T. Brown, USMC, 25 March 2015–11 January 2017), 11.

Conclusion

The Strategy of Regional Great Power Competition

The tumult and the shouting dies;
The captains and the kings depart ...
                              ~ Rudyard Kipling[1]

Then, afterwards, to order well the state,
That like events may ne'er it ruinate.
                              ~ William Shakespeare[2]

This study began with the assumption that states must compete. Though the particulars may vary from place to place and time to time, competition for regional hegemony—for a preponderance of military power and political influence over a given area—has always been a game in which states could be counted on to engage. A state can place itself in such a game by choice or by happenstance, and it is not necessary to posit reasons for the inevitability or avoidability of such games, whether based on a particular international relations theoretical school or some other rationale, to acknowledge their historical ubiquity. Whatever the case, the game can be analyzed as such: each competitor's decisions can be looked at in terms of their effect on the outcome, or, if the game is underway, their likely effect on the outcome. This being the case, classical strategic principles—the oft-styled "eternal truths" of strategy—can be applied to understanding what the competitors are doing. Similarly, the comparatively recent insights offered by certain elements of business strategy can be used for these purposes.

Conclusion

After an analysis of the concept of strategy and its evolution over the years, this study posited a framework for the analysis of state behavior in such scenarios. Its components were the analysis of strategic intent; analysis of strategic position based on the asymmetry between "alpha," the sitting hegemon of a region, and its challenger, "beta," who must create the conditions by which to dislodge it (the "finish line"); analysis of the tradeoffs among the three metrics of both state resources and state goals (power and security, finance and welfare, and intangibles) based on classical international relations theory; and finally real-time analysis of states' competitive behavior via an understanding of the interactions of their decision cycles based on U.S. Air Force colonel John R. Boyd's OODA loop theory. This framework was then tested on a real-life case: the competition between Germany and Great Britain for control of the North Sea (and to an extent its littorals) prior to World War I, known as the Dreadnought Race. This study demonstrated that this framework could be applied to such a case and could generate insights that allowed observers to understand why the states in question succeeded or failed in the competition.

Though the particulars have been exhaustively discussed in the preceding chapters, they can be recapitulated for the final time here:

1. Determine the competitive objective—what is at stake in the competition. Within the limits of this study, the competitive objective refers to regional hegemony, but exactly what that entails will vary in each case.
2. Determine which competitor is alpha, the reigning hegemon, and which is beta, the challenger.
3. Determine how that objective is manifested in terms of changes in the three metrics for each of the competitors. Determine what resources alpha and beta each have in terms of the three metrics. In so doing, also assess in particular which of the three metrics it is willing to trade for the others, and whether it has an abundance or a scarcity along

this metric, to determine whether it is utilizing a strength or a relatively weak area.
4. Assess more generally the ability of each competitor to formulate strategic intent—in particular its leadership, its ability to leverage resources and core competencies, and its overall understanding of the nature of the project it is facing. Assess in particular which competitor is more able to intelligently and ruthlessly make tradeoffs among the three metrics in pursuit of its competitive goals.
5. Assess whether either player has an endgame in mind—either to outrun the adversary (alpha) or to force a finish (beta), and whether that endgame is achievable within the context of the player's strategic intent—the decisions it knows to make and is able and willing to make.
6. Determine what core competencies each player has—what abilities it possesses that will enable it to obtain an advantage in the competition. Assess in particular each competitor's astuteness in leveraging these competencies behind strategic intent. Assess more specifically whether it has an advantage in performing the actions it will have to perform to win the competition, relative to its competitor.
7. Assess the ability of each competitor to read its adversary and make decisions that will correspondingly limit its competitor's options and situational understanding—the ability to operate "within" the adversary's decision cycle. Reassess as the competition unfolds.

The framework first asks what the competition is about, which competitor is the status-quo hegemon (alpha) and which is the challenger (beta), and how each competitor's goals manifest themselves in terms of the three metrics of security, welfare, and intangibles. The Dreadnought Race was a competition for hegemony—military and political dominance—in the North Sea and its environs, historically the exclusive province of Britain and its Royal Navy but coveted for material and immaterial reasons by Germany under Kaiser Wilhelm II. Britain was alpha; Germany was beta. Germany sought to diminish Britain's position of regional military preeminence, possibly to the point of completely dislodging it from this status, and it also sought to achieve some intangible goals—political unity at home, prestige abroad—in doing so. Germany's ultimate goals were mainly intangible, while a subsidiary set of goals involved at least maintaining its overall national security and, implicitly, enhancing it vis à vis Britain. Finance and welfare considerations did not play in, except insofar as there were limits to what Germany and German citizens were willing to pay. Finance and welfare were tradeable for a goal set that included elements of the other two metrics. For Britain, conversely, the goal was primarily one of security—because of the proximity of the North Sea to Britain's home territory, what was a lofty ideal for Wilhelm was sheer survival to a series of British governments.

The framework then inquires what resources each competitor can bring to the competition in each of the three metrics, and what core competencies each has that are relevant. It was revealed in the preceding case that Britain held core competencies that Germany lacked, both in its capacity to produce and deploy fleets and in its political acumen, competencies that Germany failed to build to its competitive detriment. Moreover, while each competitor had the financial resources to compete, with Germany retaining a nominal edge given that it was less burdened by debt, Britain ultimately showed itself much more adept at managing its resources, shrewdly maintaining its national security by a combination of political efforts and military buildup, while carefully monitoring its financial resources and incurring acceptable losses in intangibles. The outcome was far from inevitable, but the inherent difficulties of a landlubber navy in producing a fleet from scratch in the face of an adversary determined to match it ship for ship, and the conceptual problems involved in doing so, effectively ensured that Germany flailed about without accomplishing anything.

The framework then inquires as to the degree to which each competitor can formulate strategic intent: whether it can focus its resources and make use of its core competencies to gain an advantage over its adversary. This is particularly crucial for beta, which starts at a disadvantage and must put all of its available resources behind the pursuit of its competitive goals—as beta succeeds, increasingly alpha must follow suit. But in fact, Germany failed to put more than a fraction of its resources into the Dreadnought Race and balked at any greater commitment of resources, particularly financial ones. Although Britain—suitably, given its role as alpha—initially failed to focus its resources on competing with Germany, once it became clear that it had to do so, it fronted the necessary money, shed extraneous diplomatic commitments, and built ships. Britain's ability, under Royal Navy admiral Sir John Fisher, Winston Churchill, and the government of Prime Minister Herbert H. Asquith, to focus its resources on the competition ensured that Germany could not win it. The Dreadnought Race is therefore perhaps an ideal illustration of the relevance of strategic intent.

The framework then addresses the question of decision cycles, as framed by Boyd, that asks which competitor is better able in real time to make decisions more quickly and accurately and to impede its adversary's ability to do so. As was discussed in the preceding chapter, Britain repeatedly gained an edge on Germany in this regard. Although in certain elements of the OODA loop Britain and Germany might have enjoyed no clear advantage over one another, Britain repeatedly demonstrated that it retained the initiative and could render German plans moot, sometimes via outright deception as well as effort. The Dreadnought Race therefore illustrates the relevance of decision cycles and offers a case study in understanding them as they are employed by states competing for regional hegemony.

Finally, the framework demands that beta somehow create a finish line that it can cross—an endgame that will allow it to solidify and lock in its gains, or else that alpha prevent it from doing so until it can no longer sustain its own efforts. In this regard, Germany failed: as noted, the conceptual problems and outright contradictions contained within the Tirpitz Plan made it impossible for Germany even to cross the famous "danger zone," much less to achieve anything if it did so. The open question of Imperial German Navy grand admiral Alfred von Tirpitz's very intentions—mere deterrence or coercive diplomacy—behind building the fleet speaks for itself: there was no finish line in mind. This lack of conceptual clarity doomed Germany in the Dreadnought Race, but it serves here to illustrate the importance of the concept of a finish line for beta to cross. Otherwise, alpha wins by default.

No single element of the framework holds the key to predicting the result of a competition such as the Dreadnought Race, or to guiding a player in it. Nevertheless, a holistic assessment of these factors does illuminate the strengths and weaknesses not only of each competitor but of its "game"—its strategy. These factors are not only relevant in retrospective but could have been analyzed at the time in the course of attempting to predict how the Dreadnought Race might end or inform the leaders of either side. Taken together, they offer a way of understanding such a competition that can be applied to other cases.

Some caveats are in order, and they are in fact suggested by the case under study. A major insight that emerges from this study is the degree to which it is necessary to tailor one's grand strategy to fit one's competitive strategy (or vice-versa) to prevail in competition. Resources being finite and strategic focus being necessary, all else equal, the competitor that can peel away resources from other commitments to focus on the task at hand will fare better than the one that does not or, at any rate, better than it would have if it spread its resources thin. The three metrics showed, in particular, that a state had to be willing to treat its own macrogoals—the security, welfare, and intangibles that varying international relations theoretical schools posit as the drivers of state behavior—as resources as well as desired ends: a state that can trade off pursuit of one goal to focus on another will do better than a state that tries to do too much. While it must be said that no state ever makes decisions so consistently as to focus all of its resources on a fairly narrow goal set, the tradeoffs that states face are accounted for by the framework.

This, of course, raises the point that has been stated numerous times so far: a strategy can be both a plan and a process, but there is no requirement that it be made consciously or consistently, and an abstract corporate entity such as a state need not be expected to do either. Nor is such necessary to employ the framework articulated here: as this case study has shown, it is possible to evaluate the decisions made in competition without positing that they were always made by the same single person or that a consistent mind lay behind them. To have a muddled strategy, or none at all, is still in a sense to have a strategy, as assuredly as to make no conscious decisions is nevertheless in effect to decide something.

Although this study employed an historical case study for purposes of elucidation, the analysis of strategy is most useful when there is a game afoot. A logical next step, made prohibitive here by space constraints, would be to apply this framework to a competition underway in the present day. The exasperated question "how do you win a race with no finish line" could well be asked by another occupant of the White House now or in the coming years.

Subjects for such a study too readily suggest themselves as of this writing. The United States is engaged in several regional competitions at the moment. It is competing with Russia for control of Eastern Europe, and perhaps for Europe *tout court*. It is competing with China for supremacy along the first island chain, including smaller competitions on the Korean Peninsula, around Taiwan, and in the South China Sea; this larger competition is linked to competition for control of the entire Indo-Pacific, a region recently identified as being of sufficient coherence to warrant analysis as a single unit. Any or all of these competitions is susceptible to study via the framework presented here.

Space precludes such analysis here; however, this must be the beginning, not the end, of such inquiry. But it is easy to see

Conclusion

how it might be done. Alpha and beta are readily identifiable in each of these examples, and the three metrics provide insight into both means and ultimate ends. Where the United States and China are concerned, one might ask, is the Indo-Pacific competition simply a question of military balances—and, if so, is the United States or China better positioned along the security metric? It is unlikely to be about financial welfare, though a subset of the competition, control of the South China Sea does indeed have financial implications in terms of oil and other extractable resources. And what about intangibles—is there something more being sought? On this one could speak to China's determination to avenge its "century of humiliation" and to undermine Western liberalism globally. One may also note, as some U.S. analysts have argued, the comparative absence of broader goals on the American side.[3] What intelligent tradeoffs among these metrics can be made in the service of ultimate goals? All of this is up not only for debate but also for focused inquiry. One might then inquire into whether the United States or China has embedded competitive advantages that can inform the trajectory of the competition and which is better able to mobilize them. One might note that the United States, as alpha, is preoccupied elsewhere and has several alarming inefficiencies that reduce focus on the Indo-Pacific alone. One might also take account of China's shipbuilding capacity and seemingly greater efficiency of defense procurement.[4] Many other points of inquiry suggest, and would continue to suggest, themselves. And having done all this, one might examine who seems to have the initiative—who is "inside" whose decision cycle at any given time.

The same may be done for the example of the United States and Russia. One might examine whether the European competition is simply about military balances—the security metric—or whether it is driven by deep-seated pan-Western sentiment that drives competitive goals in its own right—a sought change to the intangibles metric. A sought goal in the financial metric in this area is perhaps less likely to emerge.

The Strategy of Regional Great Power Competition

One might inquire whether the United States' grand strategy elsewhere prevents the formulation of strategic intent, and on the Russian side, whether President Vladimir Putin's actions represent a long-term strategy or, as some suggest, short-term improvisation.[5] For core competencies, one might also assess the structural strengths and weaknesses of the North Atlantic Treaty Organization (NATO) alliance and, on the other side, Russia's ability to replace losses in Ukraine. One could look to the war in Ukraine for evidence in decision cycle analysis, asking such questions as whether, insofar as the competition between Russia and the United States remains such and not a full-blown military conflict, Russia is able to shape the competitive environment or whether the U.S. response has been adequate to blunt its initiative. Numerous other questions, of course—far too many to name here—might suggest themselves.

This framework could indeed inform a significant interagency or National Security Council review of these theaters of competition. But even if not formally applied, the point here is to describe a school of thought or strategic mindset—a way of calculating how competition is going and how one's own side is faring. If, at the end of the day, this study illuminates areas of the dark art of international competition in such a way that officials and analysts—or anyone of good will seeking to understand such a competitive scenario—now have at the back of their minds a list of questions to ask and a way of viewing such competition, it will have done enough.

There also remains considerable room for development in terms of how this framework may be applied. Although it is intended to inform policy as it pertains to current instances of regional great power competition, there is nothing that prevents an analysis using this framework of other past competitions similar to—or, for that matter, differing substantially from—the case study here. It may be possible to glean important insights from doing so—in particular, a comparative evaluation of multiple such analyses might yield lessons learned that could, in turn, be applied in a present day policy context. When

Conclusion

is focused strategic intent likely to emerge? Which of the three metrics is most likely to be at stake in a regional competition, and what can we learn about how great powers make tradeoffs among them? How do competitors make mistakes in their decision cycles? And so on. The relative infancy of serious strategic analysis of great power competition—as noted in chapter 1, the "how do they do it well?" and not the "why does it happen"—offers many possibilities.

As was noted in chapter 1, there exists a hierarchy of hegemony, whereby a state may control the global commons but be weak in a particular region, and whereby smaller states may, even under a global hegemon's aegis, compete for regional control and, as power transition theorists have noted, risk war at key points while doing so. Nothing prevents the application of this framework to a regional contest between subordinate powers—in fact, in such circumstances, the relationship of a larger outside power (e.g., global hegemon) to the competing states might impact any of the items of the framework, from core competencies (who has an existing relationship with the outside power), to the three metrics—Does one competitor have a higher baseline level of security due to a preexisting connection to the outside power? Are favorable terms with the outside power an intangible that drives the competition?—to decision cycles—Can one leverage diplomacy with the outside power to disrupt the adversary's plans? Further application of this framework either to historical instances of such competition or ongoing ones could yield important insights.

And equally, the work of strategic study will go on. This framework represents an attempt to integrate important strategic principles into the study of international relations, and more specifically the analysis of great power competition for control of an area of the globe—but it hopefully will not be the only such attempt. As other principles of strategy are discussed and applied, this framework may be added to or replaced by a better one. Nor should the application necessarily be limited: other strategic frameworks may be developed regarding a wider array of competitive scenarios beyond that of a dyadic competition for control over a region. Multisided competitions, competitions for global hegemony, pure arms races unrelated to territorial dominance and other scenarios deserve their own analytic frameworks, and future research may provide them.

In the end, the purpose of a framework is not to provide definitive answers but to pose questions and guide inquiry. This study has offered a framework to do just this for the analysis of a fundamental and seemingly eternal problem of international relations: how to deal with the state that wants what one has or has what one wants. Although now almost trite, Chinese military strategist Sun Tzu's famous exhortation to "know your enemy and know yourself" applies to just such types of competition. And although most states may fall far short of the mark, great leadership makes just such calculations.

Conclusion

Endnotes

1. Rudyard Kipling, "Recessional," *Spectator*, 24 July 1897.
2. William Shakespeare, *Titus Andronicus* (London: William Heinemann, 1904), 104.
3. See Elbridge Colby and Robert D. Kaplan, "The Ideology Delusion: America's Competition with China Is Not about Doctrine," *Foreign Affairs*, 4 September 2020. For an example of discussion of China's more maximalist aims, see Tanner Greer, "China's Plans to Win Control of the Global Order," *Scholar's Stage*, 17 May 2020. The point here is not to conduct an exhaustive analysis but to reference points of discussion that have been foregrounded.
4. See, for example, Ryan Pickrell, "China Is the World's Biggest Shipbuilder, and Its Ability to Rapidly Produce New Warships Would Be a 'Huge Advantage' in a Long Fight with the U.S., Experts Say," *Business Insider*, 8 September 2020.
5. See, for example, Julia Ioffe, "What Putin Really Wants," *Atlantic*, January/February 2018. The ongoing conflict in Ukraine both raises the relevance of these questions and will also no doubt continue to reframe some of the analytic debate surrounding them in real time.

Appendix
A Note on Data Sources and Methodology

Although individual data cited in this study are derived from the major historical sources consulted (and are cited as such), a significant portion of the data employed is derived from independent research. Using sources from the time period and economic research by reliable economists and social scientists, it is possible to reconstruct both Britain and Germany's public finances—and therefore their relative strategic position—at the time that the Dreadnought Race began.

This study uses a database compiled for Britain and Germany from 1875 until the outbreak of World War I—either 1914 or 1913, depending on the last year for which data were available. It chiefly employs as its source *The Statesman's Yearbook*, an almanac published from the 1860s to the present that records and preserves data on key aspects of state policy, ranging from population; to taxation, public spending, and public borrowing; to military figures; to figures on commerce and industry. This study has employed data derived from *The Statesman's Yearbook* primarily, on the theory that the data provided are essentially reliable and consistent. In practice, *The Statesman's Yearbook* must be used cautiously and judiciously, as its practices for obtaining and recording information could vary over time and were sometimes opaque. Most maddeningly, it provides two separate figures for German military spending: one of expenditures and the other for estimates (the two could vary considerably). It also provided varying information as better figures became available and sometimes provided information on a chronologically inconsistent basis as it became available (e.g., it might report the last year's figures if the current year was not yet over at the time of publication or its data were not yet available).

In compiling the database, this study has sought to abide by a few rules to remain consistent. Data that were not from the end of the relevant fiscal year would be reported as pertaining to the year to which they were closest. Gaps would be left where no data existed. Educated judgments would be made in the case of reporting data on military strength that depended on subjective assessment, such as the number of "first-rate" battleships, which in a time of technological transition could vary considerably depending on the measurement used. Where two or more differing reports were made for the same datum, such as population in a given year, the latest report given would be recorded, on the theory that it reflected the best available information. As stated in the database, population figures for Germany, which were collected only on a five-year census basis, were extrapolated for the intervening years on an average basis so as to allow for a gradation from one year to the next. Finally, data collection ended with the last edition of *The Statesman's Yearbook* published before World War I (1914), on the assumption that the outbreak of war would have wildly affected data collection.

The period 1875 to 1914 was chosen for specific reasons. It encompasses not only the Dreadnought Race but also the two decades leading up to it, allowing for reasonable points of comparison. It also begins at about the time of the consolidation of Germany as a state, minus four years and so allowing for a short period of political consolidation. It is therefore possible to build a full picture of economic, military, and fiscal changes that occurred in and between Britain and Germany not only during the Dreadnought Race but also for the preceding period in which relations were not so strained, and to compare the resulting figures.

The chief difficulty in reconstructing these states' finances has been the absence of a reference variable vital to modern

A Note on Data Sources and Methodology

budgetary analysis—namely, gross domestic product (GDP) or gross national product (GNP). The former counts all economic activity occurring within a state's borders; the latter all economic activity conducted by its citizens. Subtle differences may appear between these two measurements, and, largely for the sake of consistency, most modern governments prefer to track GDP rather than GNP for purposes of understanding their economies and public finances. Where the late nineteenth and early twentieth centuries are concerned, however, a more basic problem presents itself, in that GDP and GNP were not calculated at the time (and would not be calculated until the mid-twentieth century). This lack of a key piece of data is highly limiting where analysis of the two states' strategic positions are concerned. This study has ameliorated this problem by a rough, but hopefully consistent, calculation of GNP based on the detailed and exhaustive economic research of the late economist Angus Maddison, combined with data drawn from public sources available at the time. A key portion of Maddison's life's work focused on the assessment of key economic metrics—particularly production—during long periods of history in which data had not been collected. A spreadsheet of Maddison's calculations of adjusted per capita GNP for a large number of states from several centuries of history is available online from a webpage created to store and publish his work.[1] These figures are provided in 1990 Geary-Khamis dollars, a hypothetical currency-equivalent pegged to a specific currency, date, and purchasing power—in this case, the purchasing power of the U.S. dollar in 1990. Conversion of Maddison's per capita GNP figures to other currencies—notably the British pound sterling—was possible by using purchasing power parity (PPP) exchange rates from the relevant year: in this case, through the use of PPP exchange rates for 1990 available online through the World Bank.[2] This gives a currency conversion of $1 1990 U.S. dollar for £1.1 1990 British pounds. The National Archives of the United Kingdom provide an online service that permits one to convert historical currencies to 2005 British pounds sterling;

Appendix

in this way it was possible to convert Maddison's figures first to 1990 pounds sterling, then to 2005 pounds sterling, and finally to pounds sterling for each year under study here.[3]

Conversion to reichsmarks—the currency of Germany during the period under study—presented a further difficulty. *The Statesman's Yearbook* repeatedly gives a rough conversion figure (employed by its own editors) of 20 marks to the pound, but greater precision is in fact possible. Data presented in pounds could be converted to reichsmarks, and vice versa, through a compilation of data provided online by the Swedish Riksbank, which preserves historical exchange rates between various currencies (including the pound and reichsmark) and the Swedish krona on a monthly basis for the period in question.[4] It is in this way possible, using the krona as a bridge, and assuming that arbitrage was impossible during such a long time period as a month, to convert pounds to krona and thereafter to reichsmarks, and vice-versa, and thereby also to derive a direct exchange rate between the reichsmark and pound. This study employs such exchange rates derived from the data for December of the relevant year. All of them fluctuate quite closely around 20 reichsmarks to the pound, thereby essentially confirming *The Statesman's Yearbook* conversion (and tracking closely with it) while adding greater precision.

Through such conversions, it is possible to establish GNP (in both pounds and reichsmarks, as appropriate) for both countries by converting Maddison's per capita GNP figures into the relevant period currencies and then multiplying them by either the population given for the relevant year, or extrapolated for that year based on the growth rate between censuses, where a population figure did not exist. It was possible in turn, on the basis of this, to compare German and British financial figures (including GNP) in a single currency (the pound) by converting as necessary.

This study has not attempted to account for inflation when presenting financial figures. The reason for this is that all of the states of the time—not least Britain and Germany—were using

A Note on Data Sources and Methodology

metallic currencies whose value did not shift greatly over time and was understood by policy makers to be essentially constant.[5] The expression "sound as a pound" originally meant much more than an amusing rhyme: it reminded one of an essential fixture of British currency policy, namely, a desire to avoid fluctuation in value even at some cost to individual citizens in times of financial illiquidity.[6] Moreover, the overwhelming majority of the data involved are relative to GNP rather than a currency value, which renders inflation and deflation irrelevant in almost all cases. For this reason, it is assumed here that small fluctuations in currency value are mostly immaterial to larger questions of defense spending, debt, and overall production.

It is no doubt possible to criticize any particular part of this method; for example, this author does not consider himself expert enough to enter into substantive disputes about Maddison's economic calculations on their face. However, it is likely that consistency of method—employing the same sources over time in the same way, to the degree possible—will nullify most such disputes. Whether GNP, for example, was precisely calculated in any given year is less important than whether GNP—consistently calculated—bears a notable relationship to national debt (similarly calculated) during some period of years. Through the methods just described, it is possible to build an essentially accurate and reasonably detailed picture of the resources available and not available to the two states in question during the timeframe under study—a picture that is useful to the analysis of the strategies they employed. This picture, and the data that were obtained to create it, are cited repeatedly throughout this study as a basis for understanding the Dreadnought Race and the period leading to it.

A further note is required regarding the counting of naval strengths. There is a significant difficulty in accounting for and coding major surface combatants throughout the relevant time period, in that during this time rapid technological change and corresponding obsolescence of existing ships, combined with the allocation of some ships to coastal defense roles, made any count of German or British fleet size in particular a judgment call, since it could include or exclude various coastal defense ships, short-ranged vessels, gunboats, and obsolete craft. Accounting for each competitor's major surface combatants, using *The Statesman's Yearbook* as a source, this book has followed approximately the following rules:

- Gunboats are excluded from major combatant ship counts, along with *avisos* (dispatch) and torpedo boats;
- Coastal defense craft are not counted regardless of capability on the understanding that their relegation to that role is a reliable indicator of lack of confidence of their service in their suitability for blue-water warfare;
- "Old battleships" are counted as combatant vessels, but not obviously obsolete vessels such as sailing vessels in the 1870s.
- First- and second-class cruisers—but not third-class cruisers where identified, because they are too easily confused with smaller gunboats—are likewise counted as major surface combatants, provided that they are steamships. In instances where second- and third-class cruisers were indistinguishable—notably after 1897, the total number of cruisers was included; this, moreover, refers to a convention for identifying cruisers—in *The Statesman's Yearbook*, a parallel method was used prior to 1886 of dividing all ironclads into five classes, of which the first three were offensive and the last two suitable for coastal defense, and therefore not counted here.
- Corvettes and frigates are similarly counted, particularly as, prior to about 1890, they are difficult to distinguish in the official tables from other ironclads.
- As an accounting convention, the oddball British torpedo-ram concept (HMS *Polyphemus* [1881], notably), due to its offensive mission, is treated as a major surface combatant until (arbitrarily) 1890, by which time the limitations of the concept were apparent.

A Note on Data Sources and Methodology

For all this, much of the coding of these ships remains a judgment call based on imperfect data, particularly when ships were transitioning from one use to another. Of note: from 1877 to 1882 in particular, as well as again from 1889 to 1894, the ratios in this chart should be treated as imprecise and perhaps questionable due to a change in counting methods in the original source material, *The Statesman's Yearbook*. A similar instance happens in 1912, where it appears that obsolete cruisers are counted as such in the Royal Navy, only to be recounted as nonobsolete the following year—an effort was made here to avoid an artifact by counting said cruisers as combat-effective. Readers with more discerning judgment are invited to provide comments, but it is thought that this counting method gives a reasonably accurate picture of the relative capabilities of the German and British fleets.

Qualitatively, it may be said that relatively little of this matters, inasmuch as nobody prior to the Dreadnought Race seriously believed the Imperial German Navy to be a match for its British opposite, either in the North Sea or anywhere, and there the matter may rest.

Appendix

Endnotes

1. Angus Maddison, "Mpd_2013-01," Maddison Project, 2013.
2. "Price Level Ratio of PPP Conversion Factor (GDP) to Market Exchange Rate," World Bank, 2011.
3. "Currency Converter," United Kingdom National Archives, 2005.
4. Håkan Lobell, "Foreign Exchange Rates, 1804–1914," Sveriges Riksbank, 2011.
5. Philip Coggan, *Paper Promises: Money, Debt, and the New World Order* (London: Penguin, 2012), 72–75. As one would expect from a currency reliably pegged to gold, Britain in particular swung evenly between moderate inflation and moderate deflation on this system, effectively netting approximately zero change, with the one exception being the year 1900, at the height of the Boer War, when a small, one-off spike in inflation occurred. For simplicity's sake, this is ignored here. See Tejvan Pettinger, "History of Inflation in the UK," EconomicsHelp.org, 1 May 2022.
6. Coggan, *Paper Promises*, 84.

Selected Bibliography

Books and Monographs

Allison, Graham T. *Essence of Decision: Explaining the Cuban Missile Crisis.* Boston, MA: Little, Brown, 1971.

Andrews, Kenneth R. *The Concept of Corporate Strategy*, rev. ed. Homewood, IL: Richard D. Irwin, 1980.

Ansoff, H. Igor. *Corporate Strategy: An Analytic Approach to Business Policy for Growth and Expansion.* New York: McGraw-Hill, 1979.

Art, Robert J. *A Grand Strategy for America.* Ithaca, NY: Cornell University Press, 2003.

Bell, Christopher M. *Churchill and Sea Power.* Oxford, UK: Oxford University Press, 2013.

Berghahn, Volker R. *Germany and the Approach of War in 1914*, 2d ed. New York: St. Martin's Press, 1993.

Brands, Hal. *What Good Is Grand Strategy?: Power and Purpose in American Statecraft from Harry S. Truman to George W. Bush.* Ithaca, NY: Cornell University Press, 2014. Kindle Edition.

Brown, Ian T. *A New Conception of War: John Boyd, the U.S. Marines, and Maneuver Warfare.* Quantico, VA: Marine Corps University Press, 2018. https://doi.org/10.56686/9780997317497.

Brzezinski, Zbigniew. *Strategic Vision.* New York: Basic Books, 2012.

Cain, P. J., and A. G. Hopkins. *British Imperialism: 1688–2000*, 2d ed. London: Pearson Education Limited, 2002.

Carr, Edward Hallett. *The Twenty-Years Crisis, 1919–1939: An Introduction to the Study of International Relations.* London: Macmillan, 1958.

Chandler, Alfred D., Jr. *Strategy and Structure: Chapters in the History of the Industrial Enterprise.* Cambridge, MA: MIT Press, 1962.

Selected Bibliography

Churchill, Randolph S., ed. *The Churchill Documents*, vol. 5, *At the Admiralty, 1911–1914*. Hillsdale, MI: Hillsdale College Press, 2019.

———. *The Churchill Documents*, vol. 6, *At the Admiralty, July 1914–April 1915*. Hillsdale, MI: Hillsdale College Press, 2020.

Churchill, Winston S. *The World Crisis*, vol. 1, *1911–1914*. New York: Charles Scribner's Sons, 1923; New York: Rosetta Books, 2013.

———. *The World Crisis*, vol. 2, *1915*. New York: Charles Scribner's Sons, 1923; New York: Rosetta Books, 2013.

———. *The World Crisis*, vol. 3, *1916–18*. New York: Charles Scribner's Sons, 1927; New York: Rosetta Books, 2013.

Clark, Christopher M. *The Sleepwalkers: How Europe Went to War in 1914*. New York: HarperCollins, 2014.

Clausewitz, Carl von. *On War*. Translated by Michael Howard and Peter Paret. New York: Alfred A. Knopf, 1993.

Cline, Ray S. *World Power Assessment: A Calculus of Strategic Drift*. Washington, DC: Center for Strategic and International Studies, Georgetown University, 1975.

Coggan, Philip. *Paper Promises: Money, Debt and the New World Order*. London: Penguin, 2012.

Colaresi, Michael P., Karen Rasler, and William R. Thompson. *Strategic Rivalries in World Politics*. Cambridge, UK: Cambridge University Press, 2007. https://doi.org/10.1017/CBO9780511491283.

Corbett, Julian S. *Some Principles of Maritime Strategy*. London: Longmans, Green, 1911. Edited and republished by Project Gutenberg, 2005. Kindle edition.

Craig, Gordon A. *Germany: 1866–1945*. Oxford, UK: Clarendon Press, 1978.

Creveld, Martin van. *The Transformation of War*. New York: Free Press, 1991.

Diehl, Paul F., and Gary Goertz. *War and Peace in International Rivalry*. Ann Arbor: University of Michigan Press, 2001. Kindle Edition. https://doi.org/10.3998/mpub.16693.

Selected Bibliography

Ehlert, Hans, Michael Epkenhans, and Gerhard P. Gross, eds. *The Schlieffen Plan: International Perspectives on German Strategy for World War I*. English translation edited by MajGen David T. Zabecki, USA (Ret). Lexington: University Press of Kentucky, 2014.

Epkenhans, Michael. *Tirpitz: Architect of the German High Seas Fleet*. Washington, DC: Potomac Books, 2008. Kindle Edition.

Fehrenbach, T. R. *This Kind of War*. New York: Open Road, Integrated Media, 2001. Kindle Edition.

Fieldhouse, D. K. *The Colonial Empires: A Comparative Survey from the Eighteenth Century*. New York: Delacorte Press, 1966.

———. *Economics and Empire, 1830–1914*. London: Weidenfeld and Nicolson, 1973.

Fierke, Karin M. *Changing Games, Changing Strategies: Critical Investigations in Security*. Manchester, UK: Manchester University Press, 1998.

Fischer, Fritz. *Germany's Aims in the First World War*. New York: W. W. Norton, 1967.

———. *World Power or Decline: The Controversy over Germany's Aims in the First World War*. Translated by Lancelot L. Farrar, Robert Kimber, and Rita Kimber. New York: W. W. Norton, 1974.

Freedman, Lawrence. *Strategy: A History*. Oxford, UK: Oxford University Press, 2013.

Fuller, J. F. C. *The Reformation of War*. London: Hutchinson, 1923.

Gaddis, John Lewis. *On Grand Strategy*. New York: Penguin, 2018.

Gat, Azar. *A History of Military Thought: From the Enlightenment to the Cold War*. Oxford, UK: Oxford University Press, 2001.

Gauss, Christian. *The German Emperor as Shown in His Public Utterances*. New York: Charles Scribner's Sons, 1915. Edited and republished by Project Gutenberg, 2013.

Gilbert, Martin, ed. *The Churchill Documents*, vol. 7, "The Escaped Scapegoat," May 1915–December 1916. Hillsdale, MI: Hillsdale College Press, 2020.

Selected Bibliography

Gilpin, Robert. *War and Change in World Politics*. Cambridge, UK: Cambridge University Press, 1981. https://doi.org/10.1017/CBO9780511664267.

Gramsci, Antonio. *Selections from the Prison Notebooks*. Edited and translated by Quintin Hoare and Gregory Nowell Smith. London: Lawrence and Wishart, 1971.

Gray, Colin S. *Modern Strategy*. Oxford, UK: Oxford University Press, 1999.

Hamel, Gary, and C. K. Prahalad. *Competing for the Future*. Boston, MA: Harvard Business School Press, 1994.

Herwig, Holger H. *The German Naval Officer Corps: A Social and Political History, 1890-1918*. Oxford, UK: Clarendon Press, an imprint of Oxford University Press, 1973.

———. *"Luxury" Fleet: The Imperial German Navy, 1888-1918*. London: George Allen and Unwin, 1980; New York: Humanity Books, an imprint of Prometheus Books, 1987.

Hill, Christopher. *The Changing Politics of Foreign Policy*. Basingstoke, Hampshire: Palgrave Macmillan, 2003.

Hobbes, Thomas. *Leviathan: With Selected Variants from the Latin Edition of 1668*. Edited by Edwin M. Curley. Indianapolis, IN: Hackett, 1994.

Hobson, Rolf. *Imperialism at Sea: Naval Strategic Thought, the Ideology of Sea Power, and the Tirpitz Plan, 1875-1914*. Boston, MA: Brill Academic Publishers, 2002.

Hough, Richard. *Dreadnought: A History of the Modern Battleship*. London: Michael Joseph, 1965.

Ibrahimov, Mahir J., ed. *Great Power Competition: The Changing Landscape of Global Geopolitics*. Fort Leavenworth, KS: U.S. Army Command and General Staff College Press, an imprint of Army University Press, 2020.

Ikenberry, G. John. *After Victory: Institutions, Strategic Restraint, and the Rebuilding of Order after Major Wars*. Princeton, NJ: Princeton University Press, 2001.

Keegan, John. *A History of Warfare*. New York: Vintage Books, 1994.

Selected Bibliography

Kelly, Patrick J. *Tirpitz and the Imperial German Navy*. Bloomington: Indiana University Press, 2011.

Kennedy, Paul M. *Strategy and Diplomacy, 1870–1945: Eight Studies*. London: Fontana, 1983.

———. *The Rise of the Anglo-German Antagonism, 1860–1914*. London: Ashfield Press, 1987.

———. *The Rise and Fall of British Naval Mastery*. New York: Charles Scribner's Sons, 1976.

———. *The Rise and Fall of the Great Powers: Economic Change and Military Conflict from 1500 to 2000*. New York: Vintage Books, 1989.

Kent, Marian. *Moguls and Mandarins: Oil, Imperialism and the Middle East in British Foreign Policy, 1900–1940*. London: Frank Cass, 1993.

Keohane, Robert O., and Joseph S. Nye Jr. *Power and Interdependence*, 3d ed. New York: Longman, 2001.

Kiechel, Walter, III. *The Lords of Strategy: The Secret Intellectual History of the New Corporate World*. Boston, MA: Harvard Business Press, 2010.

King, Stephen D. *When the Money Runs Out: The End of Western Affluence*. New Haven, CT: Yale University Press, 2013.

Kissinger, Henry. *Diplomacy*. New York: Simon and Schuster, 1994.

Knutsen, Torbjørn L. *The Rise and Fall of World Orders*. Manchester, UK: Manchester University Press, 1999.

Lambert, Nicholas A. *Planning Armageddon: British Economic Warfare and the First World War*. Cambridge, MA: Harvard University Press, 2012.

———. *Sir John Fisher's Naval Revolution*. Columbia: University of South Carolina Press, 1999.

Lambi, Ivo Nikolai. *The Navy and German Power Politics, 1862–1914*. Boston, MA: Allen and Unwin, 1984.

Lemke, Douglas. *Regions of War and Peace*. Cambridge, UK: Cambridge University Press, 2002. Kindle Edition. https://doi.org/10.1017/CBO9780511491511.

Selected Bibliography

Liddell Hart, B. H. *Strategy: The Indirect Approach*, rev. ed. London: Faber and Faber, 1967.

Luttwak, Edward N. *Strategy: The Logic of War and Peace*, rev. ed. Cambridge, MA: Belknap Press, an imprint of Harvard University Press, 2001.

Lynch, Thomas F., III., ed. *Strategic Assessment 2020: Into a New Era of Great Power Competition*. Washington, DC: National Defense University Press, 2020.

Mackay, Ruddock F. *Fisher of Kilverstone*. Oxford, UK: Clarendon Press, an imprint of Oxford University Press, 1973.

MacMillan, Margaret. *The War that Ended Peace*. New York: Random House, 2013.

Mahan, Alfred T. *The Influence of Sea Power upon History: 1660–1783*. Edited by Ellen Lyle Mahan. Boston: Little, Brown, 1918; Amazon, 2011. Kindle edition.

Marder, Arthur J. *The Anatomy of British Sea Power: A History of British Naval Policy in the Pre-Dreadnought Era, 1880–1905*. Hamden, CT: Archon Books, 1940.

———. *From the Dreadnought to Scapa Flow*, vol. 1, *The Road to War, 1904–1914*. Oxford: Oxford University Press, 1961; Annapolis, MD: Naval Institute Press, 2013.

Marshall, Andy. *Long-Term Competition with the Soviets: A Framework for Strategic Analysis*. Santa Monica, CA: Rand, 1972.

Massie, Robert K. *Castles of Steel: Britain, Germany, and the Winning of the Great War at Sea*. New York: Random House, 2003.

———. *Dreadnought: Britain, Germany, and the Coming of the Great War*. London: Vintage Books, 2004.

Mearsheimer, John J. *The Tragedy of Great Power Politics*. New York: W. W. Norton, 2001.

Mintzberg, Henry. *The Rise and Fall of Strategic Planning*. New York: Free Press, 1994.

Mitchell, A. Wess. *The Grand Strategy of the Habsburg Empire*. Princeton, NJ: Princeton University Press, 2018.

Modelski, George. *Long Cycles in World Politics*. Hampshire, UK: Macmillan Press, 1987.

Selected Bibliography

Morgenthau, Hans J. *Politics among Nations: The Struggle for Power and Peace*, 7th ed. New York: McGraw-Hill, 2006.

———. *Scientific Man vs. Power Politics*. Chicago, IL: University of Chicago Press, 1974.

Ohmae, Kenichi. *The Mind of the Strategist: The Art of Japanese Business*. New York: McGraw-Hill, 1982.

Olivier, David H. *German Naval Strategy, 1856–1888: Forerunners of Tirpitz*. London: Frank Cass, 2004.

Organski, A. F. K., and Jacek Kugler. *The War Ledger*. Chicago, IL: University of Chicago Press, 1980.

Osinga, Frans P. B. *Science, Strategy, and War: The Strategic Theory of John Boyd*. London: Routledge, 2007. Kindle Edition.

Owen, David. *The Hidden Perspective: The Military Conversations, 1906–1914*. London: Haus Publishing, 2014.

Padfield, Peter. *The Great Naval Race: The Anglo-German Naval Rivalry, 1900–1914*. Edinburgh, Scotland: Birlinn, 2005.

Pakenham, Thomas. *The Scramble for Africa: The White Man's Conquest of the Dark Continent from 1876 to 1912*. New York: Avon Books, 1991.

Porter, Michael E. *Competitive Advantage: Creating and Sustaining Superior Performance*. New York: Free Press, 1985.

———. *Competitive Strategy: Techniques for Analyzing Industries and Competitors*. New York: Free Press, 1998.

Rasler, Karen, and William R. Thompson. *The Great Powers and Global Struggle, 1490–1990*. Lexington: University Press of Kentucky, 1994.

Rasler, Karen, William R. Thompson, and Sumit Ganguly. *How Rivalries End*. Philadelphia: University of Pennsylvania Press, 2013.

Robinson, Ronald, John Gallagher, and Alice Denny. *Africa and the Victorians: The Official Mind of Imperialism*. London: I. B. Tauris, 2015.

Robinson, Stephen. *The Blind Strategist: John Boyd and the American Art of War*. Dunedin, NZ: Exisle Publishing, 2021.

Rodger, N. A. M. *The Command of the Ocean: A Naval History of Britain, 1649–1815*. New York: W. W. Norton, 2004.

Selected Bibliography

——. *The Safeguard of the Sea: A Naval History of Britain, 660–1649*. New York: W. W. Norton, 1999.

Röhl, John C. G., and Nicolaus Sombart, eds. *Kaiser Wilhelm II: New Interpretations*. Cambridge, UK: Cambridge University Press, 1982.

Røksund, Arne. *The Jeune École: The Strategy of the Weak*. Boston, MA: Brill, 2007.

Rumelt, Richard. *Good Strategy, Bad Strategy: The Difference and Why It Matters*. London: Profile Books, 2011.

Seligmann, Matthew S. *The Royal Navy and the German Threat, 1901–1914: Admiralty Plans to Protect British Trade in a War against Germany*. Oxford, UK: Oxford University Press, 2012. Kindle Edition.

Seligmann, Matthew S., Frank Nägler, and Michael Epkenhans. *The Naval Route to the Abyss: The Anglo-German Naval Race, 1895–1914*. London: Routledge, 2016. Kindle Edition. (Originally published 2015).

Snyder, Jack. *Myths of Empire: Domestic Politics and International Ambition*. Ithaca, NY: Cornell University Press, 1991.

Sondhaus, Lawrence. *Preparing for Weltpolitik: German Sea Power before the Tirpitz Era*. Annapolis, MD: Naval Institute Press, 1997.

Spence, Daniel Owen. *A History of the Royal Navy: Empire and Imperialism*. New York: I. B. Tauris, 2015.

Steinberg, Jonathan. *Yesterday's Deterrent: Tirpitz and the Birth of the German Battle Fleet*. London: MacDonald, 1965.

Steiner, Zara S., and Keith Neilson. *Britain and the Origins of the First World War*. 2d ed. Hampshire, UK: Palgrave Macmillan, 2003. Kindle Edition.

Stevenson, David. *Armaments and the Coming of War: Europe, 1904–1915*. Oxford, UK: Clarendon Press, an imprint of Oxford University Press, 1996.

Stewart, Matthew. *The Management Myth: Debunking Modern Business Philosophy*. New York: W. W. Norton, 2009.

Strachan, Hew. *The Direction of War: Contemporary Strategy in Historical Perspective*. New York: Cambridge University Press, 2013. Kindle edition. https://doi.org/10.1017/CBO9781107256514.

———. *The First World War*, vol. 1, *To Arms*. Oxford, UK: Oxford University Press, 2001.

Sumida, Jon Tetsuro. *In Defence of Naval Supremacy: Finance, Technology, and British Naval Policy, 1889–1914*. London: Unwin Hyman, 1989.

Sun Tzu. "The Art of Warfare." Translated by Roger T. Ames. In *The Book of War: Sun Tzu,* The Art of Warfare, *and Karl von Clausewitz,* On War. Edited by Caleb Carr, 1–248. New York: Modern Library, 2000.

———. *The Art of War*. Translated by Samuel B. Griffith. New York: Oxford University Press, 1963.

Thucydides. *History of the Peloponnesian War*. Translated by Rex Warner. London: Penguin, 1972.

Tirpitz, Alfred von. *My Memoirs*, 2 vols. New York: Dodd, Mead, 1919; Amazon, 2019.

Trump, Donald J. *National Security Strategy of the United States of America*. Washington, DC: White House, 2017.

Tuchman, Barbara W. *The Guns of August*. New York: Macmillan, 1962.

———. *The Proud Tower: A Portrait of the World before the War, 1890–1914*. London: Hamish Hamilton, 1966.

Von Neumann, John, and Oskar Morgenstern. *Theory of Games and Economic Behavior*. Princeton, NJ: Princeton University Press, 1947.

Walt, Stephen M. *The Origins of Alliances*. Ithaca, NY: Cornell University Press, 1987.

Warfighting, Fleet Marine Force Manual 1. Washington, DC: Headquarters Marine Corps, 1989.

Wehler, Hans-Ulrich. *The German Empire: 1871–1918*. Oxford, UK: Berg, 1997.

Wendt, Alexander. *Social Theory of International Politics*. Cambridge, UK: Cambridge University Press, 1999. https://doi.org/10.1017/CBO9780511612183.

Williamson, Samuel R., Jr. *The Politics of Grand Strategy: Britain and France Prepare for War, 1904-1914*. London: Ashfield Press, 1990.

Wood, Michael. *In Search of the Trojan War*. Berkeley: University of California Press, 1996.

Zuber, Terence. *The Real German War Plan: 1904-1914*. Gloucestershire, UK: History Press, 2011.

Book Chapters

Beasley, Ryan K., and Michael T. Snarr. "Domestic and International Influences on Foreign Policy: A Comparative Perspective." In *Foreign Policy in Comparative Perspective: Domestic and International Influences on State Behavior*. Edited by Ryan K. Beasley, Juliet Kaarbo, Jeffrey S. Lantis, and Michael T. Snarr, 321-47. Washington, DC: CQ Press, 2002.

Brown, David K. "Wood, Sail, and Cannonballs to Steel, Steam, and Shells, 1815-1895." In *The Oxford Illustrated History of the Royal Navy*. Edited by J. R. Hill, 200-26. Oxford, UK: Oxford University Press, 1995.

Ehlert, Hans, Michael Epkenhans, and Gerhard P. Gross. "Introduction: The Historiography of Schlieffen and the Schlieffen Plan." In *The Schlieffen Plan: International Perspectives on the German Strategy for World War I*. Edited by Hans Ehlert, Michael Epkenhans, and Gerhard P. Gross. English translation edited by MajGen David T. Zabecki, USA (Ret), 1-16. Lexington: University Press of Kentucky, 2014.

Foley, Robert T. "The Schlieffen Plan: A War Plan." In *The Schlieffen Plan: International Perspectives on German Strategy for World War I*. Edited by Hans Ehlert, Michael Epkenhans, and Gerhard P. Gross. English translation edited by MajGen David T. Zabecki, USA (Ret), 67-83. Lexington: University Press of Kentucky, 2014.

Gross, Gerhard P. "There Was a Schlieffen Plan: New Sources on the History of German Military Planning." In *The Schlieffen Plan: International Perspectives on German Strategy for World*

War I. Edited by Hans Ehlert, Michael Epkenhans, and Gerhard P. Gross. English translation edited by MajGen David T. Zabecki, USA (Ret), 85–136. Lexington: University Press of Kentucky, 2014.

Harris, J. Paul. "Great Britain." In *The Origins of World War I*. Edited by Richard F. Hamilton and Holger H. Herwig, loc. 6701–7629. Cambridge, UK: Cambridge University Press, 2003. https://doi.org/10.1017/CBO9780511550171.008.

Kennedy, Paul M. "The Kaiser and German *Weltpolitik*: Reflexions on Wilhelm II's Place in the Making of German Foreign Policy." In *Kaiser Wilhelm II: New Interpretations*. Edited by John C. G. Röhl and Nicolaus Sombart, 143–68. Cambridge, UK: Cambridge University Press, 1982.

———. "Grand Strategy in War and Peace: Toward a Broader Definition." In *Grand Strategies in War and Peace*. Edited by Paul M. Kennedy, 1–7. New Haven, CT: Yale University Press, 1991.

Lambert, Andrew D. "The Royal Navy, 1856–1914: Deterrence and the Strategy of World Power." In *Navies and Global Defense: Theories and Strategy*. Edited by Keith Neilson and Elizabeth Jane Errington, 69–92. Westport, CT: Praeger, 1995.

Liddell Hart, B. H. "Foreword." In Sun Tzu, *The Art of War*. Translated by Samuel B. Griffith, vii–viii. New York: Oxford University Press, 1963.

Paret, Peter. "Introduction." In *Makers of Modern Strategy: From Machiavelli to the Nuclear Age*. Edited by Peter Paret, 3–8. Princeton, NJ: Princeton University Press, 1986.

Peters, Ralph. "The Seeker and the Sage." In *The Book of War: Sun Tzu,* The Art of Warfare, *and Karl von Clausewitz,* On War. Edited by Caleb Carr, vii–xxiv. New York: Modern Library, 2000.

Porter, Michael E. "What Is Strategy?." In *On Strategy*, 1–37. Boston, MA: Harvard Business Review Press, 1996.

Raymond, Gregory A. "Evaluation: A Neglected Task for the Comparative Study of Foreign Policy." In *New Directions in the Study of Foreign Policy*. Edited by Charles F. Hermann,

Charles W. Kegley Jr., and James N. Rosenau, 96–110. Boston, MA: Allen and Unwin, 1987.

Schmidt, Brian C. "Realism and Facets of Power in International Relations." In *Power in World Politics*. Edited by Felix Berenskoetter and M. J. Williams, 43–63. London: Routledge, 2007.

Steinberg, Jonathan. "The Kaiser and the British: The State Visit to Windsor, November 1907." In *Kaiser Wilhelm II: New Interpretations*. Edited by John C.G. Röhl and Nicolaus Sombart, 121–42. Cambridge, UK: Cambridge University Press, 1982.

Strachan, Hew. "The British Army, Its General Staff, and the Continental Commitment, 1904–1914." In *The Schlieffen Plan: International Perspectives on German Strategy for World War I*. Edited by Hans Ehlert, Michael Epkenhans, and Gerhard P. Gross. English translation edited by MajGen David T. Zabecki, USA (Ret), 293–317. Lexington: University Press of Kentucky, 2014.

Walt, Stephen M. "Keeping the World 'Off-Balance': Self-Restraint and U.S. Foreign Policy." In *America Unrivaled: The Future of the Balance of Power*. Edited by G. John Ikenberry, 121–54. Ithaca, NY: Cornell University Press, 2002.

Scholarly Articles

Bell, Christopher M. "Contested Waters: The Royal Navy in the Fisher Era." *War in History* 23, no. 1 (January 2016): 115–26. https://doi.org/10.1177/0968344515595330.

———. "On Standards and Scholarship: A Reply To Nicholas Lambert." *War in History* 20, no. 3 (July 2013): 381–409. https://doi.org/10.1177/0968344513483069.

———. "Sir John Fisher's Naval Revolution Reconsidered: Winston Churchill at the Admiralty, 1911–1914." *War in History* 18, no. 3 (July 2011): 333–56. https://doi.org/10.11177/0968344511401489.

Boyd, Col John R, USAF (Ret). "Destruction and Creation." *A Discourse on Winning and Losing*, 3 September 1976.

Selected Bibliography

Etzioni, Amitai. "Spheres of Influence: A Reconceptualization." *Fletcher Forum of World Affairs* 39, no. 2 (Summer 2015): 117–32.

Feaver, Peter D., Gunther Hellman, Randall L. Schweller, Jeffery W. Taliaferro, William C. Wohlforth, Jeffery W. Lergo, and Andrew Moravcsik. "Brother, Can You Spare a Paradigm? (Or Was Anybody Ever a Realist?)." *International Security* 25, no. 1 (Summer 2000): 165–93. https://doi.org/10.1162/016228800560426.

Hamel, Gary, and C. K. Prahalad. "The Core Competence of the Corporation." *Harvard Business Review* 68, no. 3 (May–June 1990): 275–92.

———. "Strategy as Stretch and Leverage." *Harvard Business Review* 71, no. 2 (March–April 1993): 75–84.

Howard, Michael. "The Forgotten Dimensions of Strategy." *Foreign Affairs* 57, no. 5 (Summer 1979): 975–86.

Lambert, Nicholas A. "On Standards: A Reply to Christopher Bell." *War in History* 19, no. 2 (April 2012): 217–40. https://doi.org/10.1177/0968344511432977.

———. " 'Our Bloody Ships' or 'Our Bloody System'?: Jutland and the Loss of the Battle Cruisers, 1916." *Journal of Military History* 62, no. 1 (January 1998): 29–55. https://doi.org/10.2307/120394.

Legro, Jeffrey W., and Andrew Moravcsik. "Is Anybody Still a Realist?." *International Security* 24, no. 2 (Fall 1999): 5–55. https://doi.org/10.1162/016228899560130.

Lissner, Rebecca Friedman. "What Is Grand Strategy?: Sweeping a Conceptual Minefield." *Texas National Security Review* 2, no. 1 (November 2018): 52–73. https://doi.org/10.26153/tsw/868.

Mansfield, Edward D., and Jack Snyder. "Democratization and the Danger of War." *International Security* 20, no. 1 (Summer 1995): 5–38. https://doi.org/10.2307/2539213.

Owen, John M. "How Liberalism Produces Democratic Peace." *International Security* 19, no. 2 (Fall 1994): 87–125. https://doi.org/10.2307/2539197.

Seligmann, Matthew S. "Naval History by Conspiracy Theory: The British Admiralty before the First World War and the Methodology of Revisionism." *Journal of Strategic Studies* 38, no. 7 (July 2015): 966–84. https://doi.org/10.1080/01402390.2015.1005443.

Silove, Nina. "Beyond the Buzzword: The Three Meanings of 'Grand Strategy'." *Security Studies* 27, no. 1 (2018): 27–57. https://doi.org/10.1080/09636412.2017.1360073.

Skold, Martin. "Book Review: *On Grand Strategy* and *The Grand Strategy of the Habsburg Empire*." *Contemporary Voices: St Andrews Journal of International Relations* 1, no. 3 (August 2019): 117–26. https://doi.org/10.15664/jtr.1512.

Waltz, Kenneth N. "The Origins of War in Neorealist Theory." *Journal of Interdisciplinary History* 18, no. 4 (Spring 1988): 615–28. https://doi.org/10.2307/204817.

Audiovisual and Multimedia Sources

Boyd, Col John R., USAF (Ret). "John Boyd Patterns of Conflict Part 1 of 7." YouTube video. Online audiovisual material recorded by Jason M. Brown and posted by Steven Shack, 14 February 2015.

———. "John Boyd Patterns of Conflict Part 2 of 7." YouTube video. Online audiovisual material recorded by Jason M. Brown and posted by Steven Shack, 8 March 2015.

———. "John Boyd Patterns of Conflict Part 4 of 7." YouTube video. Online audiovisual material recorded by Jason M. Brown and posted by Steven Shack, 8 March 2015.

———. "Patterns of Conflict." Lecture, U.S. Marine Corps Command and Staff College, Marine Corps University, Quantico, VA, 25 April and 2–3 May 1989. Transcribed by Maj Ian T. Brown, USMC, 25 March 2015–11 January 2017.

———. "Patterns of Conflict." Brief, U.S. Department of Defense, December 1986. Linked online at Belisarius.com, 2006.

Karber, Phillip. "Competitive Strategy: As an Approach to Business and Professional Life." PowerPoint presentation,

Annual Fellows Conference of the Center for the Study of the Presidency, Washington, DC, 31 October 2003.

———. "The 'Counter-Offensive' in Competitive Strategy: Lessons from the Reagan Era." PowerPoint presentation, Developing Competitive Strategies for the 21st Century Conference, U.S. Naval War College, Newport, RI, 23 August 2010.

Richards, Chet. "Boyd's OODA Loop." PowerPoint presentation published by the *Defense and the National Interest* website, 2006.

Spinney, Franklin C. "Evolutionary Epistemology Talk at [the U.S. Marine Corps Expeditionary Warfare School, 15 January 2019]." YouTube video, posted by the Warfighting Society, 4 February 2019.

Data Sets

Abbas, S. Ali, Nazim Belhocine, Asmaa ElGanainy, and Mark Horton. "A Historical Public Debt Database." IMF Working Paper no. 10/245. Washington, DC: International Monetary Fund, 2010.

———. "Debt Database Fall 2013 Vintage." International Monetary Fund, 2013.

"Currency Converter." United Kingdom National Archives, 2005.

Lobell, Håkan. "Foreign Exchange Rates, 1804–1914." Sveriges Riksbank, 2011.

Maddison, Angus. "Mpd_2013-01." Maddison Project, 2013.

Martin, Frederick, and John Scott Keltie, eds. *The Statesman's Yearbook*. London: Macmillan, 1875–1914.

Officer, Lawrence H. "Exchange Rates between the United States Dollar and Forty-One Currencies." MeasuringWorth, 2022.

"Price Level Ratio of PPP Conversion Factor (GDP) to Market Exchange Rate." World Bank, 2011.

Selected Bibliography

News/Current Events Sources

Colby, Elbridge, and Robert D. Kaplan. "The Ideology Delusion: America's Competition with China Is Not about Doctrine." *Foreign Affairs*, 4 September 2020.

Greer, Tanner. "China's Plans to Win Control of the Global Order." *Tablet*, 17 May 2020.

———. "Introducing: Asabiyah." *Scholar's Stage*, 2 May 2015.

———. "You Do Not Have the People." *Scholar's Stage*, 3 March 2018.

Ioffe, Julia. "What Putin Really Wants." *Atlantic*, January/February 2018.

Pickrell, Ryan. "China Is the World's Biggest Shipbuilder, and Its Ability to Rapidly Produce New Warships Would Be a 'Huge Advantage' in a Long Fight with the U.S., Experts Say." *Business Insider*, 8 September 2020.

Other Sources

Bülow, Bernhard von. "Bernhard von Bülow on Germany's 'Place in the Sun'." Speech to Reichstag, 1897. *German History in Documents and Images*, accessed 2 December 2021.

Commonwealth War Graves Commission Annual Report, 2013–2014. Maidenhead, UK: Commonwealth War Graves Commission, 2014.

Crowe, Eyre. "Memorandum on the Present State of British Relations with France and Germany." In *The Hidden Perspective: The Military Conversations 1906–1914*. Edited by David Owen, 216–61. London: Haus Publishing, 2014.

Dube, Claude. "The Department of National Defence and the Defence Strategies from 1945 to 1970." MBA thesis, McGill University, 1973.

Pettinger, Tejvan. "History of Inflation in the UK." EconomicsHelp.org, 1 May 2022.

Index

Admiralty (British), 197, 209, 223–24, 297, 349
Admiralty Staff (German), 159, 215–17
Afghanistan, 223
Africa, European colonialization in, 113–15, 134–42, 147, 169–70, 231, 262–63, 296–97, 324–26
Agadir Incident, 169, 220, 227, 327, 339, 351
Algeçiras crisis, 205–6, 216, 333, 351
Allison, Graham T., Jr., 60
alpha, in framework, 64–69, 99–108, 261, 268, 331–33, 336, 342–45, 366–72
American Civil War, 132
Andrews, Kenneth R., 28–30
Anglo-Russian Convention, 223
Ansoff, H. Igor, 27–30, 68–70
Arnold-Forster, H. O., 200
Art, Robert J., 44–45
Asquith, Herbert H., 212, 220–24, 228, 369
Austria, 138–39, 151
Austria-Hungary, 205, 233

Balfour, Arthur J., 199–201, 205
Baltic Sea, 217
Bath Ironworks, 158
Battenberg, Prince Louis, 229
Battle Cruiser Fleet (British), 236
battlecruiser, 209–10, 213–14, 218, 231, 236–38, 270–71, 274, 303
battleship, 113, 121–22, 126, 133, 149, 154–55, 158, 161–65, 169, 197–200, 208–14, 225, 229, 233–38, 270–71, 277, 299, 303, 306–08, 340–44, 352, 379, 383
Beatty, David, 236–37
Bebel, August, 227
Belgium, 163, 219, 266, 279, 306, 338
Beresford, Charles, 211–12, 229, 338
Berghahn, Volker, 115, 164

beta, in framework, 64–68, 99–108, 261, 268, 331, 335, 345, 366–70, 372
Bethmann Hollweg, Theobald von, 144, 226–27, 230–33, 236, 262, 349, 350–52
Bismarck, Otto von, 97, 134, 138–42, 146–47, 150–51, 158, 204, 278, 326, 344
Blücher, SMS, 236
Boer War, 103, 147, 152, 170–71, 197–200, 212–13, 263, 281–85, 288, 292, 296, 304, 311, 324, 336
Bonaparte, Napoléon, 129, 137, 202, 297, 344
Bosporus Strait, 131
Boxer Rebellion, 164
Boyd, John R., xix, xxiii, xxv, 9, 72–78, 85, 346, 351–52, 366, 369
Brands, Hal, 45
Bridgeman, Francis, 229
Britain. *See* Great Britain
British Army, 23, 218–20, 285, 288, 342
British Foreign Office, 136, 200, 219–20, 224, 227–28, 342, 351
British Parliament, 145, 222
Brzezinski, Zbigniew, 45
Bülow, Bernhard von, 115–16, 144–46, 164–66, 169, 198, 204–5, 224–26, 321, 324, 332, 335
business strategy, xviii, xix, xxiv, 5–10, 16, 21, 26, 31, 60, 68, 70, 79, 85, 365

Cain, P. J., 118
Cameroon, 140
Campbell-Bannerman, Henry, 212, 218–22
Canada, 232
Cape Colony, 141
Caprivi, Leo von, 137, 140, 150
Carr, E. H., 35
Chamberlain, Joseph, 171, 198–99
Chandler, Alfred D., Jr., 27–30

401

Index

Channel Fleet (British), 131, 207, 211, 233. *See also* Home Fleet (British)
China, xix, 5, 164, 207, 371–72
China Fleet (British), 207
Churchill, Winston S., 222–25, 228–37, 297, 341, 351–52, 369
Clark, Christopher M., 136
Clausewitz, Carl von, 21–23, 27–29, 61, 97, 269
coastal defense, 121, 160–61, 275, 334, 382–83
Colaresi, Michael P., 17–18
Cold War, xiv, xviii, 3, 37, 97–98, 221
commerce raiding, 121–27, 131, 136, 153, 165, 209–10, 215–16, 277, 303, 307, 334
Committee of Imperial Defence (British), 220, 228
competitive advantage, 9, 31, 34, 70–72, 75, 100–3, 106, 259, 330, 343, 372
competitive strategy, xiii, xv, 8, 25, 34, 42, 44–45, 62, 72, 78, 83, 86, 100–3, 113–14, 120, 145, 259, 335–38, 370
Congo, 134, 228
constructivism (international relations), 39, 41–42, 80, 99, 106, 112
Corbett, Julian S., 122–23, 127, 300
core competence, xviii, 9, 65, 71–72, 75, 85, 100–3, 106–8, 307, 330–31, 343–45, 367–69, 373–74
corvette, 383
Craig, Gordon A., 145–47, 263
Crimea, xix, 142
Crimean War, 130–31, 150
Crowe, Eyre, 148, 221–22, 324–26, 336
cruiser, 125–26, 149, 152, 157, 160–61, 165, 208–9, 214–18, 236, 308, 334, 383–84

"danger zone" concept, 165–66, 172, 204, 227, 232, 347, 370
Dardanelles, 131, 207, 235
decision cycle, 9, 72, 75–79, 85, 104, 107–8, 259, 330, 346, 350–53, 366–69, 372–74
Denny, Alice, 118
destroyer, 121, 271, 275
détente, 235

Diehl, Paul F., 18
Disraeli, Benjamin, 131
Dogger Bank Incident, 204, 236, 352
Dreadnought, HMS, 200, 206, 209–10, 212–14, 236, 271–74, 303, 337–39, 343, 347, 350
Dreadnought-class battleship, 217
Dube, Claude, 29

economic theory, 43
Edward VII, 143, 198, 201–2, 229
Egypt, 135, 140–41, 202
Eighty Years' War, 129
English Channel, 158, 297
Entente Cordiale, 201–5, 223, 227, 232, 265, 303, 339
Epkenhans, Michael, 117–19, 266, 340

Fashoda Incident, 169–70, 197, 297
Feaver, Peter D., 40–41, 79
Fehrenbach, T. R., 3
Ferdinand, Franz, 235
Fieldhouse, D. K., 118, 140–41, 148
Fischer, Fritz, 119
Fisher, John A., 116–17, 199, 205–14, 217–18, 223, 228–30, 233–38, 270–71, 294, 297, 337–46, 349–51, 369
flotilla defense, 211, 233, 275, 297–98, 341
flottenprofessoren, 168
foreign policy analysis, 42
France, 104, 123–26, 129, 133, 137–39, 142, 151, 157, 161, 169–70, 197, 199, 201–5, 218–20, 223–25, 228, 232, 235, 265–67, 277–78, 296–97, 306, 335, 342–44
Franco-Prussian War, 138, 307
Franco-Russian Military Alliance Convention, 151
Frederick II, 143
Frederick the Great, 20, 154
Freedman, Lawrence, 31–32
French Navy, 123–24
frigate, 158, 383
Fuller, J. F. C., 24

Gallagher, John, 118

Index

Gallipoli campaign, 235
Ganguly, Sumit, 18, 67, 335
George V, 229
George V-class battleship, 229
German Foreign Office, 348
Gilpin, Robert, 35–37, 59, 80–81
Gladstone, William Ewart, 131
global hegemony, 35–37, 270, 375
global reach capability, 35, 128
Goertz, Gary, 18
Grand Fleet (British), 236
grand strategy, 5, 16, 23–25, 34, 39, 43–46, 101–3, 138, 335–38, 370, 373
Gray, Colin S., 19, 23, 45, 62–63
great power competition, xiv, xviii–xix, xxv, 4–11, 26, 58, 97–99, 373–74
great power rivalry, xiii, xviii, 4. *See also* great power competition
Great War. *See* World War I
Grey, Edward, 218–22, 227–28, 231–32, 235, 342, 351
gross domestic product, 380
gross national product, 81, 129, 281, 380
Grosser Kurfürst, SMS, 158
gunboat, 135, 228, 327, 352, 383
gunboat diplomacy, 134–35, 149, 152, 207, 263

Haldane, Richard Burdon, 231
Hamel, Gary P., 63–65, 69–72, 99–103, 173, 280, 333–34, 342, 345, 353
hard power, 35–36, 100, 128, 142, 171–73, 263, 329, 332
Henderson, Bruce D., 30
Herwig, Holger, 116
High Seas Fleet (German), 161, 164, 200, 207, 212, 215–16, 225–26, 234–38, 273, 278, 300, 303, 306–7, 319–27, 332, 335–37, 350–51
Hobson, Rolf, 166, 325
Hollmann, Freidrich von, 160, 168
Home Fleet (British), 131, 207, 234. *See also* Channel Fleet (British)
Hopkins, A. G., 118

Ikenberry, G. John, 41
Imperial Admiralty Staff (German), 159

Imperial German Army, 234, 278, 306, 319, 334
Imperial German Navy, 102, 114–15, 126, 135, 144–45, 153–54, 157–60, 169–71, 200, 215–16, 224, 234, 273, 307, 312–14, 334, 350, 370, 384
Imperial Naval Cabinet (German), 159
Imperial Naval Office (German), 153, 159, 216
Indo-Pacific, 5, 371–72
Industrial Revolution, 129–30, 145
Institutionalism (international relations), 39, 59, 80
international relations theory, xix, xxv, 5, 9–10, 35, 39, 42–43, 59, 127, 366
interstate competition, 8, 69, 72, 78, 85–86, 352
Invincible, HMS, 209–10, 217–18, 237, 337, 340, 350
Invincible-class battlecruiser, 217, 238, 274
Iran, 223
Iron Duke-class battleship, 229
Italy, 139, 151, 233

Jameson, Leander Starr, 152
Jameson Raid, 170
Japan, 204
Jellicoe, John Rushworth, 236–37
Jeune École, 123–26, 130–32, 149, 157, 160, 165, 203, 215, 297, 304, 307
July Crisis, 220, 266–67, 279, 303, 338, 341
Jutland, Battle of, 209, 236–38, 270, 273, 306

Lambert, Nicholas A., 116–17, 238, 340
Lambi, Ivo, 115–19, 266
Lansdowne, Henry C.K. Petty-Fitzmaurice, 5th Marquess of, 200–1, 219
Legro, Jeffrey W., 40–41, 79
Lemke, Douglas, 37
Liddell Hart, B. H., 22, 65
Lion, HMS, 236
Lissner, Rebecca, 45
List, Freidrich, 155
Lloyd George, David, 222–25, 228, 231, 339

Index

long-cycle theory, xxiii, 35–37, 127–28
Louis XIV, 129
Luttwak, Edward N., 22, 25, 44

Kaiser Wilhelm. *See* Wilhelm II
Karber, Phillip, xxvii–xxviii, xxiii–xxiv, 33
Kayser, Paul, 147
Keegan, John, 156
Kelly, Patrick J., 116, 166
Kennan, George F., 221
Kennedy, Paul M., 25, 44, 115–16, 119, 129, 148, 157, 162–63, 201, 204, 263–64, 266–67, 298, 303, 324, 326, 336
Keohane, Robert O., 60
Kerr, Walter T., 200
Khaki Election (1900), 197
Kiderlen-Wächter, Alfred von, 227–28, 352
Kiechel, Walter, III, 27–28, 30
Kiel Canal, 217, 235
Kipling, Rudyard, 111, 136, 196, 258, 298, 365
Kissinger, Henry, 150–52, 202, 219, 265
Kitchener, Horatio H., 169
Knutsen, Torbjørn L., 35
Kondratiev, Nikolay D., 128
König Wilhelm, SMS, 158

Maddison, Angus, 380–82
Mahan, Alfred Thayer, 121–27, 131, 136, 148–49, 153, 157
Marder, Arthur J., 116, 197, 339–41
Marshall, Andrew W., xxiii, 5–6, 79, 84
Massie, Robert K., 115–19, 198, 339
Mearsheimer, John J., 35, 38
Mediterranean, 130–31, 149, 152, 211, 218, 220, 233, 297
Mediterranean Fleet (British), 131, 207, 233
merchant vessel, 125, 155, 209, 216
military strategy, xviii–xix, 19, 26, 61
Mintzberg, Henry, 28–29, 33, 260
Mitteleuropa, 233
Modelski, George, 37, 127
Moltke, Helmut von, 138
Monitor, USS, 132

Moravcsik, Andrew M., 40–41, 79
Morgenstern, Oskar, 27
Morocco, 202, 205, 227–28

Nägler, Frank, 117–19, 266, 340
Namibia, 140
Napoleonic Wars, 129, 137, 154, 200
National Security Council, U.S., 373
Naval Defense Act (1889), 133
Naval High Command (German), 154, 159–60, 216, 277, 307, 314, 333, 348
Naval Laws (German), 155, 166–70, 214, 217, 223, 231, 234, 264, 271–72, 310, 349
Naval Scare, 224, 339, 347
Neoliberalism (international relations), 39–41, 99, 106
Netherlands, 129, 279
Neumann, John von, 27
Newfoundland, 202
North Atlantic Treaty Organization (NATO), 373
North Sea, 8–10, 34, 37, 111–13, 118–19, 126, 131, 134–36, 149, 152–53, 162–64, 171–72, 207, 215, 219, 231, 234, 236–37, 261–65, 170, 277, 297–99, 303, 307, 319, 323–24, 326, 328, 331–33, 336–37, 366–68, 384
Norway, 163
Nye, Joseph S., Jr., 60

Ohmae, Kenichi, 31–33, 66, 72, 100
OODA loop, xix, xxiii, 72–78, 85, 346, 366, 369
Organski, A. F. K., 37
Orion-class battleship, 225, 274
Ottoman Empire, 131, 235

Padfield, Peter, 115–19, 197, 278, 197, 306, 339, 351
Pakenham, Thomas, 140
Palmer, William W., 200, 212
Panther, SMS, 228, 326–27
Paret, Peter, 23
Peters, Ralph, 15
Polyphemus, HMS, 383
Poplar, 158
Porter, Michael E., 7–8, 31, 71, 103

Index

Portugal, 129
power transition theory, 37, 374
Prahalad, C. K., 63–65, 69–72, 99–103, 173, 280, 333–34, 342, 345, 353
Prussia, 115, 137–38, 143, 153–54, 226
Putin, Vladimir, 373

Queen Elizabeth-class battleship, 229, 274

Rasler, Karen, 17–18, 35–37, 44, 67, 127–29, 263, 335
Rathenau, Walter, 232
Reagan, Ronald W., 3, 41
realism (international relations), 35, 39–41, 59, 80–82, 97–99, 106, 112
Reichstag, 115, 141, 145, 160, 164–67, 171, 226, 233–34, 309, 314, 321, 333
Reinsurance Treaty (1887), 139, 150–51
regional hegemony, 4–9, 16–18, 26, 34–39, 44–47, 59, 62, 66, 71, 78–79, 83–87, 96–98, 100–1, 106–8, 112–14, 137, 261, 263–64, 329, 332, 335, 353, 365–66, 369
"Risk Fleet" concept, 162, 323
Robinson, Ronald, 118
Rodger, N. A. M, 111
Røksund, Arne, 124–25, 203, 304
Royal Navy (British), 10, 103, 113, 116, 129–31, 133–35, 144, 162–64, 170, 196–98, 200, 203–9, 211–14, 217, 220, 223–25, 229–30, 234–38, 264, 273–74, 278–85, 288, 293, 297, 303, 307, 314, 328–29, 332, 336, 339–44, 348–50, 367–69, 384
Rumelt, Richard P., 70
Russia, xix, xxii, 104, 130–33, 137–39, 142, 150–51, 157, 161, 166, 199, 202–4, 207, 222–23, 232, 235–37, 267, 277–78, 335, 344, 352, 371–73
Russian Navy, 204
Russo-Japanese War, 204

Salisbury, Robert A. T. Gascoyne-Cecil, 3d Marquess of, 197–99, 219, 346, 348
Sammlungspolitik, 146, 149, 161, 168, 226, 325

Scapa Flow, 237
Scott, Percy M., 211
Seligmann, Matthew S., 116–19, 125, 209–10, 215, 266, 276–77, 340
Shakespeare, William, 365
Silove, Nina, 45
Snyder, Jack, 60
Social Darwinism, 156–57
Social Democrats (German), 114–15, 145–49, 166–68, 226–27, 321, 326, 351
soft power, 105, 173, 263
Sondhaus, Lawrence, 154, 158
South Africa, 135, 152, 170
South China Sea, 5, 371–72
Soviet Union, 79
Spain, 129, 169, 324
Spanish-American War, 169
Spence, Daniel Owen, 135
sphere of influence, 38
Stanley, Henry Morton, 134
Statesman's Yearbook, The, 378–84
Steinberg, Jonathan, 116
Stewart, Matthew, 68
Stosch, Albrecht von, 137
Strachan, Hew, xxii, 19, 24–26, 145, 216, 220, 233
Strait of Gibraltar, 202
strategic intent, 63–64, 69–72, 84, 99–100, 103, 107–8, 118, 146, 259–60, 322, 330–33, 336, 339, 342–43, 352, 366–69, 373–74
strategic rivalry, 17
strategic theory, 7
submarine, 125–26, 210–11, 216, 233, 236, 271, 275, 308, 334
Sudan, 169
Suez Canal, 141
Sun Tzu, 15, 20, 60–63, 77–78, 96–97, 100, 258, 351–52, 375
superdreadnought, 225, 229, 271, 274, 350

Taiwan, 371
Tanzania, 140
Thames Ironworks, 158
Thompson, William R., 17–18, 35–37, 44, 67, 127–29, 263, 335

Index

Tirpitz, Alfred von, 102, 115–17, 126, 132, 144, 150, 153–73, 200, 204–6, 213–18, 223–38, 262, 271–72, 277, 298–300, 307, 316, 321–25, 328, 332–35, 340, 343–44, 347–51, 370
Tirpitz Plan, 369
Togo, 140
torpedo boat, 121, 126, 133, 153, 157, 161, 211, 271–72, 275, 383
Trafalgar, Battle of, 130, 196
Transvaal Republic, 136, 152, 162, 170
Treitschke, Heinrich von, 156, 161
Triple Entente, 223, 265, 278, 295–97, 327, 333, 338, 346
Tuchman, Barbara W., 156
Two-Power Standard, 133, 158, 197–98, 205, 224, 263

Ukraine, xix, xxii, 372
United States, xvii–xix, 5, 25, 37, 44–45, 81, 371–73

Van Creveld, Martin, 20
Victoria, Queen, 143, 198

Warrior, HMS, 132
Wegener, Wolfgang, 163
Weltpolitik, 142, 146, 155, 164–66, 232, 262–63, 268, 277, 280, 323–26, 332, 344
Wilhelm II, 102, 113–18, 138, 142–61, 164–73, 200, 204–5, 215–16, 225–27, 232–34, 238, 263–64, 272, 277–78, 296–98, 316, 319–28, 332–35, 345–49, 367–68
Williamson, Samuel R., Jr., 203, 265
Wilson, Arthur K., 229
Wittelsbach, SMS, 164
World War I, 8–10, 24, 34, 87, 97, 113–14, 118–20, 124–26, 133, 137, 140–42, 151, 166, 169, 210, 217–20, 223, 264–68, 271–73, 277–79, 295–300, 306–8, 322–23, 327, 338, 341, 345, 366, 378–79
World War II, 24, 29, 119, 163

Zedong, Mao, 58

About the Author

Martin Skold is a writer and scholar on various aspects of statecraft and U.S. policy. A practitioner in the field of U.S. national security, he has served in national security and policy roles in the U.S. counterterrorism community, in the U.S. House of Representatives, and on political campaigns. He has written on U.S. national security policy in numerous publications, including *The American Interest*, *War on the Rocks*, *The Bulwark*, *Contemporary Voices* (the international relations journal of the University of St. Andrews), and the *Georgetown Security Studies Review*. He is a graduate of Georgetown University's School of Foreign Service in Washington, DC, and achieved his PhD in international relations from the University of St. Andrews in Scotland, where he also taught international relations at the undergraduate level. The views expressed in this book are his own, and not those of any employer past or present, though he hopes they may be of use. He and his wife, Christina Goodlander, live in the Washington, DC, area except when on the road and enjoy traveling, exploring places where history happened, and occasionally making some of it themselves.